Darwin

and the

General Reader

Darwin
and the
General Reader

*The Reception of Darwin's
Theory of Evolution
in the
British Periodical Press,
1859-1872*

ALVAR ELLEGÅRD

**With a new Foreword
by David L. Hull**

The University of Chicago Press

Chicago and London

Darwin and the General Reader was originally published in 1958
as volume VIII of the series Gothenburg Studies in English, edited
by Frank Behre. It was also published as Acta Universitatis
Gothoburgensis, Vol. LXIV.

The University of Chicago Press, Chicago 60637
The University of Chicago Press, Ltd., London
© 1990 by The University of Chicago
All rights reserved.
University of Chicago Press edition, 1990
Printed in the United States of America

98 97 96 95 94 93 92 91 90   54321

Library of Congress Cataloging-in-Publication Data

Ellegård, Alvar.
   Darwin and the general reader : the reception of Darwin's theory
of evolution in the British periodical press, 1859–1872 / Alvar
Ellegård ; with a new foreword by David L. Hull.
     p. cm.
   "Originally published in 1958 as volume VIII of the series
Gothenburg studies in English, edited by Frank Behre. It was also
published as Acta universitatis Gothoburgensis, Vol. LXIV"—T.p.
verso.
    ISBN 0-226-20487-1 (alk. paper)
   1. Darwin, Charles, 1809–1882. On the origin of species.
2. Journalism—Great Britain—History—19th century. I. Title.
QH365.O8E45 1990
575.01'62—dc20
                         89-78233
                         CIP

Table of Contents

By permission of the Editor of the *Journal of the History of Ideas*, parts of Chapter 2 are reprinted from my article "Public Opinion and the Press: Reactions to Darwinism," published in the *Journal* in June 1958, and parts of Chapter 9 are reprinted from my article "The Darwinian Theory and Nineteenth Century Philosophies of Science," published in the *Journal* in June 1957.

Foreword

Alvar Ellegård's *Darwin and the General Reader*, first published in 1958, is among the very few books that can actually be said to have appeared before their time. Historians both then and now comment on movements, widely held beliefs, and the tenor of the times on the basis of their own intuitions. These intuitions may well be informed by wide experience and a sensitive reading of the relevant documents, but one thing that they are not based on is extensive, statistical data. Increasingly, historians are coming to recognize the need to base their conclusions on more solid foundations. Ellegård anticipated this trend by more than two decades.

Ellegård's stated purpose for writing his book is to "describe and analyse the impact of Darwin's theory of Evolution on the British public during the first dozen years after the publication of the *Origin of Species*" (p. 5). To do so he scoured 115 British newspapers, magazines, and journals to find articles and reviews that dealt with Darwin's theory. He then grouped these publications into various categories, such as religious leanings (e.g., High Church, Low Church, Roman Catholic, Methodist, etc.), political preferences (e.g., conservative, liberal-conservative, liberal, etc.), and educational levels of their targeted readership (e.g., high, middle, and low).

Some of the results of Ellegård's investigations are not surprising. For example, the number of periodicals expressing opinions on Darwin's theory of evolution peaked in 1860, the year after he published his *Origin of Species*, in 1863 with the appearance of T. H. Huxley's *Man's Place in Nature* and Charles Lyell's *Antiquity of Man*, and again in 1871, the year that Darwin's *Descent of Man* appeared. That Darwin's book on man elicited more responses than his initial book on evolution is also not surprising. That Unitarian and Broad Church periodicals were favorably disposed toward Darwin's theory while Low Church and Methodist publications steadfastly opposed it is to be expected. However, one might not expect organs of the Catholic Church to come in a strong third after the Broad Church.

The preceding observations concern Darwin's general theory of evolution. When it came to the evolution of the human species, only Unitarians and Broad Church periodicals indicated much support, while Catholic publications joined the others in rejecting it almost totally. One of Ellegård's discoveries that has since become common knowledge among Darwin scholars is that even those authors who were favorably disposed to the evolution of species were much less enthusiastic about Darwin's mechanism—natural selection. In fact, it was those who accepted the evolution of species who were most critical of natural selection. One of the paradoxes of Darwin studies is that neither scientists nor the general public would have taken Darwin's theory so seriously had he not provided a detailed specification of his mechanism, but they did not accept the mechanism.

Ellegård discusses four main influences on the reception of Darwin's theory in Great Britain—politics, religion, philosophy of science, and scientific issues. Although Ellegård agrees with Huxley that Darwin's theory was a "veritable Whitworth gun in the armoury of liberalism" (*Westminster Review*, 1860, p. 541), he notes that conservative organs such as *John Bull* and the *English Churchman* initially came out in favor of the theory and only later came to oppose it. As far as religion is concerned, subsequent authors have attempted to play down the conflict between it and Darwin's theory, observing that deeply religious people can be found on both sides of the controversy. However, Ellegård sees religion as an important factor in the reception of evolution. In this context, he discusses the argument from design, miracles, and a literal reading of the Bible. Perhaps Darwin's theory, properly construed, did not conflict with any genuine theological tenets, but in point of fact it threatened the beliefs of millions.

Philosophers of science were faced with a somewhat different problem. Just when they were beginning to investigate the nature of science in earnest with the works of John Herschel, William Whewell, and John Stuart Mill, they were confronted with a theory that seemed to strike at the very foundations of the views of proper scientific method then being promulgated. Ellegård sees an important difference in the reactions of empiricist and idealist philosophers. Empiricists such as Herschel and Mill favored Darwin's theory while such idealists as Whewell opposed it. Subsequent investigations have cast doubt on this conclusion. Perhaps empiricists were more disposed to Darwin's theory than were idealists, but they were not so strongly disposed that they accepted it. Even Mill rejected it in favor of divine creation. The honorific term at the time was "induction," and philosophically sophisticated scientists were nearly unanimous in concluding that Darwin's theory was not sufficiently inductive.

Although Ellegård considers his book to be a social history of the reception of Darwin's theory, he cannot avoid discussing the scientific content of the controversy surrounding the theory—the sterility of hybrids, reversion to type, missing links, the fossil record, the age of the earth, useless structures, the persistence of simple forms of life, the geographical distribution of species, mimetic resemblance, etc. Strangely enough, scientific journals were a good deal less enthusiastic about Darwin's theory in the early years than were those quality organs not especially devoted to science. Ellegård's tentative explanation for this marked difference is that scientists, especially naturalists, "had to overcome a stronger resistance before being able to incorporate the new doctrines with their own established view of nature" (p. 38). In short, scientists knew too much. They realized how extensively accepted scientific beliefs would have to be modified if Darwin was right.

Ellegård addresses all the usual questions about the reception of Darwin's theory, but there is one important difference. When Ellegård ventures an opinion on the impact of some aspect of Darwin's theory, he backs it up with evidence. The evidence might not be as hard or as direct as one might wish, but it is there, and it is more extensive and dependable than has been provided since, even by those authors who think that reference to data in historical studies is worth presenting. Ellegård's appendix in which he presents his statistical analysis of the reaction of the Victorian press to Darwin's theory, his list of periodicals with an index of quotations, and his index of names with biographical notes by themselves make the book worth owning. Nowhere else can one find such an extensive list of references to Darwin's theory in Victorian periodicals.

When I began my own book on the reception of Darwin's theory by the scientific community in 1970, I tried to order a copy of Ellegård's book from Göteborgs Universitets Årsskrift to no avail and was forced to xerox a copy I borrowed through interlibrary loan. My xerox copy through the years has all but disintegrated as I have looked up references and checked citations. I am sure that Darwin scholars as well as anyone interested in Darwin's theory will be as grateful as I am to the University of Chicago Press for making this important book at long last available.

David L. Hull

Preface

The purpose of this book is to describe and analyse the impact of Darwin's theory of Evolution on the British public during the first dozen years after the publication of the *Origin of Species*. It is meant as a contribution to what may be called the social history of ideas. The aim has not been to trace the development of the ideas of Evolution and Natural Selection as scientific concepts; matters which had better be left to historians of science and philosophy. Instead, taking the Darwinian doctrine as something given, we shall study how information and opinions concerning it spread throughout the social fabric of Mid-Victorian Britain. Thus we shall seek to answer questions like the following: What did various classes of people know about the new theory? To what extent did they consider it important? What attitudes did they take towards it? In what way were their attitudes influenced by established ideas and beliefs? What parts of the theory were most vigorously controverted? What repercussions did the controversies have on the disputants' interpretation of the theory, and on their ideology? What arguments were used in the debate, and which of them were considered most important?

In order to answer these questions I have in the main relied on an investigation of the British newspaper and periodical press of the time. The controversy was of course in large part carried on in reviews and articles in the press, or in lectures and discussions which were in their turn reported in the newspapers. But the decisive argument for relying mainly on the press for the purposes of this investigation is that it can be taken to reflect, more accurately than any other source material, the climate of opinion within the various social and ideological groups of the community. Each organ appealed to a definite section of the public, great or small, and did so to the extent that its public could, by and large, accept the stand it took on the various questions of the day.

The timespan covered, 1859—1872, is short enough to allow a fairly complete coverage of the discussion. My material includes well over a hundred titles of periodicals. At the same time, the period is long enough for all the more important arguments and attitudes to find expression, and also for a definite change in the climate of opinion to take place.

Though the present work is primarily concerned with the general public's reception of Darwinism, the opinions of the contemporary scientists could not, of course, be disregarded. In the first place, the most important and most influential contributions to the Darwinian debate naturally came from men who were themselves biologists. In the second place, the scientists were not merely scientists, they were also members of the general public, and were influenced, like everybody else, by various political, religious, and ideological beliefs. It is these influences that especially interest us in the present context: they are also much better studied in the general publications, on which the present investigation is based, than in the purely scientific journals. In the latter, scientists concentrated on details; in the former, where they addressed a wider public, they treated the problems from a more general point of view.[1])

It will become evident that the scientists' attitudes towards the Darwinian theory, like the general public's, were to a large extent determined by ideological factors. On the higher intellectual levels, therefore, the Darwinian controversy inevitably led to discussions on the assumptions underlying scientific explanation. The Darwinian

---

[1]) It is significant that such a purely scientific journal as the *London, Edinburgh and Dublin Philosophical Magazine* contained no single article during the whole of the period 1859—1872 where the Darwinian theory was discussed. On the other hand, the *British and Foreign Medico-Chirurgical* reviewed fully the *Origin of Species*, as well as the *Descent of Man* and other books of Darwinian import.

theory became in fact, in the 1860's, the central point in the debate on the philosophical basis of scientific enquiry, which was a necessary preliminary to the emergence of science, in our own day, as the chief factor of change in human life and thought.

Closely bound up with these philosophical questions was the question of the relation between science and religion, which will take a large share of the discussion on the following pages, as it certainly did in the contemporary press debate. To the general public Darwinism was at least as much a religious as a scientific question. Practically all the attitudes and beliefs that prevail on these matters at the present day can be traced in the vigorous and wide-ranging Mid-Victorian debate. That is one of the reasons why a study of the impact of a theory which has revolutionized our outlook on life and on man's place in nature, and which has had repercussions in practically every sphere of human thought, should have more than historical interest.

Darwin
and the
General Reader

CHAPTER 1

The Background

To the general public in Britain before Darwin, the theory of evolution, or "transmutation", as it was usually called, was associated with a book by an anonymous author[1]) first published in 1844, called the *Vestiges of the Natural History of Creation*. It had been a popular success, and sold no less than 11 editions, or about 24,000 copies, up to 1860. But though the scientifically innocent were fascinated by the author's eloquence and imagination, the book hardly commended itself to scientific readers. There were slips on many points of detail, and as an explanation of the evolutionary process the author offered only a wholly mysterious and unanalysed "law", instituted by the Creator, and working infallibly through the ages to produce an unending series of adaptive evolutionary changes according to a providentially ordained master plan. It is hardly to be wondered at that T. H. Huxley reviewed the book savagely.[2])

The *Vestiges* was an amateurish book, and was of course regarded as such by the scientists. If they occupied themselves with it at all it was only because it made such a stir among non-scientific readers. To the scientists, the chief exponent of evolution was still Lamarck. But his theories, first published in 1802, had received such a rough handling at the hands of the great Cuvier, and in Britain from Sir Charles Lyell, that they were generally recognized as untenable. But though the Lamarckian evolution theory had practically no adherents among British scientists in the 1840's and 1850's, the transmutation theory as such was certainly not regarded as finally refuted. Lamarck had not solved the species problem, but neither had his critics: they had simply by-passed it as insoluble. They placed the origin of species outside the realm of scientific research.

---

[1]) The author was revealed in 1884 as Robert Chambers, who with his brother was editor of *Chambers's Journal*.

[2]) *British and Foreign Medico-Chirurgical Review*, 1854, 425—439.

This way out, however, was hardly satisfactory to an inquiring mind, and we have in fact direct evidence that several British biologists occupied themselves with the problem, even though they did not publish anything as long as they could not contribute anything better than Lamarck. Richard Owen, Britain's leading comparative anatomist, wrote in a quite appreciative letter to the *Vestiges* author: "The discovery of the general secondary causes concerned in the production of organized beings upon this planet would not only be received with pleasure, but is probably the chief end which the best anatomists and physiologists have in view."[3] But though the question was very much in the air, its baffling complexity had to be admitted. T. H. Huxley described the situation as follows: "I imagine that most of those of my contemporaries who thought seriously about the matter, were very much in my own state of mind — inclined to say to both Mosaists and Evolutionists, 'a plague on both your houses!' and disposed to turn aside from an interminable and apparently fruitless discussion, to labour in the fertile fields of ascertainable fact."[4]

Such a state of suspended judgment, or "thätige Skepsis" as Huxley termed it after Goethe, could not, by its very nature, be accepted as final. Inevitably scientists were feeling an inclination to adopt provisional hypotheses of one kind or another.

Now it is a remarkable fact that, at least in Britain, many and perhaps most naturalists were prepared to entertain hypotheses as to the origin of species in which a *supernatural* element entered as an integral part. Among the non-scientific public this view of the question was altogether predominant: life itself, as well as its species, was regarded as directly created by God.

It is important to realise this, since otherwise the whole of Darwin's argument in the *Origin*, and the reaction it called forth, will be inexplicable. Darwin did not contribute much really new information on the variations that occur in species, on their frequency, extent, and inheritance, nor on the causes that produced them. In fact, he repeatedly admitted that those problems remained almost completely unsolved. But in the *Origin* he was not particularly concerned with such details: his main object was to establish a solution of the species problem that would render superfluous any reference to supernatural causes in their

[3] R. Owen, *Life of Richard Owen*, I, 249—250.
[4] Darwin, *Letters*, II, 196—7.

production. A complete ejection of supernaturalism from its strong-hold within biology was necessary to ensure a rational discussion of the fundamental problems of the science. Darwin's Natural Selection theory, which was the revolutionary and radically new element in his doctrine, would achieve this.

That Darwin's contemporaries tended to look upon his theory in this light will be apparent in subsequent chapters. Darwin's own attitude is evident from several indications in the *Origin* itself. Again and again he contrasted his Evolution theory with the view that "elemental atoms have been commanded suddenly to flash into living tissues."[5] That view he took to be the prevalent one among contemporary naturalists: "Authors of the highest eminence seem to be fully satisfied with the view that each species has been independently created,"[6] he wrote in the *Origin*; and in a letter to his friend Sir Charles Lyell he said on the species question, "this branch of science is still with most people under its theological phase of development."[7] Further, at the very beginning of the *Origin* he printed quotations from Whewell and Bacon, and later from Buckland as well, to the effect that religion had nothing to fear from the substitution of a natural for a supernatural or miraculous explanation of the origin of species. Darwin no doubt primarily aimed at reconciling his non-scientific readers with the main result of his book: but it is significant that he thus implicitly represented the contrast between a natural and a supernatural explanation as the main theme of the book.

Historians have regarded this attitude of Darwin's as naïve and some-what parochial, and have spoken of his "amateurish conception of nature."[8] But if Darwin took the supernatural explanations seriously, and directed his power of argumentation against them, it must be recognized that the challenge, at the time, was real. Indeed, it is Darwin's very success that has made the pre-scientific view appear unreal to a later age.

There is no lack of independent evidence that Darwin's assessment of the climate of opinion among his contemporaries on this point was

---

[5] *Origin*, 1 ed. r., 409.
[6] *Origin*, 1 ed. r., 414.
[7] Darwin, *More Letters*, I, 194, to Lyell, 1860.
[8] E. Nordenskiöld, *Biologins historia*, II, 242. English translation, London, 1928.

correct. We have already quoted Huxley, who classed the rival schools as Mosaists and Evolutionists. T. V. Wollaston, who reviewed the *Origin* in a well-known naturalist organ, expressed himself quite definitely: "The opinion amongst naturalists that species were independently created, and have not been transmuted one from the other, has been hitherto so general that we might almost call it an axiom."[9] Should it be thought that this estimate of the relative weight of the two contrasting opinions was partly influenced by wishful thinking on the part of a writer whose opposition to Darwin was to some extent theologically inspired, we may quote in addition Darwin's friend and supporter Asa Gray, the eminent American botanist. His words have the further advantage of explicitly bringing out the supernatural element which the other writers left implicit in the word "created". Gray wrote as late as 1863: "Most naturalists believe that the origin of species is supernatural, their dispersion over a particular geographical area natural, and their extinction, when they disappear, also the result of physical causes."[10]

To the testimony of these naturalists we may add that of the leading British theorist of science of the age, William Whewell, whose *History of the Inductive Sciences* and *Philosophy of the Inductive Sciences* were standard reference books. Whewell's views were certainly not those of every natural scientist, but at least they represent an authoritative statement of the position taken by an influential portion of the scientific world, and on his particular point probably by the large majority. Both in his *History* and in his *Philosophy* Whewell had occasion to examine the transmutation theory in its pre-Darwinian, largely Lamarckian forms, and he later collected extracts from his writings on this subject into a small book, *Indications of the Creator*, specifically directed against the argument of the *Vestiges*. Whewell made no secret of his acceptance of the supernatural as an element in the explanation of nature. In his *History* he put the problem in much the same way as Darwin was to do in the *Origin*: "Either we must accept the doctrine of the transmutation of species, and must suppose that the organized species of one geological epoch were transmuted into those of another by some long-continued agency of natural causes; or else, we must believe in many successive acts of creation and extinction of species,

---

[9] *Annals and Magazine of Natural History*, 5, 1860, 133.
[10] *Annals and Magazine of Natural History*, 12, 1863, 94.

out of the common course of nature; acts which, therefore, we may properly call miraculous."[11])

What appears so remarkable to a later age is that in the mid-nineteenth century scientists could look upon a supernatural explanation as a valid alternative to a scientific one. As Whewell put it: "It may be found, that such occurrences as these are quite inexplicable by the aid of any natural causes with which we are acquainted, and thus the result of our investigations, conducted with strict regard to scientific principles, may be, that we must either contemplate supernatural influences as part of the past series of events, or declare ourselves altogether unable to form this series into a connected chain."[12])

In his discussion of the development hypothesis Whewell applied just this argument. He pointed out that no evolutionist had been able to produce any evidence of one species giving birth to another by ordinary generation, while, on the other hand, there was much evidence of the permanence of species, both at the present time, and through history. Under the circumstances, since supernatural causes were allowed, it was almost inevitable that he should prefer creation to development: "Nothing has been pointed out in the existing order of things which has any analogy or resemblance, of any valid kind, to that creative energy which must be exerted in the production of a new species . . . We are necessarily driven to assume, as the beginning of the present circle of organic nature, an event not included in the course of nature."[13])

It must be stressed, however, that the choice as it appeared to Whewell, and to the large majority of his contemporaries, was not between a purely natural explanation on one hand, and a supernatural one on the other. For none of the pre-Darwinian development hypotheses could do altogether without a supernatural element. The development, if development there had ben, had taken place in orderly fashion, leading to the continuous production of new forms which were all beautifully adapted to their environment. Such a result, one held, could not have been achieved through the chance operation of blind natural laws: one was necessarily driven to assume that the law of development had been

---

[11]) William Whewell, *History of the Inductive Sciences*, Vol. iii, 624. Also quoted in Darwin, *Letters*, II, 194.

[12]) William Whewell, *Philosophy of the Inductive Sciences*, Book 10, Chapter 2, sec. 12.

[13]) Whewell, *Philosophy*, Book 10, Chapter 3, sec. 7.

specially designed by the Creator, who had foreseen all the changes that were to take place, leading eventually to the appearance of the higher animals and man himself. This view was specifically stated by the *Vestiges* author, who quoted Buckland's *Bridgewater Treatise*: "If the properties adopted by the elements . . . adapted them beforehand to the infinity of complicated useful purposes which they have already answered . . . such an aboriginal constitution, *so far from superseding an intelligent agent, would only exalt our conception of the consummate skill and power that could comprehend such an infinity of future uses under future systems, in the original groundwork of his Creation.*"[14])

Lamarck, it is true, sought to explain evolution without any reference to an intelligent Creator, but Whewell was quite justified in insisting that all evolution theories which had been advanced when he was writing were incomplete and unsatisfying to the mind without the assumption of a supernatural intervention at some point in the chain of events. And granted that a miracle had to be admitted at any point, no good reason could be given for not admitting it at other points as well, if the facts could thus be more easily explained.

It was with views such as these that Darwin had to contend. His persistent argumentation against supernatural explanations was not due to a particularly naïve conception of nature, or to an exaggerated concern with the opinions of theologians or the general non-scientific public. The views which he combated were held by the majority of his fellow-naturalists, and by the leading British philosopher of science of the age.[15])

This background also explains why Darwin considered it essential to explain literally everything in the organic world by means of his theory. He repeatedly stated that he would have nothing to do with a law of necessary progression or development: such a law did not provide any genuine explanation, and in fact had to be referred back to the supernatural. It might seem that Darwin saddled himself with unnecessary difficulties when he said, for instance, that his theory would "absolutely break down"[16]) or be "annihilated"[17]) if any single

---

[14]) Robert Chambers, *Vestiges of the Natural History of Creation*, 6 ed. 1847, 158.

[15]) For a discussion of the relation of science to theology in Britain before Darwin, see C. C. Gillispie, *Genesis and Geology*, a book to which the present writer owes much, both for background information and stimulating discussion.

[16]) *Origin*, 1 ed. r., 162.

[17]) *Origin*, 1 ed. r., 172.

form or structure could be shown to exist, which could not be explained by Natural Selection. But Darwin took this stand because he knew that if his theory could be proved insufficient at any single point, his opponents would invoke a supernatural explanation to supply the deficiency, and once supernatural explanations were admitted, it would be impossible to set a limit to them. No natural explanation could be safely established so long as supernatural ones were allowed at all.

The Natural Selection theory was complete without any supernatural element. It could explain the adaptiveness of evolution without assuming either 'directed' individual variations, or a 'preordained' law of development. This was completely new and revolutionary. Once the possibility of such an explanation was accepted — and it was impossible not to accept its validity to some extent at least — resistance to the evolution theory was bound to disappear among the scientifically informed.

At the same time, the naturalistic character of the new theory was also bound to rouse theological opposition to Darwinism in a much higher degree than had been the case with previous evolution theories, according to which, as we have seen, God could still be regarded as directly concerned with and responsible for each new development in the organic world. In fact, as Whewell pointed out, they required that God should be so regarded. Darwin placed God altogether outside the picture. Like Laplace he could say to anybody who asked him where he fitted God into his scheme, "Sire, je n'ai pas besoin de cette hypothèse."

It is clear that Darwin's contemporaries were, in a way, prepared for an evolution theory. But they were not at all prepared for the sort of evolution theory that Darwin actually propounded. This contradiction explains one of the paradoxes of the subsequent development of opinion: though it is practically certain that the evolution theory would not have been established at all if Darwin had not been able to support it by means of the naturalistic theory of Natural Selection, yet the majority of the general public, and a good many scientists, refused to accept the Natural Selection theory, while allowing themselves to be converted to evolutionism. The ideological development was evidently more deeply influenced by traditional attitudes and beliefs than by logical considerations.

CHAPTER 2.

The Climate of Opinion

In this chapter I shall attempt a general survey of the development of opinion as regards the Darwinian theory during the years 1859—1872.[1]) My discussion here is largely based on a statistical analysis of the treatment of Darwinian questions in the contemporary press, for details of which readers are referred to Appendix I—II at the end of this book. I shall try to answer the following questions: To what extent can the press be relied upon to reflect the trend of opinion among the public? What amount of information on the Darwinian doctrine was transmitted by the press to the various sections of the public? What attitude towards Darwinism prevailed in different social groups, and how did attitudes change over the years? These questions will be treated in turn.

The Relation of Press Opinion to Public Opinion

The main assumption, when public opinion is inferred from the press, is that the views expressed in the periodicals corresponded by and large with those of their publics. It is obvious, however, that this assumption is not equally valid in all cases. The degree of correspondence depended both on the nature of the periodical, and on the question on which opinions were expressed. A periodical which was predominantly political would tend to attract readers entertaining similar political views, whereas their opinions on literary matters may have been widely different from the line (if any) taken by the paper on such questions. This implies that readers would tend to give up their

---

[1]) The year 1858, which saw the publication of Darwin's and Wallace's papers in the *Journal* of the Linnaean Society, has not been included in the survey, since the articles passed unnoticed in the general press. I have come across a mention of them in one only of the periodicals I have studied, the *Zoologist*, which reprinted the papers in Vol. 16, 1858, 6293—6308, and published comments in Vol. 17, 1859, 6357—59 and 6474—75.

allegiance to a political newspaper if they constantly found themselves at variance with its politics, but would not consider disagreements on literary topics in the same light.

The closest correlation between press and public on the question which concerns us here, the ideological one of Darwinism, was naturally to be found especially in the weekly, monthly and quarterly Reviews, where the debate on literary and philosophical questions was mainly carried on. Most of these organs tended to identify themselves with a particular ideological position, whether politico-social, philosophical, or religious, and they quite definitely appealed to a similarly minded public.

Even though the assumption that the press expressed the opinions of its readers is not always a safe one, I believe that as regards Darwinism it held reasonably well in most instances. In the first place, Darwinism was a hotly debated question, on which strong views were very often held. It would therefore tend to become a subject on which periodicals would at least be careful not to oppose the majority of their readers. We have in fact direct evidence of newspapers changing their line when their original opinion differed from that prevalent among the type of public they appealed to.[2]

In the second place, there is generally a correlation between opinions on different subjects. We know that attitudes on Darwinism were bound up with politico-ideological and religious opinions. Accordingly, if readers agreed on the whole with the periodical of their choice on those scores they would also tend to agree with it, though perhaps to a less extent, on Darwinism.

In the third place, the periodicals would have some influence on the opinions of their readers, especially if the readers had no other sources of information to go by. On the subject of Darwinism a great many readers, especially of the mass-circulation organs, would be acquainted with little besides what they could read in their own papers. Having no preconceived opinion or attitude, they would be all the more prone to accept the one presented to them by a source with which they were in general agreement on many other issues.

If the argument outlined above is accepted, the first task of the social historian of ideas will be to find out the characteristics of the rea-

---

[2] See below, p. 36.
[3] — — —

ders of the various periodicals: their number, their social status. and their opinions on various questions. Unfortunately definite evidence on this is difficult to come by. We are not, however, completely in the dark.[4]) Though circulation figures were usually kept very secret, they can be estimated with some accuracy, and the characteristics of the readers can be partly inferred from the periodicals themselves. High-priced papers could hardly reach a working-class public, and the cheapest organs were shunned in upper-class homes. Literary and scientific Reviews could appeal only to the educated, while only the uneducated appreciated the cheap weekly journals. The religious organs of various kinds were read chiefly by members of the particular group whose views they stood for. And in general conservative papers appealed to conservatives, liberal papers to liberals. This was also the opinion of contemporaries: "The Tory knew very well without our assistance, that he ought to take in the *Quarterly* — the Whig the *Edinburgh* — the "earnest and decided Reformer" (*alias* Liberal Unattached) the *Westminster* — the Papist the *Dublin* — the modern Scottish descendant of the Solemn League and Covenant-men the *North British* — the English Dissenter the *British Quarterly*, and so forth."[5])

Now even if the main assumption is granted, the inference from press opinion to public opinion would still have to be hedged around with reservations. Above all, it would be necessary to take into account the fact that the readers of the press are not now, and were even less in the 1860's, a representative cross-section of the community, since the higher and more educated classes naturally contained a much higher proportion of readers than the lower and less educated ones. On questions where there was a difference of attitude between the educated and the uneducated — and Darwinism was such a question — an opinion might possess a majority among the press readers, but only a minority among the population at large.

In our investigation, however, we are not so much concerned with the views of the majority of the whole people, as with the views that prevailed within various smaller social and ideological groups, and

---

[4]) For a discussion of the available evidence, see my paper on "The Readership of the Periodical Press in Mid-Victorian Britain". On the approach in general, see also my article on "Public Opinion and the Press".

[5]) *Globe*, 1860, Jan. 17, p. 1.

above all, the correlation between their views on Darwinism and certain other characteristics. For this purpose a knowledge the press opinion of the group is enough, provided it can be assumed to be fairly representative. The circumstance that practically every shade of opinion was in fact represented in the Mid-Victorian press is some guarantee that the periodicals can serve as a reliable basis for an estimate of the trend of thought among the public at large.

A list of the periodicals used in this investigation, together with a brief description of their general characteristics, type of readership, and coverage and attitude in regard to Darwinism, is to be found in Appendix II at the end of this book. I have aimed at selecting all the more influential organs of opinion, and also at having all the main trends of opinion represented. At the same time I have included practically all the big-circulation publications of the time, whether organs of opinion or not. The selection is therefore not a random one, but specially designed to embrace typical representatives both of the educated, articulate sections of the community, and of the less educated and less intellectually alert ones.

The main assumption underlying the whole of my procedure, namely, that the periodicals can be taken, by and large, as representative of the ideas and beliefs of their readers, and thus, with some qualifications, of the population at large, may perhaps be considered as too uncertain for any valid conclusions to be drawn. Admitting, however, the uncertainty of the inference of public opinion from press opinion, it may be of interest to consider what other possibilities there are of evaluating the public opinion of the past.

Present-day public opinion can be measured within a fairly narrow margin of error with the help of a carefully controlled sampling technique. But for the past the ordinary sampling methods are inapplicable, since the historian is limited to the actually available material, which is clearly not a random sample of the total, nor a selected sample of the kind that would satisfy a present-day public opinion pollster. The historian has to rely on written or printed sources only — diaries, letters, reports of meetings of societies, pamphlets, books, newspapers and periodicals. This limitation at once introduces a bias into the material: those people who leave literary remains are by no means a representative cross-section of the community. Even in the Victorian age, the publication of a volume of Life and Letters

implied that its subject was believed to be entitled to a certain measure
of eminence. And the eminences, however minute, are by definition
not the average.

Having recourse only to a very small sample that is not random, the
historian has to select out of it what can be accepted as typical of
the various sections which make up the community. He must be able
to project the evidence on to a wider population than that made up
of the actual writers. Now in the case of published material it ought
to be possible, as we have said, to make such a projection, namely,
from a book or periodical to its readers, or from a lecture or sermon to
its audience. But different kinds of publications are not equally useful
in these respects. Books and pamphlets, in the 1860's, had a much
smaller circulation than the most successful periodicals. Above all,
we are much better informed about the composition of a newspaper's
or periodical's readership than about the readers of a book. A book
stands comparatively alone. It has to find its own readers and win
them by the way it handles the particular subject it treats of. A book
has a restricted theme, and appeals to a public interested in that theme.
But we cannot know what other things the book's readers were interested in. We do not know, for instance, what other books they read,
what other currents of opinion they were exposed to, what other ideas
they entertained. A periodical, on the other hand, is like a personality:
it has views on a whole range of subjects, and covers a wide range of
interests. Further, it has to rely on a fairly constant public in order
to live, and to keep that public it must retain its personality, i. e.
maintain a set course or policy, at least in a community where there
is a vigorous competition between numerous organs. A hundred years
ago, as nowadays, periodicals prospered to the extent that they could
interpret, or represent, or shape, their readers' opinions. Hence the
projection from publication to public can obviously be made much
more successfully for periodicals than for other publications.

A radically different approach to the problem would be to take into
consideration not the primary material — the ideas and beliefs actually
expressed by the contemporaries — but contemporary estimates of
public opinion.

Literature, in the restricted sense of fiction, falls into this category.
Literature "holds up a mirror to the age". The implication is that

the features of the age are represented in its literature. But it is hardly necessary to insist that the mirror is a distorting medium. The artist's picture represents the world as seen through his eyes. Indeed, he may not have meant his picture to represent the real world at all. The historian, if he uses literature as source material, will therefore be faced with the problem of representativeness at one remove. Are the characters in a novel typical of their age or not? Are they typical of the kind of people they seem to portray? Even if the artist has meant his characters as typical of real people, is it certain that he has succeeded? This depends both on his own powers of observation, and on the range of his experience.

The same limitations apply of course to non-fictional writers who seek to portray the age they live in. Journalists, reviewers, essayists, and preachers often indulge in offering estimates of the opinions of their contemporaries. But such statements can seldom be accepted at their face value. In the first place, most of these people have no means of finding out the opinions of more than a small circle, and they are seldom aware of the danger of generalising from their limited experience. In the second place, they are seldom disinterested seekers after truth. The historian soon finds that writers often use assertions about the opinions of the public as weapons in the conflict of ideas in which they themselves are engaged. Pro-Darwinians almost invariably alleged that the leading scientists were favourable to the new doctrines, while anti-Darwinians asserted the opposite. In one version Darwinism was pictured as moribund, in another it appeared as triumphant.

The contemporary estimates, whether implicit, as in imaginative literature, or explicit, as in newspapers and periodicals, must be regarded by the historian as secondary sources. As such they serve as a control on the conclusions that he draws from the primary material. The imaginative literature of the age will not come under consideration in this book, while explicit contemporary estimates occurring in the press will be dealt with in a separate chapter. But in the main, it is the primary material which is considered here. The discussion above will have shown that though the uncertainties of drawing conclusions from press opinion to public opinion are great, they are less formidable than those connected with any other available source material.

Information on Darwinism in the Press

In order to study the dissemination of information on the Darwinian doctrine it is convenient to distinguish between three different parts of it: first, the Evolution idea in its general application to the whole of the organic world; second, the Natural Selection theory; and third, theory of Man's descent from the lower animals.

All through the period studied, the press gave most attention to Evolution in general, while the Natural Selection theory was usually treated very scrappily. The really popular press hardly even mentioned it. It was in fact almost only the readers of the very best quality periodicals that were at all adequately informed about the theory which formed the basis of Darwin's explanation of Evolution, and which distinguished his doctrine from those propounded by earlier evolutionists. It may be that the Natural Selection theory was somewhat out of place in the popular press, but in many organs a contributory cause of the scant attention it received seems to have been the controversies which the theory raised on theological grounds.

Discussions on the application of the evolution doctrine to Man did not go very far in the press before the publication of Darwin's *Descent of Man* in 1871. This is perhaps somewhat surprising, since there is hardly any doubt that it was this question that above all drew public attention to the new theory. In the popular press, especially, the Darwinian theory was often spoken of simply as the "ape theory." But these references were made, so to speak, with bated breath. The subject was not quite respectable, and a full and serious discussion of it was avoided. In fact the bare mention of the ape theory was an effective method for anti-Darwinians to arouse popular feelings against the new views, while pro-Darwinians often shunned the subject in order to concentrate on the less repulsive features of the theory. Relatively speaking, of course, Man's descent did figure prominently in the lowbrow press, where little else of the Darwinian theory was treated at all. In the better quality periodicals, on the other hand, where the Darwinian debate was more substantial, interest was concentrated on the other aspects of the doctrine during most of the 1860's.

Though the Darwinian question was under discussion in some form or other all through the years which we have studied, interest in it was not always equally great. It appears convenient to divide the time-span covered into three periods, the first 1859—1863, the second 1864—1869, and the third 1870—1872.

In the first sub-period, the press discussion reached the height of intensity the year after the *Origin of Species* was published on November 24, 1859. The book caused an immediate stir. The first edition of 1250 copies was sold out the day of publication, and a second edition of 3,000 had to be printed in January, 1860. Such success was indeed remarkable for a book of nearly 500 pages, packed with facts and argument. But it would be a mistake to think that the whole of Britain knew at once that a theory which was to revolutionize our outlook on man and the world had been born. The news reached the literate and educated public, but hardly the masses. The better-class periodicals occupied themselves with the book, but the popular large-circulation organs ignored it.

The newspapers of Mid-Victorian Britain usually contained only brief literature sections. In the better-class papers fairly substantial reviews of the more important books used to be published. The *Times*, for instance, had on an average 100—150 reviews per year, of which many were very brief, while others extended to 5—6 columns. The popular papers, on the other hand, when they carried a literature section at all, mainly offered brief notices only, or short extracts or episodes from the works noticed. The selection of works reviewed was decidedly trashy: attention was chiefly given to popular fiction magazines, cheap novels, and books of travel and adventure.

Under the circumstances reviews of Darwin's book could only be expected in the more expensive newspapers. In fact, the *Times*, the *Morning Post*, and the *Daily News* were alone among the morning dailies to review the *Origin*. Among the weekly newspapers, the semi-clerical *John Bull* contributed a fairly good review, the moribund (tri-weekly) *Saint James's Chronicle* reprinted the *Morning Post's*, and the *News of the World* contained a notice with an extract from the book: on slave-making ants! The most successful and popular Sunday papers, on the other hand — *Lloyd's*, *Reynolds's*, the *Weekly Times*, and the *Weekly Dispatch* — did not notice the book at all, nor did such large-circulation dailies as the *Daily Telegraph* and the *Standard*.[6])

---

[6]) A list of newspaper reviews: *Daily News*, 1859, Dec. 26, p. 2, *John Bull*, 1859, Dec. 24, 827, *Morning Post*, 1860, Jan. 10, p. 2, *News of the World*, 1860,

Some better-class magazines contained quite good articles on the *Origin*. This was a matter of course in *Macmillan's* and *Fraser's*, but even the enormously successful *Cornhill* contributed to the discussion. In most of the popular journals and magazines, on the other hand, no reference was made either to the book or to the theory during the first year. The exceptions were two journals whose intellectual standard was somewhat higher than the average of the class; Dickens' *All the Year Round*, and *Chambers's Journal*. The latter's interest is easily explained: one of its editors was R. Chambers, author of the *Vestiges*.[7])

But if reviews and notices were sparse in the big-circulation newspapers and periodicals, the Review organs, the chief vehicles of public debate of the time, did not miss the significance of the book. All the main weekly, monthly and quarterly Reviews dealt with it, often at great length. The literary world was left in no doubt as to the importance of the new theory.[8])

Jan. 8, p. 6 (slight notice), *Saint James' Chronicle*, 1860, Jan. 10, p. 3 (from *Morning Post*), *Times*, 1859, Dec. 26, 8—9 (Huxley).

[7]) Reviews in magazines and journals: *All the Year Round*, 3, 1860, 293—9, *Chambers's Journal*, 1859:2,388, *Cornhill Magazine*, 1860, 438—447 (G. H. Lewes; *John Bull*, 1859, Dec. 24), *Dublin University Magazine*, 55, 1860, 712—22 (signed D. T. A.), *Fraser's Magazine*, 61, 1860, 739—752; 62: 1860, 74—90 (W. Hopkins; Darwin, *Letters*, II, 314), *Macmillan's Magazine*, 1, 1859—60, Dec. 1859, (Huxley), *Macmillan's Magazine*, 3, 1860—61, Dec. 1860, 81—92 (Henry Fawcett; Darwin, *Letters* II, 299).

[8]) Reviews in weekly Reviews: *Athenaeum*, 1859:2, 659—60, *Critic*, 19, 1859, 528—530, *English Churchman*, 1859, Dec. 1, 1152, *Examiner*, 1859, Dec. 3, 772—3 (John Crawfurd; Darwin, *Letters*, II, 236), *Freeman*, 1860, Jan. 18, 45—6, *Gardener's Chronicle*, 1859, 1051—3, 1860, 3—4 (Hooker; Darwin, *Letters*, II, 267) *Guardian*, 1860, Feb. 8, 134—5, *Patriot*, 1860, Jan. 19, 45, *Press*, 1859, Dec. 10, 1243, *Saturday Review*, 8, 1859, Dec. 24, 775—6, *Spectator*, 1859, Nov. 26, 1210—11, 1860, Mar. 24, 285—6, Apr. 7, 334—5 (the last two by Adam Sedgwick; Darwin, *Letters*, II, 296).

Other Reviews: *Annals and Magazine of Natural History*, 5, 1860, 132—143 (Wollaston; Darwin, *Letters*, II, 275, 284), *British and Foreign Evangelical Review*, 9, 1860, 413—41, *British Quarterly Review*, 31, 1860, 398—421, *Christian Observer*, 60, 1860, 561—74, *Dublin Review*, 48, 1860, 50—81, *Ecclesiastic*, 22, 1860, 82—92, *Eclectic Review*, 3, 1860, 217—42, *Edinburgh New Philosophical Journal*, 10, 1860, 280—89, *Edinburgh Review*, 111, 1860, 487—532 (R. Owen; Darwin, *More Letters*, I, 196), *Geologist*, 3, 1860, 464—72 (F. W. Hutton; Darwin, *Letters*, II, 376), *London Quarterly Review*, 14, 1860, 281—308, *National Review*, 10, 1860, 188—214 (W. B. Carpenter; Darwin, *Letters*, II, 305), *North British Review*, 32, 1860, 455—86 (Rev. Mr. Dunns; Darwin, *Letters*, II, 311), *Quarterly Review*, 108,

It is hardly surprising that the popular press tended at first to pass by the *Origin* in silence. A book costing a pound would find few buyers below the comfortable middle class, and with its somewhat abstruse subject it demanded from its readers both knowledgeableness and intelligence. All the same it ought to be pointed out that Darwin's book received decidedly less immediate attention in the press than the theological *Essays and Reviews*, and later Bishop Colenso's *Pentateuch*.[9]) It is true that these two books led to court proceedings, and thus had additional news value, but in any case there is little doubt that science was no match for religion in the competition for public interest in Mid-Victorian Britain. Nor is there any doubt that it was largely the theological implications of the new theory that eventually won popular fame for it.

The reviews and notices of the *Origin* are a good index of the response to Darwin's new doctrine during the first year. Subsequently, the periodicals and newspapers kept their publics informed of the discussions on the theory both by their original articles and by notices and reviews. The most important of the latter concerned on the one hand Darwin's own books, and books written in direct opposition to or support of Darwin, and on the other hand, the discussions at the British Association, which were generally very fully reported in the press. The Darwinian debates at the Association's annual meetings will be dealt with in Chapter 4.

After the great stir of 1860, interest in Darwinism flagged a little during the next few years, though the extraordinary amount of notice that the popular press took of the African traveller Du Chaillu and his gorillas, in 1861, was obviously more or less consciously bound up with latent Darwinian ideas. What serious debate there was during 1861 and 1862 chiefly arose from the British Association meetings. Darwin's book on the *Fertilization of Orchids* did not cause much controversy in 1862. Reviewers often omitted any comments on the Evolution theory in connection with it.

In 1863 the press debate flared up again. That was chiefly due to

---

1860, 225—264 (Samuel Wilberforce; Darwin, *Letters*, II, 324), *Rambler*, 2, 1859 —60, 361—376 (signed R. S.), *Recreative Science*, 2. 1860—61, Jan. 1861, 268— 277 (Shirley Hibberd), *Westminster Review*, 17, 1860, 541—70 (Huxley), *Zoologist*, 19, 1861, 7577—7611 (George Maw; Darwin, *Letters* II, 276).

⁹) See below, Chapter, 5, p. 106.

two important books which were published early in that year, T. H. Huxley's *Man's Place in Nature*, and above all Sir Charles Lyell's *Antiquity of Man.*[10]) Those two books marked an important stage in the application of the Evolution theory to the human race.

The second of our sub-periods, 1864—1869, was on the whole quieter on the Darwinian front, except in 1868, during which Darwin published *Variation of Animals and Plants under Domestication*, in which was presented part of the evidence on which the discussion in the *Origin* had been built, and the theory of Pangenesis was launched. The year 1868 was also remarkable for the decidedly Darwinian tone that prevailed at the British Association meeting at Norwich.

The third period, 1870—1872, was again one of increasing controversy, caused above all, of course, by Darwin's *Descent of Man* in 1871.[11]) But there were also other substantial contributions to the Darwinian discussion during these years, notably A. R. Wallace's *Contributions to the Theory of Natural Selection*, and St. George Mivart's *Genesis of Species*. Both these books were important in switching the centre of interest from the Evolution theory as such, which both writers advocated, to

---

[10]) See below, Chapters 4 and 14 p. 73, 296.

[11]) Reviews of the *Descent of Man:* Newspapers: *Daily News*, 1871, Feb. 23, p. 2, *John Bull*, 1871, Apr 6, 234, *Observer*, 1871, Mar. 19, p. 3, *Pall Mall Gazette*, 1871, Mar. 20, 1059—60, Mar. 21, 1075—6, (Morley, Darwin, *More Letters*, I, 324.) *Times*, 1871, Apr. 7, p. 3, Apr. 8, p. 5 (Wace; *History of the Times*, II, 451),

Magazines and Journals: *All the Year Round*, 5, 1870—1, 445—450, *London Society*, 19, 1871, 371—4, *Macmillan's Magazine*, 24, 1871, 45—51.

Weekly Reviews: *Athenaeum*, 1871: 1, 275—277, *English Independent*, 1871, Mar. 23, 272—3, *Examiner*, 1871, 233—4, 256—7, *Gardener's Chronicle*, 1871, May 20, 649, *Guardian*, 1871, 935—37, *Inquirer*, 1871, May 13, 295—7, *Lancet*, 1871, Mar. 18, 381—2, *Nature*, 3, 1870—71, 442—4, 463—5 (P. H. Pye-Smith), *Nonconformist*, 1871, Mar. 8, 240—1, *Press*, 1871, Apr. 29, 262—3, *Public Opinion*, 19, 1871, Mar. 11, 297—9 (mainly *Athenaeum*), *Saturday Review*, 31, 1871, Mar. 4, 276—7, Mar. 11, 315—16, *Spectator*, 1871, Mar. 11, 288—9, Mar. 18, 319—20, *Tablet*, 5, 1871, Apr. 15, 455—6.

Other Reviews: *Academy*, 2, Mar. 15, 1871, 177—82 (A. R. Wallace), *British and Foreign Evangelical Review*, 21, 1872, 1—35 (J. R. Leebody), *British Quarterly Review*, 54, 1871, 460—85, *Contemporary Review*, 17, 1871, 274—81 (A. Grant), *Edinburgh Review*, 134, 1871, 195—235, *London Quarterly Review*, 36, 1871, 265—309, *Month*, 15, 1871, 71—101 (signed A. W.) *Popular Science Review*, 10, 1871, 292—4, *Quarterly Journal of Science*, 1, 1871, 248—254, *Quarterly Review*, 131, 1871, 47 —90 (Mivart; Darwin, *Letters*, III, 146; Wallace, *My Life*, II, 10), *Westminster Review*, 42, 1872, 378—400, *Zoologist*, 6, 1871, 2613—24 (Edward Newman).

the Natural Selection theory, in which both found difficulties. Mivart, especially, may be said to have struck the note of public opinion during this period.

Attitudes towards Darwin's theory

It has often been asserted that the first reaction to Darwin's theories was uniformly hostile. That is, however, hardly correct.[12]) Of the newspapers only the *Daily News* took a decided stand against Darwin, treating his book as a mere repetition of the fallacies of the *Vestiges*. The *Times* was favourable indeed: by an incredible stroke of luck, its review came to be entrusted to Huxley, who thus found a golden opportunity to present Darwin's work to its best advantage to a large and influential public. The *Times* was at that time not only by far the most important English newspaper, it had also the largest circulation of all. The other newspaper reviews were fair, and on the whole quite favourable.

The chief literary weekly of the time, the *Athenaeum*, was indeed violently hostile, but the majority of the weekly Reviews were fairly favourable, though cautious and often non-committal. Especially was this the case in regard to the evolution of Man. The attitude towards Natural Selection was one mixed admiration and hesitation: admiration of the ingenuity of the theory, and hesitation as to the extent of its applicability. One did not quite to grant it the power to work such wonders as Darwin claimed for it.

Two of the monthlies supported Darwin, namely, the *Cornhill*, where G. H. Lewes contributed a series of articles, and *Macmillan's*, whose first notice was written by Huxley. *Fraser's*, on the other hand, forcefully disputed the soundness of the new theory.

It was not until the influential quarterlies appeared that the real onslaught came. The Radical *Westminister*, where the indefatigable Huxley was again active, was of course very pro-Darwinian, but in the *Edinburgh*, and especially in the *Quarterly*, the heavy artillery was deployed against Darwin's doctrines. It was an open secret that the *Edinburgh* article was the work of the leading English systematic anatomist, Professor Richard Owen. The *Quarterly* article had been written by Samuel Wilberforce, Bishop of Oxford, who had been primed

---

[12]) A list of the reviews of the *Origin* was given above, footnotes 6—8.

by Owen, and who, with Greats in mathematics at the university, was considered competent to deal with scientific questions.

If the development of opinion on the Darwinian question over the years is to be adequately described, and especially, if different sections of the community are to be compared with each other as regards their attitude in regard to the new doctrines, it is necessary to introduce an ordered classification of these attitudes. Fortunately the positions taken by the disputants in the Darwinian controversy can in fact be arranged along a simple scale, so as to indicate an increasing degree of favourableness towards Darwin's theory, from total rejection to complete acceptance.

The first and lowest position on this scale may be called *Absolute Creation*. It was the fundamentalist religious position, according to which each species arose as a distinct and instantaneous creation, in the literal and naïve sense of the word. Its propounders also generally held that species had been created *en masse* at the beginning of each geological period, and not one by one, in a series exhibiting a gradual increase in complexity of structure.

The latter was the position occupying the second point on the scale, and which I shall call *Progressive Creation*. Those who held this view recognized, on geological evidence, that the further back we go in time, the simpler and the less differentiated are the forms of organic life that we find. There were molluscs before fishes, fishes before reptiles, and reptiles before mammals, with man coming at the very end of the series. On the other hand, one did not admit that the more differentiated forms were descended by ordinary generation from the earlier and less differentiated ones. They were assumed to have come into being in some wholly mysterious way, best characterized by the word creation.

The third position, which I call *Derivation*, after Owen, its chief British exponent, implied a crucial admission, namely, that the progressive evolution was to be explained by means of some Descent theory. But one insisted that the mechanism of generation was only one of the "secondary processes" employed by the Creator. The formation of the new species was still essentially a distinct and unique creative act. There was a supernatural element in it, only it consisted in changing the development of an embryo rather than in forming a whole new creature out of nothing, or out of the dust of the earth. The ordinary processes of generation came into play, but something else was superadded. There was also a tendency to regard sudden and spectacular mutations between parents and offsprings as furnishing a clue to the mystery of the origin of species.

The fourth position, to be called *Directed Selection*, allowed still wider scope to the agency of natural processes. Those who held it admitted the efficacy of Natural Selection for a considerable amount of specific differentiation, while referring to unknown factors in order to explain the more important evolutionary steps. Alternatively, unknown factors were conceived as directing the variations as between parents and offspring, whether big or small, along a pre-determined, beneficial course. The distinctive feature of the position was that it still included a teleological element as an indispensable part of the explanation of organic evolution.

The fifth and highest position, *Natural Selection*, implied acceptance of the distinctive element of Darwin's theory: a scientific, and therefore non-teleological and non-supernatural explanation of the evolution of the whole organic world.

The above five positions form, as we said, a continuous series. As we rise on the scale, less and less of the processes going into the formation of species were recognized as supernatural, or outside the range of ordinary scientific explanation. Anybody accepting a position with a higher number accepted *ipso facto* all the scientific explanations already granted by those holding a lower position. But though the five positions just described are sufficient to define the degree of acceptance of the Darwinian theory in its application to the organic world below man, a separate and additional classification must be used to cover the attitudes towards the development of Man. Some people were prepared to accept the Descent theory for the animal and vegetable kingdoms, while rejecting it for the human race. And even if man's body could be given over to the evolutionists, many, — in fact, the majority — refused to admit the same for man's immortal soul, which was to be regarded as a miraculous gift from heaven. In regard to man therefore, three independent positions could be discerned. The first, which we may call *Separate Creation*, took the whole of man, body and soul, to be created independently of the rest of the organic world. The second position, *Mental Creation*, was that man's body had evolved in the same manner as that of the lower animals, while his soul had been created separately. The third position, *Development*, admitted the gradual evolution of both body and soul.

Separate Creation for man could be combined with any of the five positions described above as regards evolution in general, whereas the second and third positions for man naturally combined only with the upper three of the general positions. Usually, however, Separate Creation for man went with Absolute Creation, Progressive Creation, or Derivation, while Mental Creation tended to combine with Directed Selection, and Development with Directed Selection or Natural Selection.

Against the classification made above the objection may be raised that it makes the issue too much one of religion versus science, of supernaturalism versus naturalism. The answer is that this was how the problem in fact appeared to the contemporaries. My classification has not arisen out of *a priori* considerations; it is based on the actual positions taken in the contemporary press.

When this classification is applied to the actual material, an undramatic but distinct change in the climate of opinion can be discerned during the years covered by our survey. What happened may be briefly described by saying that the majority of the general public were in the end prepared to accept the Evolution part of Darwin's doctrine, at least for the organic world below man, while they rejected Darwin's explanation of it, namely, the theory of Natural Selection. In our first period, up to 1863, most people seem to have clung to the lowest two positions, Absolute Creation or Progressive Creation. In the last period, on the other hand, the most popular positions were the fourth and third: Directed Selection an Derivation. They were both middle-of-the-road positions, paying homage both to science and to religion. Moreover, they could be easily assimilated with the old *Vestiges* views, for which the ground had been long prepared among the broad public. Position five, Natural Selection, implying full acceptance of the Darwinian theory, was all the time granted by a small minority only. It evidently met with strong resistance. But it took time for the resistance to become organised and articulate, and during our second period, 1864—1869, Natural Selection clearly gained ground among the scientifically informed, who tended to give at least a provisional assent to it. In the last period, however, the number of adherents to position 5 seem to have declined. This was probably

chiefly due to the influence of Mivart's *Genesis of Species*, where the
arguments against Natural Selection were put forth ably and effectively. His influence was acknowledged by Darwin himself. "It quite
delights me," he wrote to Huxley, "that you are going to some extent
to answer and attack Mivart. His book, as you say, has produced a
great effect; yesterday I perceived the reverberations from it, even
from Italy . . . The pendulum is now swinging against our side, but I
feel positive it will soon swing the other way; and no mortal man will
do half as much as you in giving it a start in the right direction, as
you did at the first commencement . . . It will be a long battle, after
we are dead and gone."[13]) The swing of the pendulum, however,
concerned only the Natural Selection theory: Evolution as such was
steadily gaining ground. Among the educated, in fact, the battle for
it was virtually over by 1870.

In regard to man, resistance to Evolution had also decreased
considerably from 1859 to 1871, though naturally not as much as in
regard to the lower organic world. Nor did the development follow
quite the same pattern. Among the broad public, especially, complete
rejection of the theory of Man's descent remained by far the most
common position until the very end of our period of survey, in spite
of the advance towards general evolutionary views going on all the
time. Even at the end, only a small minority were prepared to accept
the gradual development of man's soul.

Seeing that the general development of the climate of opinion was
one of gradual advance towards a more pro-Darwinian stand, the
question arises whether any significant differences existed between
the various sections of the public in regard to their attitude on Darwinism. The answer is quite definitely yes. There were differences, and
they were bound up with such factors as education, ideology and
religion.

Educational Factors

By and large, the lower the educational standard, the less was the
inclination to accept the Darwinian doctrines. The periodicals appealing
to educated readers were fairly consistently more pro-Darwinian (or
less anti-Darwinian) than the mass circulation organs with a less
educated readership. It is true that the mass circulation organs, then

---

[13]) Darwin, *Letters*, III, 148—149.

as now, had a tendency to play safe in controversial matters. This tendency would naturally lead them to prefer established, traditional views. As one of the most successful of them said in an advertisement, "There is not a line, not a word, in the columns of the *London Journal* which unfits it for the perusal of the young and innocent."[14] Such a set policy undoubtedly partly explains why many of these organs avoided discussing Darwinism altogether. But it does not wholly invalidate the conclusion from press opinion to public opinion in this instance. In the first place, the difference between educated and uneducated does not only come out when the mass circulation organs are compared with the quality press, but also, though naturally not so markedly, when "highbrow" and "middlebrow" periodicals are compared. In the second place, the public of the mass circulation organs had few other sources of information open to them on these matters, so that to the extent that they cared to have any opinion on them at all, they would have little reason to differ from the one that was offered them. It is also a noteworthy fact that the periodicals which contributed most to the Darwinian discussions were also in general more favourably disposed towards the theory. Towards the end of the 1860's, in fact, it was practically impossible to be adequately informed of the doctrine and yet fail to admit its truth at least to some extent.

It may perhaps cause some surprise that among the quality organs the general periodicals were at first on an average more accommodating towards Darwin than those which specially emphasized science and natural history. The difference in this respect was quite marked during the first years. The explanation may be that the scientists — or rather, naturalists — had to overcome a stronger resistance before being able to incorporate the new doctrines with their own established view of nature. A contributory factor may have been that naturalists, when addressing the general public — and we are here only concerned with productions of this sort — had long been in the habit of mixing up their science with a conventional sort of natural religion. That religious prepossessions had some share in determining their attitude is further supported by the fact that the lag of the scientific group of periodicals was consistently greater for man than for Evolution in general. The lag disappeared, however, after the first few years. Once

---

[14] *London Journal*, 1858, preface.

the scientists had overcome their initial conservatism in favour of their established views, they could recognize the scientific advantages of the Darwinian doctrines.

Politico-Ideological Factors

It has been said that Darwinism has been "a myth of the left."[15] This was also the opinion of many of Darwin's contemporaries. G. H. Lewes, for instance, wrote in 1861: "The hypothesis is clamorously rejected by the conservative minds, because it is thought to be revolutionary, and not less eagerly accepted by insurgent minds, because it is thought destructive of the old doctrines."[16] Similarly, Huxley held that "every philosophical thinker hails it as a veritable Whitworth gun in the armoury of liberalism."[17] It seems that these generalisations contained a large measure of truth. Judging from the press, the conservative and the liberal sections of the public differed strikingly in their attitude towards Darwinism, though the difference was not at all so marked in the highest educational group as in the middle and lower ones. This is perhaps not surprising. The political bias could be given greater scope when the need to take factual evidence into consideration was less. In the best quality Reviews, appealing to an educated and intellectually alert public, the Darwinian theory could be treated mainly on its scientific merits. But in the large-circulation middlebrow organs the doctrine was largely used as a weapon in the political and ideological warfare.

Darwin's theory undoubtedly had a dissolving effect on established beliefs. Liberals and radicals, standing in principle for innovation and experiment, did not object to this tendency, whereas conservatives naturally both feared and detested it, especially when considering its possible effect on the lower classes. Illustrations of this way of reasoning will be found in subsequent chapters.

One remarkable feature is the adverse trend in the development of opinion on Darwinism in part at least of the conservative section of the public, a trend which was opposed to the general tendency of the age, and to the development in practically all other sections of

---

[15] E. g. Max Rouché, "Herder précurseur de Darwin", *Publications de la faculté des Lettres de l'Université de Strasbourg*, Fasc. 94, 1940, p. 9.

[16] *Blackwood's Magazine*, 89, 1861, 166.

[17] *Westminster Review*, 17, 1860, 541.

the public. What happened was that several conservative or liberal-conservative organs, which at first had received the *Origin* favourably, later changed their views to conform with what was prevalent among leading conservative circles, as represented above all by the *Quarterly Review* and by several High Church organs. There are several instances. The *Times* could hardly be expected to recognize as its own the views Huxley had propounded in its columns. Gradually the normal spirit of the paper asserted itself in later pronouncements on Darwinian questions, and when the *Descent of Man* appeared, the *Times* review was violently hostile.[18A] A similar development occurs in the *Morning Post, John Bull*, the *Press, St. James's Chronicle* and the *English Churchman*.[18B] All these conservative organs had started by reviewing Darwin favourably, but ended up among his most decided opponents.

Religious Factors

The connection between Darwinism and religious opinions was clearly recognized by contemporaries. The theory, said one commentator "has been vehemently abused, and not less extravagantly commended, by illogical and intemperate partisans on both sides, who supposed it could affect the truths of the Christian religion."[19]

The nature of the religious opposition to Darwinism will be fully discussed in subsequent chapters: we are here only concerned with the general difference between the religious section and the rest of the public in their attitude towards Darwinism. There was such a differance; but it is necessary to distinguish between the various religious groups. There were consistent differences between them as well.

The obvious line of division in the religious community was between Church of England and Dissent. But further subdivision is desirable and possible. It was customary at the time to distinguish between three Church of England groups, namely, High, Low and Broad. The majority of the High Church group represented a conventional, middle of the road position in theology, whereas an extreme fringe was made up of the Anglo-Catholic, Ritualistic sections. The Low Church group was much more decidedly Protestant and Evangelical in theology,

---

[18 A] *Times*, 1871, Apr. 8, p. 5.
[18 B] See under those names in Appendix II.
[19] *Illustrated London News*, 1871, Mar. 11, 243.

giving more emphasis to the individual Christian's direct study of the Bible than to the theology of the Church. In this respect, as well as in its social position, the Low Church group was approaching to the Dissenting sects.

Both High Church and Low Church were conservative in religious matters, and also in several other respects. Broad Church stood for neology and even to some extens for rationalism in theology. It was also associated with political liberalism. Disraeli's determined attack on the Broad Church in his famous Oxford Speech on November 25, 1864, is only one indication of this.[20]) Among the clergy, Broad Church views seem to have been less common than among the religiously interested laymen: there was no purely clerical Broad Church journal. On the other hand, the High Church group included a fair proportion who were willing to follow the road which the leading Broad Churchmen were exploring. The success of the weekly *Guardian*, where latitudinarian views were treated with respect, is evidence of this.

Among the Dissenting groups a distinction has to be made in the first place between Roman Catholics and Protestants. The Roman Catholics consisted of two fairly distinct strata: an old, upper class English Roman Catholicism, and a newer, working class Roman Catholicism among the Irish immigrants in the North, and to some extent in London. Theologically there was an obvious similarity between Roman Catholicism and High Church Anglo-Catholicism.

As for the Protestant Dissenters we may place on one side the small Unitarian group, whose main characteristic was a thoroughly rationalistic approach to theological questions. Indeed, they were hardly recognized as Christians by the other sects. At the other end there were the Methodists, who were the largest Dissenting group in the country, connected both historically, theologically, and socially with the Evangelical section of the Church of England.

The remaining, older Dissenting sects — Congregationalists, Baptists, Presbyterians, and Quakers — may be brought together into one omnibus group. Theologically as well as politically, they seem to have been somewhat more liberal than the Methodists. The Congregationalists, the biggest of these old Dissenting groups, were sometimes referred to as the Broad Church of Dissent.

The attitudes towards Darwinism in the various sects reflects with

---

[20]) See below, Chapter 14, p. 295—6.

remarkable accuracy their ideological position. As we should expect, Unitarians and Broad Church people were much less anti-Darwinian than the other religious groups. At the other end of the scale were the Methodists, where Evangelical fundamentalism, political conservatism, and a fairly moderate educational standard all combined to produce a strongly anti-Darwinian attitude. The Low Church group had many of the same characteristics, and in relation to Darwinism they were equally hostile. The other Dissenting groups took a less adverse stand, though their attitude was much nearer to that of the other Evangelical Protestants than to the latitudinarian sections. The two ecclesiastical groups, High Church and Roman Catholics, both took a middle position. Temperamentally they were averse to innovations; on the other hand, relying on a much more developed theology than the others, on the experience of centuries of ideological argument, and on the ability of the Church rather than the individual to solve Scriptural perplexities, they could afford to give greater scope to new physical theories than their Evangelical brethren.

All the religious groups were considerably less accommodating in their attitude towards the descent of Man than in regard to Evolution in general, which is hardly surprising. The discrepancy was especially great in the Roman Catholic group. That was probably at least partly due to the distinction, which many Fathers of the Church had laid down as absolute, between men and animals — culminating in the Cartesian doctrine of animals as automata. A good exponent of this line of reasoning was Mivart, himself a Roman Catholic convert, who expressly differentiated between man and the lower animals in regard to evolution.

CHAPTER 3

The Press on the Progress of Darwinism

The previous chapter, together with Appendix I—II, have provided the groundwork for an estimate of the development of opinion with regard to Darwinism in the '60s and early '70s. In this chapter we shall approach the same subject from another angle, and consider, not the information actually provided, nor the opinions actually expressed, but contemporary *estimates* of these matters.

As contemporaries can rely on both written and oral evidence in order to assess the state of opinion within their community, it might seem that they ought to have an advantage over the historian, who must needs confine himself to written sources. But this advantage is in reality more than compensated for by the fact that contemporaries seldom go to the same lengths as the historian in order to obtain truly representative evidence. In the present instance it is quite clear that they did not: the estimates which we are going to discuss on the following pages are obviously impressionistic. They were not, and did not claim to be, based on extensive research. We must in fact always suspect them of being influenced by the writer's necessarily limited experience. When they purported to refer to the state of opinion within the community at large they were likely to be illegitimate generalisations.

Even more important than the limitations due to carelessness, ignorance or naïveté, is the fact that in most cases contemporary estimates of public opinion were made in order to serve a purpose. Darwinism was a highly controversial subject, and all sorts of arguments were used in the ideological battle around it. One of these arguments— perhaps the most effective among non-specialists—was to represent one's own side as strong and triumphant, and the opposing side as weak and divided. Truth, like victory, is on the side of the big battalions.

It is in fact chiefly when considered from this point of view that these contemporary estimates are of interest. Taken by and large, letting contrary opinions cancel each other, they yield indeed the same picture of the progress of Darwinism as we arrived at in the preceding chapter, and so provide some confirmation of its correctness. But the material presented here serves above all to exemplify the *bias* of the various writers. Assuming that the direct evidence already discussed gives a tolerably good estimate of the true state of opinion in the various social groups, it is easy to see how far, and in what way, each writer's assessment diverges from the truth. In this manner we can judge the nature and extent of his bias. Not unnaturally the bias proves to be systematic. Pro-Darwinians tended to exaggerate the strength of the Darwinian faction, the anti-Darwinians to play it down.

As before, we shall treat separately estimates of the extent of information, and estimates of attitudes. It is above all as regards the latter that the systematic bias reveals itself, and therefore it will naturally be given most attention. We shall distinguish between estimates of public opinion in general, and the opinions of the scientific world. The names of prominent individual scientists were of course wielded as weapons in the propaganda battle: this subject also deserves separate treatment.

Information on Darwinism

Darwin had no reason to complain of the interest the *Origin* aroused. On the contrary, his belief, expressed in a letter to a friend,[1]) that it would be "popular to a certain extent", proved an understatement. The fairly extensive sale of the *Origin of Species* itself, as well as the amount of attention given in the press to it, and to discussions arising out of it, is proof enough that his book was a literary sensation. Reviewing the *Origin* in the *Westminster*, in April, 1860, Huxley described the position as follows:

> "Overflowing the narrow bounds of purely scientific circles, the 'species question' divides with Italy and the Volunteers the attention of general society. Everybody has read Mr. Darwin's book, or, at least, has given an opinion on its merits or demerits; pietists, whether lay or ecclesiastic, decry it with the mild railing which sounds so charitable; bigots denounce it with ignorant invective; old ladies, of both sexes, consider it a decidedly dangerous book, and even savans, who have no better mud to throw, quote antiquated writers to show that its author is no better than an ape himself; while every philosophical thinker hails it as a veritable Whitworth gun in the armoury

---

[1]) Darwin *Letters*, II, 153, to J. D. Hooker, April 1859.

of liberalism, and all competent naturalists and physiologists, whatever their opinion as to the ultimate fate of the doctrines put forth, acknowledge that the work ... is a solid contribution to knowledge which inaugurates a new epoch in natural history."[2])

Similar views were expressed by several other organs. The *Saturday Review* wrote that the controversy excited by Darwin's book had "passed beyond the bounds of the study and lecture-room into the drawing-room and the public street."[3]) And in the new and successful *Cornhill* G. H. Lewes asserted that "Darwin's book is in everybody's hands."[4])

These expressions obviously overstate the case. Four thousand copies was indeed a big sale for a scientific book, but it naturally did not reach the broad public, nor more than a small proportion even of the educated classes. These latter, it is true, knew of the book through reviews and notices in the better-class periodicals: readers of the mass-circulation organs did not get even that. The book was indeed a sensation, but it must be kept in mind that it was a literary and scientific sensation, not a popular one. The periodicals from which the above quotations were taken chiefly addressed educated readers: the writers' generalisations hold at most for their own public. In these matters, of course, one might admit that theirs was the public that counted. The readers would of course be flattered by that implication.

Other writers in the press were more careful to indicate just what portion of the public they had in mind. Several early reviewers ventured to forecast the sort of reaction that the *Origin* was to meet with. The *Critic* found it likely to "create more stir in the world of science than anything which has appeared since 'The Vestiges'",[5]) and the *Examiner* also thought that the book was "sure to make a mighty stir among the philosophers".[6]) The *News of the World* added, "perhaps even among theologians".[7]) These writers clearly did not expect the question to be of any concern to the man in the street. On the whole, journals which appealed chiefly to the less educated portions of the public

[2]) *Westminster Review*, 17, 1860, 541.

[3]) *Saturday Review*, 9, 1860, 573.

[4]) *Cornhill*, 1, 1860, 603. See also: *Press*, 1860, 1207; *Nonconformist*, 1860, Dec. 26, 1035, *Daily Telegraph*, 1860, July 26, p. 3.

[5]) *Critic*, 1859, Nov. 26.

[6]) *Examiner*, 1859, Dec. 3, 772.

[7]) *News of the World*, 1860, Jan. 8.

reported the sensation caused by the *Origin* as taking place in another world than their own. It belonged to the world of "science", "philosophers", "thinkers".[8]) It is a significant fact that *Punch*, a faithful mirror or ordinary middle-class culture, did not take up the matter of Darwinism until November 10, 1860, and then only in passing.

The above only confirms what we found in the previous chapter, namely, that the low-brow press in general ignored the Darwinian debate, while the highbrow press devoted much space to it. The *Spectator* summed up the position fairly accurately when it said that "Mr. Darwin has made the origin and succession of living species the leading philosophical question of our day."[9]) That puts the matter in the right perspective. The class of people who took an interest in such questions was well defined by the *Edinburgh Review* when it stated that the *Origin* had been "perused with avidity, not only by the professed naturalist, but by that far wider intellectual class which now takes interest in the higher generalisations of all the sciences."[10]) Another writer referred to them as "the scientific and semi-scientific circles"[11]).

It is evident, however, that interest in the Darwinian question was not only scientific, it was in many cases primarily theological or semi-theological. The religious communities had the new theory thrust upon them very much against their own will, but it was impossible to adopt an isolationist attitude, for, as one religious organ said, "so far as relates .. to the more educated sections of our congregations, to calculate upon their ignorance of these modern heresies would be most unwise."[12]) And it is natural that they should adopt a somewhat acrid tone in commenting on the spread of interest in the new theory: "If notoriety be any proof of successful authorship, Mr. Darwin has had his reward."[13])

In the *Origin* Darwin said next to nothing about the descent of man, and though that subject was usually brought up in reviews and other

---

[8]) *Chamber's Journal*, 1859, Dec. 31, 431, *Illustrated London News*, 1860, Feb. 11, 139, Feb. 18, 163, Nov. 10 449, *Temple Bar*, I, 1861, 541; *John Bull*, 1861, Jan. 5, p. 11.

[9]) *Spectator*, 1860, 1172.

[10]) *Edinburgh Review*, III, 1860, 488.

[11]) *Blackwood's Magazine*, 1861, 165.

[12]) *Christian Observer*, 1861, 332.

[13]) *North British Review*, 32, 1860, 455.

discussions, it was on the whole kept in the background at first. But it is clear that it was this aspect of the theory that all along chiefly interested the broad public. "The Darwinian theory would lose half its interest with the public if it did not culminate in a doctrine on the origin of the human species", said one religious organ,[14]) and a scientific journal admitted that this implication had "enlisted the sympathies of many who otherwise take little interest in natural history"[15]). And it is this which explains why the low-brow press gave almost as much attention to Darwin in 1861 as in 1860: that year saw the publication of Du Chaillus's book of African travels, where the existence of the gorilla was for the first time brought to the notice of the general public in England. Every newspaper and magazine carried stories of what was called man's nearest relation, and though both Du Chaillu and most of the press expressly repudiated the Descent theory, there is no doubt that precisely as Darwin had prepared the way for Du Chaillu's success, so Du Chaillu's book drew fresh attention to Darwin's doctrine, which was by now currently described as the 'ape theory'. Again we may use *Punch* as an indicator: while the journal had almost wholly ignored Darwin in 1860, it made the gorilla and its relationship to man one of the chief features of 1861.[16])

Contemporary estimates of the state of information grow markedly sparser in the years after 1860-1: the novelty had worn off, and some knowledge and interest was taken for granted. What little material there is confirms the conclusions we drew in the previous chapter about the fluctuations in the interest given to Darwinism. Both the peak in 1863[17]) and the flagging interest in 1864-5[18]) are attested by contemporaries.

By the end of the 'sixties Darwin's name, and the terms of Evolution and Natural Selection had been established firmly in the minds of the public at large. In 1868 the *Saturday Review* was able to claim about the *Origin* that "so rapid has been the hold that it has taken on the public mind, that the language incident to the explanation of the 'struggle for life' and the gradual evolution of new forms consequent

[14]) *Literary Churchman*, 6, 1860, 393.
[15]) *Edinburgh New Philosophical Journal*, 14, 1861, 129.
[16]) *Punch*, 40, 1861, Preface, 206; 41, 1861, 64, 244, 245, 257. Below, p. 295.
[17]) *Geologist*, 1863, 77.
[18]) *Temple Bar*, 15, 1865, 237; *Punch*, 47, 1864, 89.

thereon, has passed into the phraseology of everyday conversation"[19]). It is true, of course, that knowledge of the theory did not, even among educated readers, go much beyond mere words and phrases, and the leading Church weekly Review could rightly ask those who criticised the clergy for ignoring the latest developments in science—"Is it true that clergymen are so much behind the rest of the world in scientific knowledge? Are laymen in general so well informed on these topics?"[20]) Very few indeed were really well informed about the Darwinian theory of Natural Selection: the general public tended to equate Darwinism with Evolution pure and simple. But Evolution, however it was interpreted, had taken a hold on people's mind thanks to Darwin. The *Edinburgh* summed up the position in 1871:

> "Since the publication of the 'Origin of Species' in 1859, no book of science has excited a keener interest than Mr. Darwin's new work on the 'Descent of Man.' In the drawing-room it is competing with the last new novel, and in the study it is troubling alike the man of science, the moralist, and the theologian."[21]).

Darwin had come to stay.

Attitudes on Darwinism
The battle of authorities

Contemporary estimates of the amount of interest given to the Darwinian question were seldom very wide of the mark. When attitudes were reported, on the other hand, the bias at once became marked. It was common form to allege roundly that the chief authorities on the the subject were on one's own side: we shall here first deal with those writers who mentioned by name the authorities they considered of most weight in the context.

This game of claiming the support eminent authorities seems to have been played more often on the anti-Darwinian than on the Darwinian side. This is easily explained. Evolutionism before Darwin had meant chiefly Lamarckian "Vestigianism", and almost every British naturalist of any note during the period 1830—1859 was on record against it. Only a few of these older authorities were converted to Darwin's views; those who did not publicly acknowledge their change of opinion could

---

[19]) *Saturday Review*, 1868, April 11, 491.
[20]) *Guardian*, 1868, Dec. 23, 1425.
[21]) *Edinburgh Review*, 134, 1871, 195.

therefore be freely cited against him. Thus the pro-Darwinians had at first a much less extensive choice of authoritative names: they would have to wait until the new generation had established itself.

The best that the pro-Darwinian organs could do under the circumstances was to represent Darwin's supporters as more or less equal in strength with his opponenets. *Macmillan's*, for instance, late in 1860 held that "A Darwinite and an anti-Darwinite are now the badges of opposed scientific parties. Each side is ably representend".[22 A] And it cited four names from each camp: for Darwin were T. H. Huxley, J. S. Henslow, J. D. Hooker and Sir Charles Lyell. Against Darwin were Richard Owen, W. Hopkins, Sir Benjamin Brodie and Adam Segwick, Darwin's old teacher.[22 B]. But this balance-sheet was not left unchallenged. Henslow wrote a letter to the journal, protesting against his inclusion among the Darwinians, though he admitted that he believed Darwin's theory was "a stumble in the right direction."[23] His letter caused one anti-Darwinian organ to comment: "We believe that many names have been made use of in different quarters, as supporting the theory, quite irrespectively of the consent of the owners."[24] Mistakes surely sometimes occurred—on both sides—but in general writers naturally avoided citing names without good evidence: those who could not find enough names to support their claims chose instead to make unspecific assertions, as we shall see below.

Anti-Darwinian organs ordinarily cited only their own authorities: Darwinism was representend as supported by cranks only. Thus the Roman Catholic *Rambler*, in a brief notice of Darwin's forthcoming book, wrote,"This is the theory of the author of the Vestiges of Creation, and before him of Lord Monboddo. It has hitherto been discredited by the authority of Humboldt, Professor Owen, Forbes, and others."[25] Many organs, especially religious ones, continued in this vein for a long time. The *Leisure Hour*, in 1863, was indignant that such "rubbish" as Darwinism could still be propounded in lecture rooms "after all that

[22 A] *Macmillan's Magazine*, 3, 1860—1, 81.

[22 B] For information on the scientists and other authorities mentioned, in this chapter and elsewhere, see Index.

[23] *Macmillan's Magazine*, 3, 1860—1, 336.

[24] *Literary Churchman*, 7, 1861, 74.

[25] *Rambler*, 1859, 110.

[26] *Leisure Hour*, 1863, 192.

[27] *British Quarterly Review*, 41, 1865, 143.

has been written on the subject by Brewster, Owen, Hugh Miller, and other men of science."[26]) And as late as 1865 the *British Quarterly Review* declared that "it is only here and there that a second-rate naturalist will sympathize at all with such dreamy views."[27]) The journal cited a whole list of authorities against Evolution, including not only such names as Cuvier, Owen, Agassiz, Sedgwick, and Hugh Miller, but also W.R. Carpenter and Lyell, both of whom were in fact by that time favourably inclined towards Evolution.

Organs appealing to a better-informed public had to weigh their words more carefully. From 1861 onwards, it became increasingly difficult to pretend that the world of science in general regarded Darwin's theory as a mere flash in the pan. Reviewing a book by the geologist John Phillips, the *Annals and Magazine of Natural History* wrote that "considering that the Darwinian chariot has been accompanied on its course by such shouts of triumph from the supporters of the theory contained in it, it may be some consolation to the more sober-minded of our readers to learn that one at least of our naturalists ... conspicuous for his extensive and accurate knowledge ... does not hesitate to come forward in support of ... old-established views."[28]) Darwin's supporters could not be completely ignored, and it was difficult to impugn their scientific ability. When a Dublin professor of geology had alleged that Darwinians must be "untrained in the use of the logical faculties"[29]) A.R. Wallace answered, "This is the judgment of the Rev. S. Haughton on such men as Lyell, Hooker, Lubbock, Huxley, and Asa Gray."[30])

Towards the end of the 'sixties pro-Darwinian writers in the press usually claimed that Darwin's victory was complete among the experts. *Nature* said:"Since the publication of the 'Origin of Species', we may say that almost the whole body of the younger naturalists of this country and of Germany —Von Baer, Huxley, and Haeckel leading the way after Darwin and Wallace—have given in their adhesion to the doctrine of Evolution."[31]) The *Examiner* wrote in a similar vein: "Anti-Darwinianism has been very unfortunate with the experts. Our best botanists, Hooker and Bentham, the greatest living geologist, Sir Charles Lyell, and a crowd of great names in physiology and zoology, might be quoted.

[28]) *Annals and Magazine of Natural History*, 7, 1861, 404.
[29]) *Annals and Magazine of Natural History*, 11, 1863, 428.
[30]) *Annals and Magazine of Natural History*, 12, 1863, 308.
[31]) *Nature*, 3, 1870—1, 270.

The best testimony to the success of Mr. Darwin's views is Dr. Bree's [an anti-Darwinian] statement that our learned societies have 'had the tone of their publications gradually changed into the phraseology and teachings of an unproved hypothesis'."[32]

The change in the climate of opinion was undeniable, and the tone in the anti-Darwinian organs was modified. Hence in 1869 the *Athenaeum* did not venture to suggest outright that the majority of the experts were against Darwin, but in criticising the historical sketch that Darwin had appended to the later editions of the *Origin*, the journal did what it could to leave the reader with that impression: "No account is given [by Darwin] of any hostile opinions. This fact is very significant. The refutations of Cuvier and De Blainville, Agassiz and Flourens, and other notabilities, are . . . omitted by Mr. Charles Darwin . . . In his ballot, most of the Noes are left out; the Ayes are carefully counted, and more are counted than can bear a scrutiny."[33]) A religious organ, noticing the trend of opinion, exclaimed in horror, "Were Lucretius, Horace, Hume, Monboddo, followed by the author of the 'Vestiges', and by Darwin and Huxley, right? Were Moses, Plato, Lord Bacon and Hugh Miller wrong?"[34])

But though the authorities that could be cited against the development theory were becoming fewer, Darwin's opponents in the press had one consolation. The evolutionism of the experts, more often than not, was not Darwinian. Accordingly, in the better-class organs at least, one chose to acknowledge Darwin's victory on the score of Evolution, while carefully emphasizing the criticism that many experts directed against his Natural Selection theory. A leading Church of England organ admitted that "very few naturalists of any eminence are bold enough now to claim for every species an absolute separation from every other", but on the other hand, it found that "the tendency of scientific thought is to limit and circumscribe the power of Natural Selection", and gave as instances such names as A. R. Wallace, St. George Mivart, J. J. Murphy, and R. Owen.[35]) Thanks to this distinction between Evolution pure and simple, and Darwinian Natural Selection, it was possible even for the *Edinburgh Review*, in 1872, to include Darwin

---

[32]) *Examiner*, 1872, June 8, 576.
[33]) *Athenaeum*, 1869, Aug. 14, 210.
[34]) *Methodist Recorder*, 1863, May 15, 155.
[35]) *Guardian*, 1871, 936.

with Lyell, Huxley, Wallace and Lubbock — of whom Darwin, Huxley and Lubbock were Darwinians, while the other two did not go the whole way with Natural Selection — among Britain's "ablest scientific thinkers".[36])

Owen

Two key figures in the Darwinian controversy were Richard Owen and Sir Charles Lyell. They were by common consent the leading authorities in their respective fields — comparative anatomy and geology — and they both made substantial contributions to the Darwinian debates of the 1860s.

Owen was a disciple of Cuvier, and his contempt for the Lamarckian development theory was apparently as profound as his master's. On the other hand, as a comparative anatomist he could not but be impressed by the close correspondences in the structure of the great groups of animals and plants. Like many idealistic Nature Philosophers he tried to explain such correspondences as due not to a real, but to an ideal development. All vertebrate animals, he held, were formed according to one common plan, represented by the "archetype", and each separate species represented but one of the infinite number of modifications which the Creator had introduced into this plan, in order to fit each organism to its peculiar station in the economy of nature. Thus the organisms were certainly related: only they were not related by ordinary descent, but by the fact of being connected in the mind of God.

When stripped of their metaphysical superstructure Owen's views hardly meant more than a confession of ignorance as regards the origin of specific differences, though his pompous and involved mode of expression concealed the barrenness of his theory to most of his readers. But however obscure his own theory may have been, his contempt for the Darwinian one was apparent to everybody: it was widely known that Owen had written the anti-Darwinian review of the *Origin* in the *Edinburgh*, and that he had inspired the Bishop of Oxford's attacks in the *Quarterly* and at the British Association meeting in 1860. The *Saturday Review*, for instance, wrote about the author of the *Edinburgh* article, "We will not attempt to decide whether he sat at the feet, or stood in the very shoes, of the author of

---

[36]) *Edinburgh Review*, 135, 1872, 89.

Palaeontology".[37]) Accordingly, when Owen's *Palaeontology* was published in 1860, it was searched for a refutation of Darwin, and not wholly in vain. One religious organ sighed: "How consolatory to the hearts of such as are alive to the scientific validity and higher relations of natural theology . . . how cheering to listen to words such as these of Britain's greatest anatomist, which are not more lofty in their wisdom than exemplary by their humility!"[38])

In the non-scientific press the actual arguments advanced by Owen were hardly important: neither the writers themselves, nor their readers, were in a position to judge of the value of the argumentation. It was the bare fact that Owen controverted Darwin that was essential. A straightforward appeal to authority was more effective than facts and logic. The *John Bull* expressed this attitude clearly and unmistakeably in a review of Huxley's *Man's Place in Nature:* "Mr. Owen and Mr. Huxley . . . place a very different value upon the truth of Mr. Darwin's theory. Had they agreed, we should in all humility, in spite of our own suspicions, have been compelled to accept Mr. Darwin. As it is, we can adopt one side or the other, with the comfortable assurance that a champion whose blows all must respect is fighting with us."[39])

But many were disappointed by the vagueness of the views Owen propounded in opposition to Darwin. The *Athenaeum*, quoting Owen's "law of the ordained becoming of organized beings" complained: "And is this all? might the natural theologian ask as he lays down the book . . . Do you only conclude that there is a 'great First Cause, which is certainly not mechanical?' . . . We have enough written on the other side to freeze us or to ossify us; we have men fearless in feigning hypotheses: let then an accomplished successor of Cuvier fully declare all that is within his knowledge."[40]) As a matter of fact, however, Owen was never to supply what his anti-Darwinian friends required. On the contrary, he was steadily moving towards acceptance

[37]) *Saturday Review*, 9, 1860, 573—4. (By Mivart?) Darwin noticed this article in a letter to Huxley (*Letters*, II, 311, note) Huxley worked for the S. R. at this time. (Huxley, *Life* (1. ed.) I, 302).
[38]) *British and Foreign Evangelical Review*, 9, 1860, 431. See also: *Rambler*, 3, 1860, 128.
[39]) *John Bull*, 1863, Feb. 21, 123.
[40]) *Athenaeum*, 1860: 1, Apr. 7, 479.

of the Evolution theory. To begin with, he acknowledged the effectiveness of Natural Selection in *destroying* species.[41]) This caused one of his reviewers to comment, "we would ask whether it is not actually an admission of the Darwinian theory."[42]) Owen at once wrote a letter of protest to the journal, upon which the reviewer drily remarked: "So far as we can gather from his communication, he denies the Darwinian doctrine, admits the accuracy of its basis, and claims to be the first to point out the truth of the principle on which it is founded."[43]) And in 1872 the *Popular Science Review*, which was by no means a pro-Darwinian organ, went as far as stating that "we may not call Professor Owen a Darwinian, but if we could take Mr. Darwin's name from the doctrine, we doubt not Professor Owen would hold to it."[44]) This, however, was hardly fair to Owen, who was undoubtedly consistent in refusing to entertain the full Natural Selection theory, which was the distinguishing mark of Darwin's Evolution theory. But in the non-scientific press such nice distinctions could not easily be upheld, and hence Owen's name could no longer be used to such good advantage as before by anti-Darwinians. Moreover, his reputation as a scientist had suffered severely as a result of a sharp controversy with Huxley in the early 'sixties. At the British Association meeting in 1860 Owen had made some statements as regards the anatomical differences between man and the monkeys, statements which Huxley had directly and flatly contradicted, pledging to support his contradiction "elsewhere". He did so in the *Natural History Review* for January, 1861. Owen, however, reasserted his views at the British Association meeting in 1861, and Huxley, who was not present at the meeting, wrote a letter to the President, again contradicting Owen. The famous *hippocampus minor* debate was in full swing. The press gave much space to it; it even gave an echo in *Punch*.[45]) The controversy continued into 1863, until a note in Huxley's *Man's Place in Nature* put an effective stop to it: it had to be recognized that Huxley had proved his case.

The line taken by Owen's many supporters in the press was to avoid

---

[41]) E. g. *Critic*, 1862, June 7, 558, *Popular Science Review*, 1866, April, 212.
[42]) *London Review*, 12, 1866, 482.
[43]) *London Review*, 12, 1866, 516.
[44]) *Popular Science Review*, 1872, July, 291.
[45]) *Punch*, 40, 1861, Preface; May 18, 206.

discussing the facts, while deploring the bad taste of Huxley's attack. One writer said, "Here, then, we have the highest authority in England — the so-called British Cuvier — ... persistently and publicly contradicted by his juniors in age and inferiors in fame."[46]) The *Lancet* recommended Huxley to "try to imitate the calm and philosophical tone of the man whom he assails"[47]), while a religious organ found it "infinitely amusing to see how Mr. Huxley crows and struts, and goes quite beside himself with inflation and self-delight, in the hope to be able to fasten the charge of want of learning upon the venerable name of Owen."[48]) The *Edinburgh Review*, somewhat sourly, declared that it was "a matter of perfect indifference"[49]) whether Owen or Huxley was right on the point of fact. In the light of such comments, the summing-up of the pro-Darwinian *Reader* seems no more than fair: "Singularly enough, Professor Huxley has been twitted by many ... for his pertinacity in the defence of truth, while but a solitary voice here and there has been raised to condemn Professor Owen's chivalric devotion to error."[50])

The *hippocampus* controversy only served to underline the eclipse of Owen in the world of science. Even the *Intellectual Observer*, which was not at all favourable to Darwinian views, admitted in 1865 that Owen had been "distanced in science as well as convicted of error"[51]) by Huxley. Therefore his theory of Derivation, an Evolution theory with a strong religious colouring, made little impression when it was published in 1868. Scientists were impatient with Owen's metaphysics, and theologians were shocked that he should have come to embrace evolutionism, not even stopping short of Spontaneous Generation. To the anti-Darwinians Owen was by now a spent force.

Lyell

The other key authority in the Darwinian debate was Sir Charles Lyell. In his *Principles of Geology*, first published in 1830—33 and reissued all through the following decades, Lyell had thoroughly

---

[46]) *Patriot*, 1862, Oct. 9, 669.
[47]) *Lancet*, 1862: 2, 487.
[48]) *Eclectic Review*, 4, 1863, 340.
[49]) *Edinburgh Review*, 117, 1863, 564.
[50]) *Reader*, 1863, Mar. 7, 234.
[51]) *Intellectual Observer*, 6, 1864—5, 372.

discussed the Lamarckian theory, ending by rejecting it. Before Darwin published, Lyell had therefore been widely cited as a supporter of the theory of specific permanence. But in the first edition of the *Origin* Darwin stated that he had "reason to believe" that he had at least made his friend entertain grave doubts on the subject.[52]) Lyell himself, however, was slow in stating his position publicly, though he sponsored Darwin's and Wallace's early contributions to the Linnaean Society[53]), and mentioned the forthcoming *Origin* in favourable terms at the British Association meeting in 1859. Informed people therefore usually counted Lyell among Darwin's supporters, but few knew just how far he was prepared to go. His book on the *Antiquity of Man* was naturally expected with great eagerness in 1863. But in it Lyell expressed himself very cautiously; at one place, for instance, he wrote, "If it should ever become highly probable" that species change by variation and natural selection,[54]) which surely must be interpreted as meaning that he did not consider it probable at the time. Darwin was understandably greatly disappointed by Lyell's half-hearted stand, and wrote in a letter to Asa Gray, "If such a man dare not or will not speak out his mind, how can we who are ignorant form even a guess on the subject?"[55])

The reaction in the press is interesting. Pro-Darwinian organs deplored Lyell's extreme caution, while pointing out that "the whole tenor of his argument is in favour of Mr. Darwin's theory."[56]) The majority of the commentators agreed in this view of Lyell's attitude,[57]) while several, especially anti-Darwinian ones, omitted to say anything about the matter, although the book carried the subtitle "with remarks on the Darwinian Theory as to the Origin of Species."[58 A]) Apparently one did not wish to leave readers with the impression that Lyell gave even a qualified support to Darwin. Those who did mention Lyell's stand attempted instead to play down Lyell's authority: "It has

---

[52]) *Origin* 1 ed. r., 264.

[53]) *Journal* of the Linnaean Society, 1858.

[54]) Sir Charles Lyell, *Antiquity of Man*, p. 469.

[55]) Darwin, Letters, III, 10, to Asa Gray, 1863.

[56]) *Natural History Review*, 1863, 213.

[57]) *Spectator*, 1863, 1726, *John Bull*, 1863, Mar. 21, 187, *Edinburgh Review*, 118, 1863, 294—5, *London Quarterly Review*, 20, 1863, 302, *Critic*, 25, 1863, 297.

[58 A]) *Times*, 1863, Apr. 9—10, *Athenaeum*, 1863: 1, 221, *Examiner*, 1863, Feb. 28, 132—3, Mar. 21, 180—1.

occasioned sincere grief to any man who loves both science and the
Bible to see the course adopted by Sir C. Lyell . . . He shows so evident
a *bias* in favour of Darwinian speculations as must cause his doctrines
henceforth to be received with great distrust."[58 B])

Lyell's caution made it possible for the obscurer organs to continue
even to cite him *against* Darwin for some years still.[59]) His definite
acceptance of Evolution did not come until 1868, in the 10th edition
of the *Principles of Geology*. The *Westminster* immediately drew attention to this, stressing that his adoption of the continuity doctrine for
biology was a logical and necessary extension of the views which had
made him famous in geology.[60]) But Darwin's gain was of course at
least as great as Lyell's: Wallace, writing in the *Quarterly*, emphasized
this: "For more than thirty years . . . Sir Charles Lyell had been
constantly quoted as the greatest and most authoritative opponent of
"transmutation" . . . [now he] gives his adhesion to a theory which
to superficial readers will appear hardly distinguishable from
[Lamarck's] . . . If for no other reason than that Sir Charles Lyell in
his tenth edition has adopted it, the theory of Mr. Darwin deserves
an attentive and respectful consideration from every earnest seeker
after truth." [61]) The *Saturday Review* put this point even more forcibly:
"Followers of Darwin may therefore so far claim the authority of
Sir Charles Lyell as having more than half abandoned his old weapons
and adopted the newer style of fence, and to them, considering the
well-deserved weight of his writings, this is a great gain . . . the force
of his example will be felt."[62])

Anti-Darwinian writers naturally did not regard Lyell's change in
this light. The indefatigable *Athenaeum* regretted that Lyell had "so
completely adopted the Darwinian hypothesis", in view of "the complete
confutation which many think inevitably awaits his theory."[63]) But
this was prophecy; the present fact, that Darwin had the full support of
Lyell as regards Evolution, if not altogether as regards Natural Selection, could not be denied. Thus the year 1868 was an important one in

[58 B]) *Freeman*, 1863, Apr. 8, 222.
[59]) *British Quarterly Review*, 41, 1865, 143.
[60]) *Westminster Review*, 34, 1868, 549.
[61]) *Quarterly Review*, 126, 1869, 381.
[62]) *Saturday Review*, 1868, Apr. 11.
[63]) *Athenaeum*, 1868, Mar 28, 455

the crystallisation of opinion the Evolution theory. Darwin's most decided opponents definitively lost Owen, and Darwin himself gained Lyell.

Scientific and Public Opinion in General

The references to contemporary opinion that we have dealt with in the preceding sections were supported by factual evidence, namely, the names of the authorities appealed to. Thereby writers were prevented from straying too far from reality, and in fact we found that the trend of opinion on Darwin was fairly well mirrored in the material. The estimates that we are now going to consider are vaguer, and may therefore be expected to be more markedly biassed. It is easier to assert roundly that "everybody" says so and so, than to say — without too great a risk of being controverted — that A, B and C say so and so. The nature of the bias, however, is the same as we have already become familiar with. Claims were generally made that the majority, or the majority of those who really counted, were on the writer's own side, or else that the opposition was due to irrelevant, sinister, or perverse considerations.

1. *1859—1863.*

Nobody in 1860 ventured to suggest that Darwinian views were adopted by a majority of scientific men. On the contrary, many alleged that Darwin had failed to make any impression at all on his colleagues. It is hardly surprising to find the *Athenaeum* taking this view,[64]) but even generally pro-Darwinian organs like *Chambers's Journal*[65]) and the *Leader* were inclined to agree: Darwin's theory seemed "destined to receive a disapproving notice from nearly every literary and scientific periodical of the day."[66]) This however, was an overstatement, and other writers were soon to introduce the necessary qualifications. One of them found that the theory was being accepted "among a limited group of scientific men"[67]) while another stated that "men eminent in science . . . gave at least a provisional and partial adhesion to the hypothesis."[68]) Pro-Darwinian writers were of course quick to point out

[64]) *Athenaeum,* 1859: 2, 659.
[65]) *Chambers's Journal,* 1860: 1, 415.
[66]) *Leader,* 1860, Apr. 21, 378.
[67]) *All the Year Round,* 3, 1860, 175.
[68]) *Press,* 1860, 656.

that this provisional assent was above all forthcoming from "the best thinkers of the day",[69]) while anti-Darwinians represented the converts as misled by their inexperience: "several, perhaps the majority, of our younger naturalists have been seduced into the acceptance of the homoeopathic form of the transmutative hypothesis now presented to them by Mr. Darwin."[70])

From 1861 onwards, Darwinians began to claim a larger share of converts. A writer in the *Gardener's Chronicle* stated as a fact that among Darwin's opponents there were "few who have any *status* in science" and concluded that "there has been little avowed opposition."[71]) Asa Gray, writing in the *Annals and Magazine of Natural History*, a journal which certainly did not regard Darwin with any favour, summed up the position in 1863 as follows: "The general drift of opinion . . . carries along naturalists of widely differing views and prepossessions, some faster and further than others, but all in one way. The tendency is, we may say, to extend the law of continuity."[72])

The progress of Darwinian views was recognized by several anti-Darwinians as well, and they reacted in various ways. Some simply shook their heads,[73]) others sought to explain away the deplorable fact in various ways. One writer in the *Annals and Magazine* thought it an "astonishing circumstance" that such a theory as that contained in the *Origin* had met with acceptance from "several of the leading naturalists of this country", but ascribed it to the esteem in which the name of Charles Darwin was held.[74]) Others put the blame on the "scepticism" of the age.[75]) But in general, when anti-Darwinians admitted the progress of Darwinism at all, they declared that the converts were not competent judges. Samuel Haughton, professor of geology at Dublin, found it curious to observe how Darwin, like Lamarck, appealed "from the judgment of their peers to the young, the enthusiastic, and the inexperienced."[76])

---

[69]) *Macmillan's Magazine*, 1859, Dec. The writer was T. H. Huxley.

[70]) *Edinburgh Review*, 111, 1860, 488. The writer was R. Owen.

[71]) *Gardener's Chronicle*, 1861, 219.

[72]) *Annals and Magazine of Natural History*, 12, 1863, 88.

[73]) E. g. *Friend*, 1861, 212.

[74]) *Annals and Magazine of Natural History*, 7, 1861, 399.

[75]) *Recreative Science*, 3, 1861—2, 219 (C. R. Bree).

[76]) *Annals and Magazine of Natural History*, 11, 1863, 425.

It was practically only in the non-scientific, and chiefly in the religious press, that one still continued to argue that the scientific world had overwhelmingly rejected Darwin's views. The *Family Herald* found the evolutionists "reasoned out of their position, defeated by the arguments of the physiologist and anatomist,"[77]) and the *Leisure Hour* thought in 1863 that "almost all scientific men" were convinced of the permanence of species.[78]) Even an old scientific hero like Sir David Brewster, writing in *Good Words* in 1862, declared that Darwin's hypothesis had been "long ago refuted by the most distinguished of our naturalists."[79])

In the better-class organs these extreme views were not endorsed, but one did claim, and probably rightly, that Darwin had not yet won over the majority of his colleagues.[80]) But the trend of opinion was obviously setting towards Darwin. The *Spectator* expressed this somewhat cryptically; the descent theory, it said, "can scarcely fail to obtain a position which naturalists are at present far from willing to concede to it."[81]) Even more oracular was the Duke of Argyll, writing in the *Edinburgh:* "The present state of accurate scientific thought in England, with reference to the theory of Mr. Darwin, falls exceedingly short of entire and unmixed assent."[82])

The spread of Darwinism among the general public during these early years also attracted the attention of commentators. The close correspondence between acceptance of Darwinism and certain political and religious views did not pass unnoticed: we have already quoted G. H. Lewes, writing in *Blackwood's*, to this effect.[83]) The anti-Darwinian Haughton expressed a similar view in a rather different tone; Darwin's arguments, he said, "will no doubt find acceptance with those political economists and pseudo-philosophers who reduce all the laws of action and human thought habitually to the lowest and most sordid motives."[84])

How far had the broad public accepted the theory? In 1861, the *Athenaeum* held that the traditional view of the permanence of species

---

[77]) *Family Herald*, 1861—2, 157.
[78]) *Leisure Hour*, 1863, 192.
[79]) *Good Words*, 3, 1862, 4.
[80]) *Friend*, 1861, 10.
[81]) *Spectator*, 1862, 780—1.
[82]) *Edinburgh Review*, 117, 1863, 547.
[83]) See above, Chapter 2, p. 35.
[84]) *Annals and Magazine of Natural History*, 11, 1863, 423.

was still "if not the most philosophical, at least the most prevalent belief."[85]) This cannot but be interpreted as an admission that the new theory was gaining ground, among the informed public at least. Thus even in the anti-Darwinian press there was apparently little inclination to underestimate the number of converts that Darwin was making. "Alas, their name is legion,"[86]) said one religious organ, and another went as far as claiming that the theory had been "all but universally accepted."[87]) This was of course going too far, and in fact the last-quoted writer only made his admission in an article which was intended as a complete refutation of Darwin: it was a case of *reculer pour mieux sauter.* But though it was an exaggeration, it was not far wrong. Darwin had already a strong position, both among scientists and among the general public, by the end of our first period.

2. *1864—1869.*

In our second subperiod, almost everybody, Darwinian or not, was ready to attest that the scientific world had in general come to accept evolutionary views. The popular *Temple Bar* magazine asserted in 1865 that "naturalists, who best should know . . . mostly uphold the hypothesis."[88]) *Fraser's* put it even more strongly in 1868: "nearly all men qualified to form an opinion are convinced of its substantial truth."[89]) Many similar statements could be quoted.[90])

Anti-Darwinian organs, though expressing themselves more cautiously, often seem to have held a similar view. The *North British Review* acknowledged that Darwin's theory had been "received as probable, and even as certainly true by many who . . . are competent to form an intelligent opinion."[91]) Such a development was of course looked upon with great misgivings in religious circles, where it was regarded as just another instance of the increasing "Scepticism of

[85]) *Athenaeum*, 1861: 1, 866, quoting David Page.

[86]) *Friend*, 1861, 212.

[87]) *British Quarterly Review*, 38, 1863, 492.

[88]) *Temple Bar*, 15, 1865, 238.

[89]) *Fraser's Magazine*, 78, 1868, 353.

[90]) See also: *Westminster Review*, 32, 1867, 2, *London Review*, 16, 1868, 178, *Student*, 1, 1868, 180, *Quarterly Journal of Science*, 2, 1865, 194, *Popular Science Review*, 1867, Jan., *Westminster Review*, 35, 1869, 207.

[91]) *North British Review*, 46, 1867, 277.

Science."[92]) That explanation seems to have been to some extent
endorsed by the *Edinburgh Review*, which thought that the advance
of evolutionary views was not so much due to Darwin's theory having
been accepted, as to a general movement of scientific opinion towards
"disbelief in special creations." This was, said the journal somewhat
ungenerously, "not so much the product of the investigations of one
man, as the necessary result of the progress of natural history during
the last fifty years."[93 A])

Writers in the non-scientific press seldom made any clear distinction
between Evolution pure and simple, and the peculiarly Darwinian
doctrine of Natural Selection. Darwinism and Development were more
or less synonymous to the general reader. But when the two concepts
were distinguished, it was generally in order to point out that most
experts did not go as far as Darwin, especially as regards Natural
Selection.[93 B]) Inevitably some anti-Darwinians would interpret this
resistance against Natural Selection as a hopeful sign that the *whole*
of Darwin's doctrine was losing favour in the scientific world. The
British Quarterly, for instance, asserted in 1865 that "such is the united
voice of all the men most eminent in geology and its kindred sciences."[94])
The *Athenaeum*, expressing itself more cautiously, noted in 1868 that
Natural Selection was seen "rapidly to decline in scientific favour", and
asked, "what sober naturalist will insure the life of Natural Selection,
or the main theories of Darwinism, for ten years only?"[95]) The Roman
Catholic *Tablet*, reviewing the Duke of Argyll's book on *Primeval Man*,
was equally optimistic: "That so close a reasoner as the Duke has been
led to reject [Darwin's] theory, is a satisfactory sign that its hold on
the so-called scientific mind is diminishing, and perhaps will soon be
lost altogether."[96])

As regards the state of opinion among the general public, most writers
now believed that Darwin was gaining ground, and there is no reason to
doubt that they were right. The *Spectator* said in 1868: "We are certainly

---

[92]) *Dublin University Magazine*, 65, 1865, 593.

[93 A]) *Edinburgh Review*, 128, 1868, 414.

[93 B]) *Intellectual Observer*, 6, 1864—5, 12; *Lancet*, 1868: 1, 501.

[94]) *British Quarterly Review*, 41, 1865, 143.

[95]) *Athenaeum*, 1868, May 28, 455.

[96]) *Tablet*, 1869, Apr. 10, 780. See also: *Athenaeum*, 1869, Mar. 27, 431,
Family Herald, 1867—8, 380.

not exceeding the mark when we assert that a large majority both of the general public and, in this country, of scientific naturalists, have given in their adhesion, more or less, to Darwinian principles."[97]) As in the previous period, the dependence of the attitude towards Darwin on people's philosophy of life was noticed: the *Pall Mall Gazette* commented, in 1868: "Mr. Darwin's 'Origin of Species' . . . exasperated many theological philosophers, who saw in it the advent of a 'dreary materialism' (not otherwise specified); and it delighted freethinkers, because it exasperated their opponents . . . Everywhere 'Darwinism' has become a byword, which has gone far to replace 'materialism'."[98])

3. *1870—1872.*

After 1870, contemporaries were even more emphatic than before about the general acceptance of evolutionary views among the scientists. The estimate of the new scientific journal, *Nature*, may be accepted as well-informed. It said: "The fascinating hypothesis of Darwinism has; within the last few years, so completely taken hold of the scientific mind, both in this country and in Germany, that almost the whole of our rising men of science may be classed as belonging to this school of thought. Probably since the time of Newton no man has had so great an influence over the development of scientific thought as Mr. Darwin."[99]) Similar verdicts were given in many other journals.[102]) Even the *Athenaeum*, which was beginning to come round, admitted that Darwinian views were accepted by "an increasingly powerful body of the younger men of science."[100]) And a religious organ like the *Inquirer* informed its readers that "no scientific journal of any standing ventures to impugn the theory".[101]) Several anti-Darwinians also reluctantly admitted the fact, though not without adding sour comments; the *Quarterly*, for instance, ostensively adopting an attitude of impartiality, found that "many persons, at first violently opposed through ignorance or prejudice to Mr. Darwin's views, are now, with

---

[97]) *Spectator*, 1868, 318.

[98]) *Pall Mall Gazette*, 1868, Feb. 10, 555.

[99]) *Nature*, 3, 1870—1, Nov. 10, p. 30.

[100]) *Athenaeum*, 1870, July 23, 109.

[101]) *Inquirer*, 1871, May 13, 295.

[102]) See also: *Annals and Magazine of Natural History*, 6, 1870, 34; *Examiner*, 1870, May 28, 341, *Saturday Review*, 31, 1871, 276. *Academy*, 2, 1870—1, 13, *Daily News*, 1871, Feb. 23.

scarcely less ignorance and prejudice, as strongly inclined in their favour."[103]) Others insisted that the acceptance was by no means unanimous,[104]) but met with the rejoinder that the opposition consisted almost only of the old guard — the "honoured but older chiefs in science"[105 A]) — while all the coming men ranged themselves on Darwin's side.

The trump card of the anti-Darwinians was now, even more than in the previous period, the distinction between Evolution and Natural Selection. The success of Mivart's book on the *Genesis of Species* focussed attention on this distinction, and *Nature*, reviewing the book, admitted that "a reaction has been setting in" against the Natural Selection theory.[105 B]) There is no doubt that this was true — Darwin himself, as we noted above, was also conscious of it — and anti-Darwinian organs naturally seized upon this trend. The *Guardian*, commenting on the British Association meeting in 1871, concluded that it had given prominence to two facts, namely, "that Evolution, and specifically the evolution of life from life only, is now the dominant biological doctrine, and that Natural Selection is becoming more and more discredited as an adequate rendering of that doctrine."[106]) In the following year the *British and Foreign Evangelical Review* was emboldened to assert that "the number of scientific men who hold the theory of evolution as Mr. Darwin states it, constitute but a small minority of our leaders in scientific thought."[107]) These are two instances out of many.[108])

Few commentators had any means of really assessing the penetration of Darwinian views among the general public. *Fraser's* ventured to suggest that Darwin had "ceased to shock, not only the ignorant prejudices of the multitude, but even the delicate susceptibility of divines."[109]) That, however, was an exaggeration. The *Edinburgh* would go no farther than admit that "many theologians" had given up

---

[103]) *Quarterly Review*, 131, 1871, 48. See also: *Zoologist*, 6, 1871, 2624, *London Quarterly Review*, 36, 1871, 281.

[104]) E. g. *Edinburgh Review*, 133, 1871, 175, *Once a Week*, 9, 1872, 522.

[105 A]) *Observer*, 1871, Mar. 19, p. 3.

[105 B]) *Nature*, 3, 1870—1, Feb. 2, 271, (A. W. Bennett).

[106]) *Guardian*, 1871, 1020.

[107]) *British and Foreign Evangelical Review*, 21, 1872, 30.

[108]) E. g. *Guardian*, 1870, Sep. 28, 1137, *Tablet*, 1871, Feb. 25, 233.

[109]) *Fraser's Magazine*, 2, 1870, 251.

the old view of species as specially created,[110]) and the *Illustrated London News* thought that Darwin had "not yet convinced all the world that he is right."[111]) The *Westminster Review* even found that the people who rejected Darwin on the basis of "imperfect details which they can gather from newspaper articles and popular lectures ... form a very large class in our present society."[112])

The writer in *Fraser's* believed that the divines were beginning to give up their resistance to Darwinism. That was true of the Broad Church clergy whose views *Fraser's* knew so well. It was not true of the British clergy in general, though the Roman Catholic *Tablet*, looking at the matter from a diametrically opposed angle, seems to have implied as much: "Protestantism, having completely broken down under the pressure of men's minds, Rationalism, in its mitigated or absolute form, has taken possession of the universities, of the mechanics' institutes, and of a large proportion of the literature of the day; and therefore Christianity is loosening its hold on the mind and heart of a continually increasing number. Science takes the place of dogmatic religion, and the worship of nature the place of the worship of God. Mr. Huxley is the favourite and popular apostle of this new creed."[113]) A writer in the *Academy*, who said that "the attacks on Darwinism come chiefly from the side of religious orthodoxy"[114]) was considerably nearer the mark.

---

[110]) *Edinburgh Review*, 134, 1871, 199.
[111]) *Illustrated London News*, 60, 1872, Nov. 23, 491.
[112]) *Westminster Review*, 40, 1871, 256.
[113]) *Tablet*, 6, 1871, Nov. 4, 582.
[114]) *Academy*, 2, 1870—1, 560.

CHAPTER 4

Darwinism at the British Association

It is no part of the object of the present work to investigate in detail the impact of Darwin on biological theory and practice. However, as the expressed opinions of the scientists were obviously of great importance for determining the attitudes of the public at large, it appears desirable to go somewhat further into the development of opinion within the scientific world than has been possible on the somewhat narrow basis of evidence provided by the general periodicals. The fresh material that we shall now consider is provided by the debates at the meetings of the British Association for the Advancement of Science. I have chosen to take into account at the same time the comments appearing in the press on these debates: in this manner it becomes possible to discuss not only the views of the scientists themselves, as represented through the papers read at the meetings, but also the public reactions to these views, represented by the discussions at the meetings an subsequently in the press. It would in fact hardly be possible to keep these two things apart, since most of what we know of the meetings has to be gathered from the press reports of them. The official *Reports*[1]) often contain nothing but the titles of the papers read, and give no idea of the subsequent discussions.

It has long been customary to look upon the dramatic clash between T. H. Huxley and the Bishop of Oxford at the British Association meeting at Oxford in 1860 as the event that set the tone of the Darwinian controversy among the general public. Is that a correct view? Were the contemporaries as much impressed by Huxley's skilful repartee as posterity has undoubtedly been? Did the public take much interest in the British Association meetings generally? What did the Association as a whole stand for? Can it be relied upon as a fair indicator of the trend of opinion within the scientific world? What

---

[1]) *Report* of the British Association for the Advancement of Science, 1831, in progress (*Report*).

did the public, as represented by the press, think of it and its doings? These are questions that need to be considered.

The British Association was founded in 1831, and stated its objects as follows:

> To give a stronger impulse and a more systematic direction to scientific inquiry, — to promote the intercourse of those who cultivate Science in different parts of the British Empire, with one another and with foreign philosophers, — to obtain a more general attention to the objects of Science, and a removal of any disadvantages of a public kind which impedes its progress.[2])

The Association sponsored much serious scientific research, the results of which were published in its annual Reports. But it is probably true to say that its main function, besides the social one of enabling scientists to meet each other during a pleasant holiday week, was what may be called a propagandistic one. By visiting a new provincial town each year — London was avoided — and by being fairly liberal in allowing the general public to attend, the Association was hoping to spread interest in science all over the country.

The meetings themselves followed a set pattern. At the opening of each meeting the President for the year delivered an address, which usually, at least from the 'forties onwards, took the form of a survey of progress in the different branches of science during the year. On the following days the meeting divided into sections, in order to hear and discuss such papers as had been approved by a committe for each section. During the 'sixties it became common practice for the sectional presidents as well to start the proceedings in their own sections by delivering an opening address. Evening lectures and conversaziones intended for the whole meeting were also held, and from 1867 on there was also a "Lecture to the Operative Classes".

The association was treated somewhat superciliously at first in many quarters, especially, it seems, in the ancient seats of learning. This opposition found effective expression in the *Times*, which poured scorn and contempt on the "gipsies of science".

But by the 'fifties the Association had become an established and respected institution in the national life: the Prince Consort attended the meetings in 1846 and 1851, and was President in 1859. Interest in its proceedings was great, and increased very markedly in the

---

[2]) *Report*, beginning of each volume.

'sixties, the decade of the great Darwinian debates. This increase is apparent in many ways. The average attendance at each meeting had been c. 1300 in the 'thirties, c. 1100 in the 'forties, and c. 1400 in the 'fifties. In the 'sixties it rose dramatically to c. 2300, and to 2500 in the period 1870—72.[3]) Another indication of the increasing interest in the Association, and presumably in science generally, is afforded by the number of corresponding societies affiliated to it. In the first three decades, 1831—1860, the number of new affiliations were 8, 5, and 16, respectively, as against 19 in the 'sixties and 21 in the 'seventies, after which the numbers dropped again.[4]) Last but not least, the amount of space given by the press to the meetings increased considerably during the 'sixties. The *Times*, for instance, had afforded only an average of 7 columns of reports per meeting during the 'fifties: in the next decade the average was 25 columns per meeting. In the other newspapers the pattern was similar.

In the 'sixties most of the daily papers, especially the dearer ones, reported the meetings quite fully. From 1863 onwards it was common practice to print the whole, or almost the whole of the presidential address,[5]) which usually covered some five columns in close print. Leading articles often occurred in connection with the address. The weekly press usually commented on the meetings in editorials, but did not as a rule offer much in the way of straight reports: the *Saturday Review* and the *Spectator* are instances in point. The *Athenaeum*, and later on *Nature*, on the other hand, gave a very ample coverage of the proceedings, including very often the full text of some papers.

In the cheap Sunday papers, naturally, much less attention was given to the Association. If readers of *Lloyd's Weekly*, or the *Weekly Times*, for instance, heard anything about the meetings, it was in connection with such things as Du Chaillu's gorilla hunts, or Stanley's quest for Dr. Livingstone. Occasional papers of social or political interest were also sometimes taken up in these popular organs.

The newspaper press could give so much attention as it did to the

---

[3]) *Report*, 1873, xlv.

[4]) O. J. R. Howarth, *The British Association*, 95.

[5]) The *Daily News* and the *Morning Post* had done so in 1861 and 1862 as well. These two, and the *Times*, gave most space to the address in the following years; the *Star*, the *Standard* and the *Morning Advertiser* were somewhat more restrictive, and the *Daily Telegraph* seldom exceeded one or two columns for the address.

meetings of the Association because they were held during the politically "dull" season. The *Illustrated London News* explained: "It is a great convenience to newspapers that our men of science hold their annual festival in this, the dullest month of the year... By meeting in August, moreover, the British Association more effectually obtains the public ear than could be the case in the busier periods of the year. The papers are only too glad to report its meetings and their readers have leisure to skim the cream of scientific knowledge put to them."[6] At other times of the year the press took very little interest indeed in scientific matters. The Victorian newspaper press was almost exclusively political. But the British Association meetings were used as occasions for a stock-taking of scientific progress during the year. The editorial comments on the meetings — and especially on the presidential addresses — provide us in fact with almost the only information we have on the opinions of an important and influential section of the public — the newspaper editors — in regard to science. To these men, and to that large majority of their readers who did not read scientific periodicals, the British Association meeting was the scientific event of the year. "The presidential address finds as many readers as a statement from Lord Palmerston or a budget speech by Mr. Gladstone"[7] wrote one newspaper, and another called it "a sort of Queen's Speech of Science for the year."[8 A] Similar views were expressed on every hand.[8 B]

The increasing public interest in the meetings during the 'sixties was also noticed by contemporaries. "The Association has risen rapidly in public favour of late years", wrote the *Manchester Guardian* in 1863,[9] and six years later the *Guardian* found that "the meetings of the British Association are watched year by year with a keener interest by a larger public."[10]

I would suggest that part at least of this increase of interest was due to the stir caused by the Darwinian debates. It is significant that though Darwinian questions only cropped up in a few of the couple

---

[6] *Illustrated London News*, 1869, Aug. 28, 197.

[7] *Manchester Guardian*, 1863, Aug. 31, p. 2.

[8 A] *Guardian*, 1871, Aug. 9, 950.

[8 B] *Morning Post*, 1863, Aug. 28, p. 4, *Methodist Recorder*, 1863, Aug. 28, 299, *Daily News*, 1871, Aug. 4 p. 5.

[9] *Manchester Guardian*, 1869, Aug. 31, p. 2.

[10] *Guardian*, 1867, Aug. 25, 948.

of hundred papers read to the meeting each year, these papers, and the discussions they gave rise to, were usually among those selected for reports and comments in the press. The same is true of the presidential addresses. Only three of them during our period, namely, 1866, 1868, and 1870, were to any great extent concerned with Darwinian topics. But reports and comments seldom omitted those brief references to Darwin or Evolution which normally occurred. To some extent this may be due to the fact that most presidents, in their surveys of scientific work during the year, started with astronomy and ended with biology: in this way the references to the organic world often became involved in the final rhetorical display.

As usual, it is also possible to get contemporary testimonies as to the importance of the Darwinian debates for stimulating public interest in the meetings themselves. The *Daily News* expressed this in a general way when it said, "It is where science necessarily takes a controversial turn that the greatest popular interest is shown."[11] The great scientific controversy of the time was of course the Darwinian one, and one is hardly surprised to see one contemporary comparing the two protagonists in the debate, Owen and Huxley, with the two leading prizefighters of the day.[12]

The way in which Darwin tended to monopolize public interest in the proceedings at the Association was naturally deplored in the anti-Darwinian press. One religious organ expressed its dissatisfaction in a leader, entitled "Science astray": "The admiration of the select few will be given to those who count no toil too great to accumulate irrefragable evidence of truths which may have no immediate bearing on practical life, and patiently to investigate facts which are incapable of being utilized now . . . many of the notables of science have become enamoured of their own notability . . . To what other cause can the birth of such ridiculous theories as the Darwinian hypothesis be ascribed?"[13] But whatever the select few might think, the general public persisted in paying attention to the Association meetings to the extent that they were dramatized by the clashes of men and ideas. "They like to see the scientific 'lions' of the day, and to hear them roar",[14] wrote the

---

[11] *Daily News*, 1869, Aug. 19, p. 4.
[12] *Pall Mall Gazette*, 1865, Sep. 8, 346.
[13] *Patriot*, 1863, Sep. 10, 594.
[14] *Manchester Guardian*, 1867, Sep. 10, p. 4.

Manchester Guardian. And the *Daily News* said, "Among the thousands who become members of the British Association, only a small proportion would care to watch the progress of really scientific work ... We are sure to hear of 'wars and rumours of wars' over the question of the origin of man, though Darwin himself avoids very wisely the arena of dispute."[15])

It is thus quite clear that the meetings of the British Association were very much in the public eye at the time we are studying. What was said at the meetings, and especially the presidential addresses, was in fact delivered to a national audience. That audience, consisting as it did chiefly of people with a non-scientific education, were above all interested in the question of the relation between science and the general problem of life, and especially between science and religion. They were therefore keenly sensitive to every reference to Darwinian matters. The speakers at the meetings must obviously have been aware of the attention given to their utterances on these subjects.

In the eyes of the general public the British Association was therefore one of the chief arenas of the Darwinian battle. A study of the development of opinion at the annual meetings, and of the comments of the press upon them, ought therefore to help substantially to elucidate the question of the spread of the Darwinian doctrines, both in the scientific community, and among the public at large.

Oxford, 1860

If the Oxford meeting of the British Association in 1860 was an historical event, the press missed a great opportunity. Partly this seems to have been the fault of a defective organisation, as no special facilities were provided for the journalists.[16]) Such practical difficulties, however, would hardly have been insuperable, if one had wished to overcome them. But apparently the press did not realize that the session would be as exciting as it was. One had not yet got used to regarding the British Association as a forum for Darwinian controversies. In any case the Oxford meeting was exceptionally meagrely reported: in the *Times* hardly at all.

---

[15]) *Daily News*, 1872, Aug. 14, p. 5. Similar views were expressed e. g. in *John Bull*, 1862, Oct. 18, 672, *Morning Post*, 1863, Aug. 28, p. 2, *Daily Telegraph*, 1869, Aug. 21, p. 6, *Observer*, 1871, Aug. 6, p. 5, *Daily Telegraph*, 1872, Aug. 16, p. 3.
[16]) *Daily News*, 1860, June 30, p. 3.

Thus it came about that the clash between Professor Huxley and the Bishop of Oxford had little press publicity. It will be remembered that the Bishop, towards the end of a long speech where he denounced the Darwinian doctrines, had turned to Huxley and mockingly asked him whether he reckoned his descent from an ape on his grandfather's or on his grandmother's side? — to which Huxley retorted: "If the question is put to me, would I rather have a miserable ape for a grandfather or a man highly endowed by nature and possessing great means and influence, and yet who employs those faculties and that influence for the mere purpose of introducing ridicule into a grave scientific discussion — I unhesitatingly affirm my preference for the ape." On this, says Huxley himself, "there was inextinguishable laughter among the people, and they listened to the rest of my argument with great attention."[17]

It is interesting to see how this episode was reported in the press. The daily papers, as already indicated, almost ignored the whole meeting, and I have found a mention of the clash only in the *Star*, which spoke of the Bishop's speech as "of great power and eloquence", and Huxley's as "an argumentative speech, which was loudly applauded."[18 A] In the weeklies, references were more numerous, but few brought out the force of Huxley's repartee,[18 B] though *Chambers's Journal*, without giving any details, called his reply "smart and somewhat silencing."[19] The *Guardian* on the other hand, wondered what would become of the British Association "when Professors lose their tempers and solemnly avow they would rather be descended from apes than Bishops; and when pretentious sciolists seriously enunciate follies and platitudes of the most wonderful absurdity and draw upon their heads crushing refutations from the truly learned."[20] The *Athenaeum* slanted its report more subtly: "The Bishop of Oxford came out strongly against a theory which holds it possible that a man may be descended from an ape, — in which protest he is sustained by Prof. Owen, Sir Benjamin Brodie, Dr. Daubeny, and the most eminent naturalists assembled at Oxford.

---

[17] Letter to Dr. Dyster: rep. from *Nature*, 1953, Nov. 14, 920. in the *Manchester Guardian*, 1953, Nov. 14.

[18 A] *Evening Star*, 1860, July 2, p. 3.

[18 B] *Press*, 1860, July 7, 656, has a fairly full report.

[19] *Chambers's Journal*, 1860: 2, 64.

[20] *Guardian*, 1860, July 4, 593.

But others, — conspicuous among these, Prof. Huxley — have expressed their willingness to accept for themselves, as well as for their friends and enemies, all actual truths, even the last humiliating truth of a pedigree not registered in the Herald's College. The dispute has at least made Oxford uncommonly lively during the week."[21]

The clash between Huxley and the Bishop occurred in what should have been the discussion of a paper by an American scientist, Dr. J. W. Draper, "On the Intellectual Development of Europe Considered with reference to the views of Mr. Darwin and others." Yet another paper in which Darwinism was touched upon was read at the meeting: Professor Charles Daubeny offered "Remarks on the Final Causes of the Sexuality of Plants, with particular reference to Mr. Darwin's Work 'On the Origin of Species'." The author cautiously defended Evolution as a not impossible hypothesis, while stressing that such evolution could not be due to chance, but to "the watchful super-intendence of an intelligent Cause."[22]

The presidential address by Lord Wrottesly[23] contained nothing on the Darwinian question. The following devout passage represented biology:

> [Consider] the beauties and prodigies of contrivance which the animal and vegetable world display, from mankind to the lowest zoophyte ... Marvels indeed they are, but they are also mysteries, the unravelling of some of which tasks to the utmost the highest order of human intelligence. Let us ever apply ourselves seriously to the task, feeling assured that the more we thus exercising improve our intellectual faculties, the more worthy shall we be, the better shall we be fitted to come nearer to our God."[24]

Manchester, 1861

At Manchester William Fairbairn in the chair delivered the opening address, which was concerned mainly with technology. Darwin's name was mentioned only in passing in a somewhat ambiguous sentence:

---

[21] *Athenaeum*, 1860: 2, 19. For other brief mentions of the incident, see e. g. *Inquirer*, 1860, July 7, 566, *John Bull*, 1860, July 7, 422.

[22] Reprinted as a pamphlet: preserved in the British Museum.

[23] Reported in the *Athenaeum*, 1860: 1, 890, and briefly in the *Evening Star*, 1860, June 28, p. 1.

[24] *Report*, 1860, lxxv.

How interesting is the organization of animals and plants, how admirably adapted to their different functions and spheres of life! They want nothing, yet have nothing superfluous. Every organ is adapted perfectly to its functions; and the researches of Owen, Agassiz, Darwin, Hooker, Daubeny, Babington, and Jardine fully illustrate the perfection of the animal and vegetable economy of nature."[25]

It must be kept in mind that "the perfection of nature" was a pregnant phrase in those days: Darwinians tended to deny it, while anti-Darwinians like most of those named by Fairbairn, affirmed it.

Among the papers read at the meeting there was one by Henry Fawcett, "On the Method of Mr. Darwin in his Treatise on the Origin of Species". It was apparently received somewhat coolly, and the press largely ignored it.[26] A paper by Professor Owen, who was very active at this meeting, also touched Darwinian topics. It was on Du Chaillu's gorillas: Du Chaillu himself, the African explorer, also made an appearance at the meeting.

The press apparently found Owen's paper rather technical, and hardly any reports occur. On the other hand, practically every newspaper and periodical which covered the meeting at all gave publicity to some remarks which Owen made in the subsequent discussion. One of the scientists present had reminded the audience of the Darwinian theory "brought forward so often" at the Oxford meeting, and asked Owen to point out the difference between man and the great apes.[27] As Huxley was not present at the 1861 meeting, Owen could be taken as representing the voice of science in general, and his devout treatment of the question of Man's origin was gratefully acknowledged in the religious press.[28] Some of his statements on the differences between the brains of men and apes, however, were not left unchallenged by Huxley, who wrote a letter to the president of the section which was read at the meeting. The letter was also published in the *Athenaeum*,[29] and marks one of the stages in the famous *hippocampus minor* controversy.[30]

---

[25] *Report*, 1861, lvi.

[26] Reports only in *Manchester Guardian*, 1861, Sep. 9, p. 6, *Nonconformist*, 1861, Sep. 18, 752. Brief adverse comments in *Watchman*, 1861, Sep. 18, 309.

[27] See report in the *Daily News*, 1861, Sep. 7, p. 6.

[28] *Nonconformist*, 1861, 732—3.

[29] *Athenaeum*, 1861: 2, 378.

[30] See above, p. 50.

Leading articles on the meeting in the daily papers still held aloof from the Darwinian question. Nor did the presidential address offer any natural opportunity. The *Times*, however, skirted the subject by discussing the origin of languages and the antiquity of man, in connection with a paper read by John Crawfurd, who was president of the Geography and Ethnology section. Though non-committal, the *Times* appears to come down on the side of orthodoxy on the point of the antiquity of the human race, while being prepared to give up, like Crawfurd, the unity of the human race.[31]) One was by now at the start of the American Civil War, and the question of the Negro's place in nature was beginning to loom large, especially among those favouring the Confederate side — and they included almost all the important English press organs.

Cambridge, 1862

At the Cambridge session Professor R. Willis's address was unusually thin. He did not venture at all into the controversial Darwinian field. Though undoubtedly the doctrine was beginning to revolutionize biological thinking, the subject was apparently still too hot to be even mentioned by the elected leader of the scientific world.

In the sections, on the other hand, discussions were lively. The protagonists on either side, Owen and Huxley, were both present. Huxley was president of the Zoology section, and in his opening address he alluded to the Darwinian theory. The *Times* reported: "He passed a graceful encomium on the labours of Mr. Darwin, whose name was received with a burst of applause. The Professor emphatically affirmed that Mr. Darwin's work was as perfect in its logical method as it was accurate in its scientific facts."[32])

Owen offered two papers, and everybody expected controversies to arise. One reporter wrote that "an eager crowd of listeners was to be expected when Professors Owen and Huxley battled so strenuously concerning the mysteries of being, the differences of species, and the tremendous issues of life."[33]) Many of the eager crowd, however, seem to have been disappointed, and one writer in a weekly Review declared that "the performance . . . was quite unequal to the expectation; the two champions seemed shy of putting forth all their strength. . . It is to be

---

[31]) *Times*, 1861, Sep. 16, p. 6.
[32]) *Times*, 1862, Oct. 3, p. 5.
[33]) *John Bull*, 1862, Oct. 18, 672. (quoting the *Cambridge Chronicle*).

regretted that the sense of public duty was not strong enough in the two professors to force them into the battle which so many had come to see."[34] There was, however, a lively discussion after one of the papers, and it was widely reported in the press. It turned on the significance of the difference between man's brain and the ape's, and Huxley, who declared that the difference was one of degree only, seems to have been supported by most of the scientists present. The chairman on the occasion, W. N. Molesworth, when trying to mediate between the opposing sides, hardly helped Owen by declaring that "as there was admitted to be a moral and psychical gulf between man and the higher apes, there must also be a structural gulf, although Professor Owen might not have succeeded in exactly ascertaining it, and although it might never be discovered."[35]

Yet another paper which was widely reported was Molesworth's "On the Influence of Change in the Conditions of Existence in Modifying Species and Varieties." The speaker cautiously supported the Evolution theory, which he did not consider irreconcilable with Scripture, while contending that Natural Selection was insufficient to account for the changes. Molesworth held that external conditions "impress on the variations a definite direction."[36] Molesworth's paper is important because it indicates the lines which discussion on Darwin's theories were henceforth to follow in the Association, and among informed people generally. The Evolution theory was to be accepted, or at any rate not directly rejected, while a large number of objections were to be raised against Natural Selection as an explanation of Evolution. It is also significant that though Molesworth did not accept Darwin's theories he referred to Darwin in respectful terms. The same tone prevailed in the discussions that followed upon the paper. The President of the section, Professor Daubeny — whose paper in 1860 had outlined views similar to Molesworth's — recommended to the audience to read Darwin's 'second book' (i. e, the *Orchids*) "as it would dispel many notions which had been wrongly entertained with regard to the tendency of his writings."[37] It seems reasonable to conclude that 1862 marked a turning-point for Darwin at the Association. He was perhaps not yet quite

---

[34] *London Review*, 5, 1862, 314.

[35] *Daily Telegraph*, 1862, Oct. 4, p. 2.

[36] *Report*, 1862, 113.

[37] *Daily News*, 1862, Oct. 3.

respectable in the eyes of the majority, but he was definitely gathering strength in the Biology section.

The prevalence of Darwinian, or near-Darwinian views at the 1862 meeting, was a constant theme in leaders in the press. In the religious press this gave rise to grave apprehensions. The *English Churchman* wrote: "We regret to observe that Messrs Crauford and Darwin's essentially antiscriptural notions with regard to the origin of Man have been again brought forward at this meeting. Professor Huxley was the champion of Mr. Darwin's mischievous theory — that Man is a development from brutes — and Mr. Crauford . . . that mankind did not spring from a single pair of human beings."[38]) The *Times* expressed its sympathy with this point of view: "We must not be surprised . . . at the indignation which the 'Darwinian theory' represented by the Professor [i. e., Huxley], never fails to provoke."[39])

Newcastle, 1863

The Darwinian question was very much in people's minds when the 1863 meeting convened at Newcastle. Sir Charles Lyell had published on the *Antiquity of Man*, and Huxley on *Man's Place in Nature*. The application of Darwin's theory to man had begun in earnest.

This is the first time that there occurs a direct reference to Darwin's doctrines in the presidential address. Sir William Armstrong, who chiefly concerned himself with technological problems, said in the part devoted to biology:

> "Neither is there any lack of bold speculation contemporaneously with this painstaking spirit of inquiry. The remarkable work of Mr. Darwin promulgating the doctrine of natural selection has produced a profound sensation. The novelty of this ingenious theory, the eminence of its author, and his masterly treatment of the subject have perhaps combined to excite more enthusiasm in its favour than is consistent with that dispassionate spirit which it is so necessary to preserve in the pursuit of truth. Mr. Darwin's views have not passed unchallenged, and the arguments both for and against have been urged with great vigour by the supporters and opponents of the theory. Where good reasons can be shown on both sides of a question, the truth is generally to be found between the two extremes. In the present instance we may without difficulty suppose it to have been part of

---

[38]) *English Churchman*, 1862, Oct. 9, 987. See also: *Patriot*, 1862, Oct. 9, 668 —9, *Record*, 1862, Oct. 8, p. 2.

[39]) *Times*, 1862, Oct. 7, p. 7.

the great scheme of creation that natural selection should be permitted to determine variations amounting even to specific differences where those differences were matters of degree; but when natural selection is adduced as a cause adequate to explain the production of a new organ not provided for in original creation, the hypothesis must appear, to common apprehensions, to be pushed beyond the limits of reasonable conjecture."[40]

This mode of treating the Darwinian theory was becoming increasingly common among the informed public. Evolution was granted, at least to a large extent, but the necessity of supplementing Natural Selection was stressed. Armstrong ended his reflections on these matters by expressing his belief in a "Great presiding Intelligence",[41] thus continuing the line of thought already indicated by Daubeny in 1860 and Molesworth in 1862, and around which practically all well-informed anti-Darwinians were to rally towards the end of the 'sixties.[42]

The section of Zoology contained nothing of Darwinian interest this year. In the subsection of Physiology, on the other hand, Darwinism seems to have had a stronghold. Its president, Professor Rolleston, praised Huxley's *Man's Place* in his opening address, and a paper by one Dr. Embleton on the chimpanzee supported Huxley's views on this subject. Huxley himself, however, was not present at the meeting, and perhaps this explains why some of the audience seem to have been dissatisfied with the discussions: the *Guardian*, somewhat surprisingly, thought that the subject of Man's relationship with the Ape had been "buried beneath abstruse anatomical details, and ... hardly emerged into the light of open day".[43] Several papers, however, reported Embleton's paper.[44] The *Star* concluded: "The foot of the ape being proved to be the same as man's it is of no use now to show that the feet of the ape and the negro are identical."[45]

These comments show how the Darwinian question was bound up with one of the burning political questions of the day: that of the place of the negro. This question was hotly debated in the Geography and

---

[40] *Report*, 1863, lxiii.

[41] *Report*, 1863, lxiv.

[42] See above, p. 28, 33.

[43] *Guardian*, 1863, Sep. 9, 841.

[44] Full report in *British Medical Journal*, 1863: 2, Sep. 12, 305. Brief report e. g. in *Daily News*, 1863, Aug. 29, p. 2.

[45] *Evening Star*, 1863, Sep. 3, p. 3.

Ethnology section, where two speakers especially, John Crawfurd and James Hunt, actively defended the thesis that the negro was a different species of man from the European. As they both rejected the Darwinian theory, the implication was that they denied any connection between the white man and the negro. Such a doctrine was obviously very convenient to the defenders of white supremacy: the negro was represented as intermediate, from a systematic point of view, between the white man and the ape, while at the same time the absolute distinction between man — or at least European man — and the brutes was upheld. The political implications of the doctrine were evident: one speaker, Carter Blake, was introduced in the *Star* as one "who appears to act the part of Confederate physiologist, in company with Dr. Hunt."[46]) But the *Star* was a paper with radical leanings. Poking fun at the Confederates was not popular in the rest of the press, in which Crawfurd's paper, especially, was reported faithfully. The *Times* gave him three whole columns,[48]) but unlike most of the other papers it omitted to report the subsequent discussion, where Crawfurd's views were controverted by one Mr. Craft, who was described as "a gentleman of colour". The *Daily News* reported cheers for all those who spoke for the unity of the human race, and the equality of all men.[49])

The editorial statements on the meeting turned largely on the questions raised by Crawfurd and Hunt. It is interesting to see how many religious papers, especially, made no distinction between Darwin and Crawfurd, in spite of the fact that Crawfurd explicitly rejected the Darwinian theory. What was considered relevant was that they both disregarded the evidence of the Bible. The *Inquirer* held that "those who adopt the Darwinian hypothesis, and believe that we are first cousins of the ape, may perhaps with some consistency go on to represent the African as not only a different but an essentially inferior race."[50]) The *Guardian*, however, made a distinction: "There can be no truce between those who hold the theory of Development, and those who deny the Unity of the Human race," and drew the comforting

---

[46]) *Evening Star*, 1863, Sep. 2, p. 3.
[47]) — — —
[48]) *Times*, 1863, Aug. 31, p. 7.
[49]) *Daily News*, 1863, Aug. 31, p. 2.
[50]) *Inquirer*, 1863, Sep. 5, 566. See also *Patriot*, 1863, Sep. 10, 594, and *Morning Advertiser*, 1863, Sep. 4, p. 4.

conclusion that therefore "those who entertain their own doubts on both may fairly require the opposing champions to fight out their battle in open lists, before they are required to give their adhesion to either."[51]

Bath, 1864

Sir Charles Lyell's election to the presidency of the British Association at its Bath meeting marks a stage in the attitude of British science towards Darwin. Many people connected Lyell's name with Darwin's, and in fact their views coincided on many points. Lyell's election may thus be seen as a sort of recognition of the Darwinian theory. In his address, however, Lyell was extremely cautious, and did not at all mention his friend's name, except to say that he agreed with him in regarding the geological record as extremely fragmentary. The address, which was reported in full in all the chief dailies, was received favourably on every hand; in some papers, it seems, even with relief. The *Morning Advertiser* thought that "Sir Charles Lyell has, to a great extent, neutralised the poison which his own pen had lately infused into the scientific literature of our age."[52] Almost everybody praised the cautious tone of the address,[53] or, as one paper called it, its freedom from "dogmatism."[54]

No Darwinian discussion seems to have occurred in the Zoological section this year, and in the section of Ethnology also the activity seems to have roused less interest than the year before. The presidential address in the Geology section, by Professor John Phillips, did, however, skirt the controversial subject of gradual evolution. Without mentioning Darwin's name, Phillips made clear his opposition to the theory, and to Uniformitarianism in general:

> "If in some thousands of years of human experience no very material change has happened in our wild plants or wild animals, or in cultivated grains, or domestic birds and quadrupeds, it is evident that no considerable changes of this kind can arise from such causes as are now in action without the aid of periods of time not contemplated in our chronology."[55 A]

---

[51] *Guardian*, 1863, Sep. 9, 841.

[52] *Morning Advertiser*, 1864, Sep. 17, p. 4.

[53] *London Review*, 9, 1864, 320, *Daily Telegraph*, 1864, Sep. 15, p. 4.

[54] *Evening Star*, 1864, Sep. 15, p. 4.

[55 A] *Report*, 1864, 46.

Phillips' was reported lengthily in the conservative *Globe* and *Standard*,[55 B]) but was ignored in the other newspapers.

Birmingham, 1865

A curious feature of the British Association meetings from 1864 to 1870 is the regular alternation of Darwinians and anti-Darwinians in the presidential chair, the former occupying it in even years, the latter in odd years. After Lyell came Phillips, who, as an exponent of catastrophist views in geology, was not prepared to accept Darwin's evolution theory. In his presidential address, he referred to the *Origin of Species* as an "elegant treatise", and though he did not specifically reject the hypothesis, he stressed above all the need for more evidence: "That such evidence will be gathered and rightly interpreted, I for one neither doubt nor fear, nor will any be too hasty in adopting extreme opinions or too fearful of the final result, who remember how often that which is true has been found very different from that which was plausible." The utmost that he himself would grant as a justifiable generalisation from the available evidence was that each species "has been submitted to, or is now undergoing, the pressure of a general law, by which its duration is limited in geological time."[56]) This was the position of the idealistic Nature philosophers of the early nineteenth century Continental school. Phillips's adoption of the idea of a providentially ordered evolution is also clear from his treatment of the question of man: "We have ... sufficient proof of the late arrival of man upon the Earth, after it had undergone many changes, and had become adapted to his physical, intellectual, and moral nature."[57])

The difference between Phillips's and Lyell's views was commented on in several leaders on the address.[58]) Many writers praised the President for his "freedom from dogmatism",[59]) and the religious organs, especially, were highly appreciative.[60])

Little else of Darwinian interest was reported in the press from this meeting, except some brief references to discussions in the section of

[55 B]) *Globe*, 1864, Sep. 16, p. 4, *Standard*, 1864, Sep. 16, p. 6.

[56]) *Report*, 1865, lxiii.

[57]) *Report*, 1865, lix.

[58]) *Nonconformist*, 1865, Sep. 13, 743. *Manchester Guardian*, 1865, Sep. 11, p. 2, *Evening Star*, 1865, Sep. 7, p. 2, *Daily News*, 1865, Sep. 7, p. 4.

[59]) *Daily News*, 1865, Sep. 7, p. 4, *Standard*, 1865, Sep. 8, p. 4.

[60]) *Guardian*, 1865, Sep. 13, 928—9, *Patriot*, 1865, Sep. 14, 593.

Geography and Ethnology, where Rawlinson, the president, contended that man was created in a state of civilisation[61]) and Crawfurd, as was his wont, advocated his views of the "plurality" of the human race.[62 A])

Nottingham, 1866

At the Nottingham meeting in 1866 Darwin's theory was for the first time one of the main themes of the presidential address. W. R. Grove, who had been primed by J. D. Hooker on the biological topics he treated, [62 B]) discoursed at length on the uniformitarian doctrine of Continuity and gradual change:

> I am not going to put forward any theory of my own, I am not going to argue in support of any special theory, but having endeavoured to show how, as science advances, the continuity of natural phenomena becomes more apparent, it would be cowardice not to present some of the main arguments for and against continuity as applied to the history of organic beings . . . If I appear to lean to the view that the successive changes in organic beings do not take place by sudden leaps, it is, I believe, from no want of an impartial feeling, but if the facts are stronger in favour of one theory than another, it would be an affectation of impartiality to make the balance appear equipoised."[63])

It is evident that Grove felt he must adopt a defensive attitude. He knew, of course, that he was the first to advocate these views from the presidential chair. It is true that Grove said nothing on the Natural Selection theory: but the general tenor of his argument, which was forceful and strictly scientific, was clearly pro-Darwinian, and was so interpreted in the press. Liberal organs tended to praise the address, though several of them do not seem to have understood the continuity doctrine in the same way as Grove: they wished to give it a metaphysical interpretation, as a providential law of development.[64])

The *Times* was less satisfied with Grove: "He makes a transition from the experimental field of science to the speculative, and enters upon the much-debated question of the Origin of Species . . . Science . . . can observe facts, but the origin is out of her reach."[65]) And most

---

[61]) Reports in *Daily Telegraph*, 1865, Sep. 11, p. 3, *Pall Mall Gazette*, 1865, Sep. 12, 381.

[62 A]) Report in *Evening Star*, 1865, Sep. 11, p. 3.

[62 B]) Leonard Huxley, *Life and Letters of Sir J. D. Hooker*, Vol. II, 105.

[63]) *Report*, 1866, lxxviii.

[64]) *Daily Telegraph*, 1866, Aug. 24, p. 4, *Daily News*, 1866, Aug. 24, p. 2.

[65]) *Times*, 1866, Aug. 24, p. 8.

religious organs were severe:[66]) the *Methodist Recorder* was "surprised and grieved to find that an accomplished philosopher like Mr. Grove should commit himself so decisively to the derivative hypothesis of Mr. Darwin."[67])

In the Sections also Darwinism was strongly represented at this meeting, so that the *Guardian* reported that "it was impossible to pass from Section to Section without seeing how deeply those views [of continuity and gradual development] have leavened the scientific mind of the day . . . Some form or other of the Darwinian theory . . . was everywhere in the ascendant."[68])

Huxley was this year president of the section of Biology, and Hooker was also present at the meeting. There were, however, anti-Darwinians as well. One of them, Professor G. M. Humphry, who was president of the sub-section of physiology, entered in his address upon a detailed discussion of the Natural Selection theory, advocating instead an evolution theory where changes took place, "not by slow gradations, but by sudden start, by something resembling a new creation."[69]) Humphry's paper, which was fully reported in the *Morning Post* and the *Guardian*[70]) is of interest because he foreshadows a line of argument which was later to be developed more fully by St. George Mivart, in his *Genesis of Species*. Humphry's was the first close criticism of the Natural Selection theory to be heard at the Association.

In the new sub-section of Anthropology, A. R. Wallace was president, but Carter Blake — and Crawfurd in the Geography section — were still very active promulgating their old views: their anti-Darwinian pronouncements were carefully reported, above all in the anti-Darwinian press.[71]) The most anti-Darwinian paper of all, however, was read in the section of Geography, by one J. M. Reddie, "On the various Theories of Man's Past and Present." The *Daily News*, which briefly reported the paper, gave a vivid picture of its reception: "The opinion of the scientific men present was expressed by Mr. Huxley, who said that the paper really belonged to the ethnological section, but would

[66]) E. g. *Record*, 1866, Aug. 27, p. 2, *Patriot*, 1866, Aug. 30, 572.

[67]) *Methodist Recorder*, 1866, Aug. 31, 308.

[68]) *Guardian*, 1866, Sep. 5, 917.

[69]) *Report*, 1866, 88.

[70]) *Morning Post*, 1866, Aug. 24, p. 2, *Guardian*, 1866, Sep. 5, 925—8.

[71]) E. g. *Standard*, 1866, Aug. 27, p. 6, *Record*, 1866, Aug. 27, p. 3.

perhaps not have passed muster with the committee there."[72]) In the anti-Darwinian press, Reddie was of course praised: the *Standard* devoted three columns to his paper, but did not report Huxley's sarcastic remarks. The *Record* commended Reddie in a leader in the following words: "How refreshing it is to turn from the inaugural address to the manly and faithful paper read by Mr. Reddie!"[73])

Dundee, 1867

The presidential address of the Duke of Buccleugh at Dundee was not on traditional lines. Instead of discoursing on the state of scientific progress, the Duke served the audience with an ordinary speech: the official Report, contrary to custom, did not print it.[74]) The press were evidently taken aback, and few rendered a full version of the ducal platitudes. Praise for the Duke was almost wholly confined to such Tory organs as the *Morning Post*, which commended the Association for the choice of a man of "social position"[75]) as president, and the *Globe*, which was gratified that the Duke "ventured to tell the meeting that the study of science ought to be closely connected with true religion."[76]) The *Churchman*, a High Church and Tory organ, thought that the president delivered a "manly, unpretending address."[77])

In organs of liberal opinions the leader writers were usually severe,[78]) and the *Morning Advertiser* managed to hit at both Buccleugh and the scientific spirit which it found prevalent in the Association: "we regard the selection of a nobleman . . . as a proof that some misdirection of its scientific advantages has rendered necessary adventitious aid."[79])

From the meeting itself the most important feature of Darwinian interest was Sir John Lubbock's paper "On the Origin of Civilization

---

[72]) *Daily News*, 1866, Aug. 28, p. 2. For "ethnological", we should read "anthropological". Traditionally, Ethnology formed part of the Geography section; from 1869 on, it was included as a sub-section of Biology, and was renamed Anthropology from 1870 onwards. The 1866 experiment of a separate sub-section of Anthropology was discontinued in 1867 and 1868.

[73]) *Record*, 1866, Aug. 27, p. 3, *Standard* 1866, Aug. 27, p. 6.

[74]) *Report*, 1867.

[75]) *Morning Post*, 1867, Sep. 10, p. 4.

[76]) *Globe*, 1867, Sep. 6, p. 2.

[77]) *Churchman*, 1867, Sep. 12, 584—5.

[78]) *Daily Telegraph*, 1867, Sep. 7, p. 4, *Guardian*, 1867, Sep. 11, 973. *English Independent*, 1867, Sep. 12, 1196—7.

[79]) *Morning Advertiser*, 1867, Sep. 11, p. 4.

and the Early Condition of Man", in which he argued that the original condition of all mankind was one of utter barbarism, and that an advance to civilization had occurred independently at several places. The opposite doctrine, associated at the time with the name of Archbishop Whateley, to the effect that man was created in a state of civilization, was defended by several anti-Darwinians, but Lubbock seems to have had the support of the majority of the speakers. The paper and the discussion were widely reported in the press.[80])

In the section of Geography, Crawfurd still defended his old views, while Hunt seems to be moving towards Darwin: "I have felt for some time that enough has now been said by us against the views of Mr. Darwin and his English disciples."[81]) The emancipation of the negro slaves in America was now an accomplished fact.

The most widely reported speech of the session was that of John Tyndall to the Operative Classes, on "Matter and Force." Tyndall maintained the right of scientists to carry on their research and draw their conclusions regardless of what he bluntly called religious superstitions. He stressed that scientific explanations were concerned with facts only, and that special interpositions of Supernatural power had to be absolutely excluded from the scientist's account of nature. At the same time, Tyndall also admitted that the scientist, because he was concerned with facts only, was altogether unable to answer the question *whence?* All this had an evident bearing on the Darwinian questions, and leader-writers on the meeting often pointed out the connection. Some drew comfort from the fact that Tyndall had at least left the question of the origin of the universe outside the range of science: "It is here that Revelation steps in to lift the veil."[82]) Others criticised him for his "scornful denunciation of belief in special dispensations of Providence."[83])

Norwich, 1868

At Norwich the Association again saw a Darwinian in the presidential chair. Much of J. D. Hooker's address consisted of an exposition of the Darwinian doctrine. Among other things he asserted that the theory

---

[80]) E. g. *Daily Telegraph*, 1867, Sep. 10, p. 3, *Daily News*, 1867, Sep. 11, p. 3.

[81]) *Daily News*, 1867, Sep. 9, p. 3.

[82]) *Nonconformist*, 1867, 767.

[83]) *English Independent*, 1867, Sep. 12, 1196.

of Natural Selection was "an accepted doctrine with every philosophical naturalist," though he admitted that he then included "a considerable proportion who are not prepared to admit that it accounts for all Mr. Darwin assigns to it."[84]) He also discussed various objections raised against the theory, including what he termed "metaphysical" ones, and ended with a warning directed to both scientists and theologians not to mix up the two domains:

> "Both parties must beware how they fence with that most danger-
> ous of all two-edged weapons, Natural Theology; a science, falsely
> so called, when, not content with trustfully accepting truths hostile
> to any presumptuous standard it may set up, it seeks to weigh the
> infinite in the balance of the finite, and shifts its ground to meet the
> requirements of every new fact that science establishes, and every
> old error that science exposes. Thus pursued, Natural Theology is
> to the scientific man a delusion, and to the religious man a snare,
> leading too often to disordered intellects and to atheism."[85])

As was to be expected, Hooker had a rather bad press in most con-servative and religious organs. The *John Bull* called his address "a melancholy exhibition of verbose mediocrity in excelsis", which only gave Hooker "an opportunity of puffing Mr. Darwin's latest hallu-cinations".[86]) The *English Churchman* looked upon it as just another sign that "rank infidelity" was the "predominating hue"[87]) of the Association. Not all, however, resorted to this blind denunciation,[88]) and in the more liberal religious organs one used the occasion for a rational discussion of the Darwinian doctrines.[89]) In the *Daily Tele-graph*, *Daily News*, *Manchester Guardian* and the *Examiner* Hooker was given high praise.[90])

The "Lecture to the Operative Classes" this year was given by Huxley, who spoke "On a piece of Chalk."[91]) The lecture contained little that

---

[84]) *Report*, 1868, lxx.

[85]) *Report*, 1868, lxxiv.

[86]) *John Bull*, 1868, Aug. 22, 571.

[87]) *English Churchman*, 1868, Aug. 27, 525. See also *Morning Advertiser*, 1868, Aug. 21, p. 4.

[88]) *Morning Post*, 1868, Aug. 29, p. 4, *Standard*, 1868, Aug. 21, p. 4.

[89]) *Guardian*, 1868, Sep. 2, 977; *Inquirer*, 1868, Aug. 29, 549—50.

[90]) *Daily Telegraph*, 1868, Aug. 21, p. 6, *Daily News*, 1868, Aug. 21, p. 4, *Manchester Guardian*, 1868, Aug. 24, p. 2, *Examiner*, 1868, Aug. 22, 531, 538.

[91]) Subsequently published in *Macmillan's Magazine*, 1868, and in *Lay Ser-mons*, 1870.

was controversial, and few papers reported it. The anti-Darwinians
at the meeting were represented in the person of the Rev. F. O. Morris,
whose lecture "On the Difficulties of Darwinism" got publicity in the
Times and *Morning Post*.[92]) The impression it made was summed up
by the *Guardian:* it "served only to bring out the fact that nobody
else in the Section found any difficulties at all in it."[93]) This may be
accepted as an apt commentary on the whole Norwich meeting: the
Guardian recognized the "predominance of Darwinism" throughout:
"its reign was triumphant and almost unopposed."[94])

Exeter, 1869

At the Exeter meeting the following year the balance was restored.
G. G. Stokes, the physicist, when dealing with biology in his presidential
address, stressed the absolute distinction between dead and living
matter, the mysteriousness of life and even more of mind, and gave
his support to the argument from Design. All these points were normally
insisted on by Darwin's opponents.

"Admitting to the full as highly probable, though not completely
demonstrated, the applicability to living beings of the laws which
have been ascertained with reference to dead matter, I feel constrained
at the same time to admit the existence of a mysterious *something*
lying beyond, — a something *sui generis*, which I regard, not as balancing
and suspending the ordinary physical laws, but as working with them
to the attainment of a designed end.

What this *something*, which we call life, may be, is a profound mystery.
We know not how many links in the chain of secondary causation may
yet remain behind: we know not how few. It would be presumptuous
indeed to assume in any case that we had already reached the last
link, and to charge with irreverence a fellow-worker who attempted
to push his investigation yet one step further back. On the other hand,
if a thick darkness enshrouds all beyond, we have no right to assume
it to be impossible that we should have reached even the last link of
the chain; a stage where further progress is unattainable, and we
can only refer the highest law at which we stopped to the fiat of an
Almighty Power . . .

When from the Phenomena of life we pass on to those of mind,
we enter a region still more profoundly mysterious . . . Science can be
expected to do but little to aid us here, since the instrument of research

---

[92]) *Times*, 1868, Aug. 27, p. 8, *Morning Post*, 1868, Aug. 27, p. 2.
[93]) *Guardian*, 1868, Sep. 2, 977.
[94]) *Guardian*, 1868, Sep. 2, 977.

is itself the object of investigation. It can but enlighten us as to the depth of our ignorance, and lead us to look to a higher aid for that which most nearly concerns our wellbeing."[95]

The devout tone of this address, ending in a Prayer Book phrase, was in marked contrast to Hooker's. The press received it very favourably: the religious organs were delighted, and so was the *Times*.[96] Only the *Manchester Guardian* indicated some impatience.[97]

Among the features of the 1869 meeting were the papers of three clergymen, who attacked different aspects of the Darwinian theory. The papers, as well as the subsequent discussions, were briefly reported in most of the newspapers. The consensus of opinion was that the anti-Darwinians "got the worst of it in argument."[98]

At this meeting also Sir John Lubbock continued his dissertation on early man, started at the Dundee meeting in 1867. Lubbock now defended his own views against those of the Duke of Argyll, who had in the meantime published a book on the subject, *Primeval Man*. The discussion — widely reported in the press — was remarkable, among other things, because A. R. Wallace here declared his belief that Natural Selection was inadequate to account for the origin of man. This was a view that he defended at the same time in the columns of the *Quarterly Review*,[99] and later in his *Contributions to the Theory of Natural Selection*. This partial conversion of a prominent Darwinian naturally pleased the anti-Darwinians, who did not fail to give prominence to the news.[100]

Liverpool, 1870

After Stokes, there was to be a Darwinian in the chair again, and a Darwinian there was, with a vengeance. After some manoeuvring, which is not without its amusing points, Huxley was elected President for the Liverpool meeting. The *Times* reports the president of the Council of the Association as saying that he "was quite aware that if they went into individual opinions it might have rendered it difficult

---

[95] *Report*, 1869, civ—cv.

[96] *Standard*, 1869, Aug. 20, p. 4. *Daily News*, 1869, Aug. 19, *Star*, 1869, Aug. 19, p. 4, *Times*, 1869, Aug. 19, p. 6.

[97] *Manchester Guardian*, 1869, Aug. 20, p. 2—3.

[98] *Pall Mall Gazette*, 1869, Aug. 24, 739. See also *Guardian*, 1869, Aug. 25, 960—1.

[99] *Quarterly Review*, 126, 1869, 359—394.

[100] *Record*, 1869, Aug. 25, p. 3.

for him to introduce the gentleman he had mentioned." But he was convinced that Huxley would not misuse his position, but would "put aside individual feelings for the feelings of those for whom he spoke." The reporter added: "There seems to be a very general feeling that Professor Huxley in the chair of the British Association will be in as difficult position as Mr. Bright in the Ministry. He is the champion of views to which large classes of people entertain very strong objections; and however discreet he may be in the absence of opposition, his best friends tremble for him if those views should be impugned."[101])

From reports in other, less discreet organs it is clear that there was very nearly a split in the Council over the election: the opposition to Huxley had wished to have Lord Stanley as president, but that nobleman refused to stand when he understood that he would not be unopposed, and that "the chances were two to one that [Huxley] would have been the victor."[103]) As it was, Huxley was unopposed, but the acting president of the Council, Sir Stafford Northcote, seems to have taken the extraordinary step of openly dissociating himself from the vote for Huxley.[104])

In several organs — chiefly liberal ones — Huxley's election was acclaimed with satisfaction. The *Spectator* severely condemned the attitude expressed in the *Times:* "The *Times* exactly represents, in this instance, the idea of the majority of Englishmen, and we cannot conceive of any idea at once more unwise and more ignoble ... But, says the *Times* — it is not merely a reporter who says it, though the words appear in a report, for the *Times* does not allow its reporters to lecture in that style — Mr. Huxley is so indiscreet ... If we had only an 'indiscreet' Archbishop! — but that being impossible, let us be thankful that we shall next year have an indiscreet President of the British Association."[105]) The *Star* expressed similar feelings. It condemned Northcote's "faltering, apologetic, patronising, warning, hoping, fearing phrases."[102])

In his presidential address Huxley discoursed on a single subject: the question whether living organisms can be produced out of dead

[101]) *Times*, 1869, Aug. 25, p. 6.

[102]) *Star*, 1869 Aug. 26, p. 4.

[103]) *Daily Telegraph*, 1869, Aug. 24.

[104]) *Spectator*, 1869, Aug. 28, 1008.

[105]) *Spectator*, 1869, Aug. 28, 1010.

matter. After considering the evidence carefully, and putting great weight on Pasteur's experiments, Huxley concluded that the feat had not been achieved. As far as present scientific knowledge went, *abiogenesis* did not occur, and *biogenesis* was the rule. Further, Huxley rejected *xenogenesis* — the theory that organisms of one species can suddenly give birth to organisms of a quite distinct species: this had an immediate bearing on the Darwinian discussions at the time.[106] But though Huxley rejected the experiments of those who thought they had proved Spontaneous Generation, he did not stop at that point:

> But though I cannot express this conviction of mine too strongly, I must carefully guard myself against the supposition that I intend to suggest that no such thing as abiogenesis has ever taken place in the past, or ever will take place in the future. With organic chemistry, molecular physics, and physiology yet in their infancy, and every day making prodigious strides, I think it would be the height of presumption for any man to say that the conditions under which matter assumes the properties we call "vital" may not, some day, be artificially brought together. All I feel justified in affirming is, that I see no reason for believing that the feat has been performed yet . . . if it were given me to look beyond the abyss of geologically recorded time to the still more remote period when the earth was passing through physical and chemical conditions, which it can no more see again than a man can recall his infancy, I should expect to be witness of the evolution of living protoplasm from not living matter. I should expect to see it appear under forms of great simplicity.[107]

"This bold, but not unphilosophical, statement," wrote the *Manchester Guardian*, "was listened to without a single token of applause. Rather, the audience seemed for a moment appalled by the cool, unflinching audacity of the speaker."[108] But the general reaction to the address seems to have been almost disappointment: "it was not naughty enough."[109] And in fact, the "bold statement" where Huxley introduced the important qualifications to his general acceptance of the "biogenesis" doctrine, was either overlooked or brushed aside. The religious press was apparently both astonished and relieved at finding Huxley defending, in effect, a doctrine which seemed to them to be incompatible

---

[106] See below, p. 265—7.
[107] *Report*, 1870, lxxxiii—lxxxiv.
[108] *Manchester Guardian*, 1870, Sep. 16, p. 3.
[109] Ibid.

with his Darwinian views.[110]) "We trust that the great question of biogenesis or abiogenesis is now settled,"[111]) wrote the *English Church-man*, and many expressed similar convictions.[112])

In the sections there were several papers relative to the Darwinian theories, the most important being one by A. W. Bennett "On the Theory of Natural Selection looked at from a Mathematical point of View," in which the author maintained that small *random* variations would never accumulate in any single direction, since the incipient steps of a modification of an organ would be entirely useless to the individual.[113]) The paper is symptomatic of the increasing emphasis that Darwin's opponents were at this time placing on the difficulties of Natural Selection. Bennett himself gave plain adherence to the evolution part of the theory. The change of emphasis, however, was taking the theory out of the realm of popular controversy. Natural Selection was too complicated for the common man to fathom. Reports of Bennett's paper were very sparse.

Another feature of Darwinian interest — though hardly reported in the press — was Professor Rolleston's address to the section of Physiology, in which he spoke of the relation between science and religion, and mentioned Darwin's theory as an instance. He exhorted theologians to consider the possibility that trespasses into the field belonging to the other party did not only occur from the side of the scientists.

The most widely reported lecture of the session, besides Huxley's address, was a lecture by Tyndall at one of the general meetings, "On the Scientific Use of the Imagination." What Tyndall advocated was complete freedom for the scientists to try out any hypotheses they might imagine, unfettered by the prejudices of the multitude: and he cited Darwin as an instance of a scientist to whom such freedom should be granted. Many defenders of traditional views were outraged. The *Times* commented sarcastically: "We had been under the impression that Natural Philosophers drew no bills. We do not presume to say

---

[110]) *Guardian*, 1870, Sep. 21, 1109, *English Independent*, 1870, Sep. 22, 922, *Record*, 1870, Sep. 16, p. 2.

[111]) *English Churchman*, 1870, Sep. 29, 485.

[112]) *Methodist Recorder*, 1870, Sep. 23, 538, *Morning Advertiser*, 1870, Sep. 20, p. 4.

[113]) The whole paper was published in *Nature*, 3, 1870, 30—33.

one word about the Evolution Hypothesis . . . The greater part of the
opposition [to that theory] is provoked . . . not by Science, but by the
imagination of men of Science . . . we look to men of Science rather for
observation than for imagination."[114]) The *English Churchman* observed
that "Mr. Darwin . . . appears to us as an instance of a philosopher
whose imagination has run away with him."[115])

Edinburgh, 1871

Like all scientifically informed people at this time, Sir William Thomson, who occupied the Chair at the Edinburgh meeting, was an Evolutionist. But he was not a Darwinian: In his address he declared that
the doctrine that life must spring from life seemed to him "as sure a
teaching as the law of gravitation."[116]) He then launched himself
upon expanding the famous hypothesis that life had come to this earth
from outer space:

> "We must regard it as probable in the highest degree, that there
> are countless seed-bearing meteoric stones moving about through
> space. If at the present instant no life existed upon this Earth, one
> such stone falling upon it might by what we blindly call *natural* causes,
> lead to its becoming covered with vegetation."[117])

This, said Thomson, "may seem wild and visionary; all I maintain is
that it is not unscientific."[117]) But Thomson's excursion into the biological field did not stop at that point. He also explicitly criticized
Darwin's Natural Selection theory:

> I have always felt that this hypothesis does not contain the true
> theory of evolution, if evolution there has been, in biology. Sir John
> Herschel, in expressing a favourable judgment on the hypothesis of
> zoological evolution (with, however, some reservation in respect to
> the origin of man), objected to the doctrine of natural selection, that
> it was too like the 'Laputan' method of making books, and that it
> did not sufficiently take into account a continually guiding and controlling intelligence. This seems to me a most valuable and instructive
> criticism. I feel profoundly convinced that the argument of design
> has been greatly too much lost sight of in recent zoological speculations."[118 A])

---

114) *Times*, 1870, Sep. 19, p. 9.

115) *English Churchman*, 1870, Sep. 29, 485. See also: *Methodist Recorder*, 1870,
Sep. 23, 538, *Record*, 1870, Sep. 23, p. 2, *English Independent*, 1870, Sep. 22, 922.

116) *Report*, 1871, Cii.

117) *Report*, 1871, Cv.

118 A) *Report*, 1871, Cv.

These passages were included in practically all the reports of the address, whether long or short. Editorial reactions to the comments on Natural Selection were favourable almost everywhere,[118 B]) while the 'dispersion' theory caused bewilderment. One reporter from the meeting described its reception by the audience: "This theory took everybody by surprise. There was a little laughter, and many people looked as if they expected it was the preliminary to some good joke."[119]) The *Times* leader commented sarcastically on the different and contradictory theories of the successive Presidents, and ridiculed the dispersion theory, but admitted, finally, that "after all, it only suggests a mode of Creation; it does not get rid of Creation or a CREATOR, and therefore leaves the great question of all where it finds it."[120])

The more extreme religious press were shocked by the dispersion theory: many would not believe that Thomson could have intended it seriously. The *Record* believed Thomson's position "to have been simply argumentative and controversial, and to have been directed throughout against the crude speculations of Darwin."[121]) The *Morning Advertiser* expressed itself in a similar vein: "Why should not Sir William Thompson enjoy and ventilate his craze?",[122]) and the *Daily News* wondered, "Is Sir William Thomson poking fun at some of his colleagues, or showing in his own person what the imagination can do when it leaves the common earth behind it, and soars forth into space?"[123])

I have found only one criticism of Thomson from a Darwinian point of view. The *Examiner* favourably contrasted the Huxleian hypothesis of abiogenesis with Thomson's dispersion theory, which was "its own refutation."[124 A]) The journal also published a letter from a correspondent, in which Thomson's use of the argument of design was criticized.[124 B])

In the sections this year there was little of Darwinian interest, except a discussion, where Huxley took part, of a paper by one Staniland Wake "On Man and the Ape", where the author maintained the view

---

[118 B]) Also religious organs: *Methodist Recorder*, 1871, Aug. 8, 438, *Inquirer*, 1871, Aug. 12, 505, *Nonconformist*, 1871, Aug. 9, 785.

[119]) *Manchester Guardian*, 1871, Aug. 4, p. 3.

[120]) *Times*, 1871, Aug. 7, p. 9.

[121]) *Record*, 1871, Aug. 9, p. 2.

[122]) *Morning Advertiser*, 1871, Aug. 8, p. 4.

[123]) *Daily News*, 1871, Aug. 4, p. 5.

[124 A]) *Examiner*, 1871, Aug. 5, 776.

[124 B]) *Examiner*, 1871, Aug. 12, 801.

advocated by Wallace, in 1869, that Man could not be due to the operation of Natural Selection alone. The interest in just these questions
had naturally been further increased by the appearance, earlier in
the year, of Darwin's *Descent of Man.* Almost all the papers reported
the discussion: the *Times* pouring contempt on the section where it
was read: "anything more utterly 'unscientific' than the tone of this
new department . . . it could never enter into the mind of man to conceive. The questions at issue seem to be essentially insoluble in the
present state of knowledge about facts, and they were at once carried
into the region of the emotions."[125])

In the section of Biology the president, Allen Thompson, rather
pointedly supported Darwinian views against Sir William, endorsing
Huxley's belief in the probability of "abiogenesis" at some time in
the earth's history.[126]) The press, however, hardly reported his paper.

Brighton, 1872

The president at Brighton was William B. Carpenter. He discoursed
in his address on the nature of science, and on its relation to philosophy
and religion. Science, said Carpenter, was concerned with finding natural
laws. The most fundamental of these laws — indeed, the only ones
properly so called — are those describing the action of a force. However, all laws are only human conceptions:

> They are Human conceptions, subject to Human fallibility; and . . .
> they *may* or *may not* express the ideas of the Great Author of Nature.
> To set up these Laws as self-acting, and as either excluding or rendering
> unnecessary the Power which alone can give them effect, appears to
> me as arrogant as it is unphilosophical. To speak of *any* Law as "regulat-
> ing" or "governing" phenomena, is only permissible on the assumption
> that the Law is the expression of the *modus operandi* of a Governing
> Power.[127])
> But when Science, passing beyond its own limits, assumes to take
> the place of Theology, and sets its own conception of the *Order* of
> Nature as a sufficient account of its *Cause*, it is invading a province
> of thought to which it has no claim, and not unreasonably provokes
> the hostility of those who ought to be its best friends.[128])

[125]) *Times*, 1871, Aug. 11, p. 4. See also Globe, 1871, Aug. 8, p. 3, *Daily
Telegraph*, 1871, Aug. 8, p. 6.
[126]) Brief mention in the *Manchester Guardian*, 1871, Aug. 5, p. 8.
[127]) *Report*, 1872, lxxxiii.
[128]) *Report*, 1872, lxxxiv.

It appears that in Carpenter's view Science was subservient to Theology. If natural laws were to be conceived as expressing the *modus operandi* of God, it could be maintained from the side of Theology (though Carpenter, to be sure, would not allow this) that God might well suspend their operation, or alter the conditions under which they are valid, in order to achieve specially designed ends. In other words, scientists would have to admit the legitimacy of miraculous explanations.

It was clear that these ideas could be applied to the Darwinian theory, which so often came up against appeals to the miraculous. In religious organs Carpenter was therefore reviewed very favourably. The *Nonconformist* wrote: "At length the extreme pressure which has been put on our reason by materialism has broken down. The hard, unbending ideal of a mechanical universe originating itself, without a Maker, and evolving itself without an Evolver, is becoming distrusted by Evolutionists."[129] Other religious organs tended to share these views,[130] while the *Morning Advertiser*, true to its narrow Evangelicalism, found Carpenter deviating too far in the direction of Berkeleianism, which might lead to scepticism.[131] In general, conservative organs were appreciative. The *Standard* wrote that if the address "tends rather more than might be thought desirable in the direction of metaphysics, it does so in order to encounter some of the wilder speculations of the day,"[132] and the *Morning Post* found it undeniable that "an authoritative check was required to the bold theorizing of a certain school, at all events, of our philosophers."[133] Some liberal papers, on the other hand, were critical. The *Daily Telegraph* thought Carpenter's criticism "barren of results,"[134] and the *Examiner* wrote almost caustically: "A profession of philosophical faith eminently conservative. The isolation of the speaker — the feeling that he represented a school of thought whose dwindling adherents are almost wholly made up of very old men, that his words would find few echoes in the lecture room though many in the pulpit — gave a dramatic point to his eloquence . . . this latest plea for Scoto-German transcendentalism."[135]

---

[129] *Nonconformist*, 1872, Aug. 28, 889.
[130] *English Independent*, 1872, Aug. 22, 885, *Inquirer*, 1872, Aug. 24, 541.
[131] *Morning Advertiser*, 1872, Aug. 17, p. 4.
[132] *Standard*, 1872, Aug. 15, p. 4.
[133] *Morning Post*, 1872, Aug. 16, p. 4. See also *John Bull*, 1872, Aug. 17, 572.
[134] *Daily Telegraph*, 1872, Aug. 16, p. 4.
[135] *Examiner*, 1872, Aug. 17, 815.

Carpenter also contributed to the Darwinian discussion elsewhere. In the section of Biology, Sir John Lubbock maintained in his presidential address, that Darwin's theory of Natural Selection had to be recognized as a *vera causa*.[136]) In the discussion Carpenter did not agree. "He thought that the cause really lay in the developmental powers which gave origin to the advances of type and varieties of form; and that 'natural selection' producing the 'survival of the fittest,' limited and directed its operation."[137])

To judge from press reports, there was little else of Darwinian interest at the meeting, and the discussion just quoted was not widely reported. When the theory of development had become accepted almost everywhere, and had almost ceased to shock, the nonscientific public lost interest in the discussions. The reporter of the *Daily Telegraph* seems to have taken this view: "[The section of Biology] possibly disappointed the ladies in striking no new Darwinian vein. It rigorously followed out the old one, however, and established with great copiousness of illustration, the doctrine of animal development."[138])

Summary

The British Association was not a purely scientific forum. As its audience consisted not only of scientists, but also of scientifically interested laymen, and as, moreover, it was recognized that the meetings also served a propagandistic function, papers had to be selected not only on the strength of their scientific quality, but also for their appeal to the general public. Above all, the discussions arising out of the papers often had another character than the one to be expected in purely scientific contexts. This was probably especially true of the Darwinian discussions: it is significant that Huxley, in his first clash with Owen in 1860, declined to enter into a discussion of facts at the meeting, but promised to present them elsewhere.[139])

For these reasons it would be rash to take the performers at the British Association, especially in the matter of Darwinism, as an accurately representative sample of the British scientific community. They do, however, certainly provide a valuable indication of the trend of

---

[136]) *Report*, 1872, 124.
[137]) Report in the *Times*, 1872, Aug. 17, p. 10.
[138]) *Daily Telegraph*, 1872, Aug. 16, p. 3.
[139]) Leonard Huxley, *The Life and Letters of T. H. Huxley*, I, 261.

thought within that community, and the development does indeed correspond fairly well with what we know from other evidence.

The presidential addresses reflect by and large the ideas of what may be called the governing class of scientific men: not, perhaps, the leading scientists, but those eminent scientists who had reached a position of social authority. They were the "older and honoured chiefs in science", whose opinions naturally tended towards the traditional and conservative. It was to such men in the inner councils of the Association that the job fell of electing the President for each successive year. And we know, from the case of Huxley, that the general opinions of the candidates were taken into consideration in these elections.

It is therefore hardly surprising to find the opinions on Darwin set forth in the presidential addresses to be on the whole less advanced than in the discussions in the sections, and in the scientific press. The Darwinian theory was not mentioned in any address until 1863, and then only to be rather severely judged. The next time it was taken up, in 1865, no judgment was passed upon it from the presidential chair, but a critical attitude was clearly conveyed. It was only in 1866 that a clearly Darwinian point of view was expressed by a British Association president: and after this date, with Darwinians and anti-Darwinians alternating in the chair for several years, the question of Evolution was a constantly recurring theme in the addresses. Pro-Darwinianism reached its culmination point in 1868, when Hooker was president: after that year, though Evolution was taken more or less for granted, criticism was increasingly directed against the Natural Selection theory.

In the sectional discussions Darwinian topics were debated all the time. In 1860 and 1861 it appears that the opponents of the new theory were still in the ascendant, but in 1862 the climate of opinion was already changing, and we find how anti-Darwinian speakers chiefly focus attention on Natural Selection rather than on Evolution. Henceforth, opposition against the Evolution theory no longer came from scientific biologists and physiologists, but mainly from clerical amateurs of natural history, and also from a certain group of anthropologists. Again, the later years of the period we are discussing were marked by an increasingly critical attitude towards the theory of Natural Selection. This is altogether the same sort of development as the one which emerged from our study of the press discussions in Chapter 3. Darwin had come to stay, but his battle was far from won.

As regards the press coverage of the Association debates, two facts stand out: first, that the press clearly tended to give prominence to Darwinian questions whenever they were brought up at the meetings; and second, that both comments and reports on these questions were marked by the same sort of bias as we have been accustomed to in other contexts. Finally, there is some evidence that the Darwinian debates were largely responsible for the increasing interest that the public bestowed on the Association's meetings. In this respect as well, the British Association serves well to illustrate the impact of Darwin on Mid-Victorian Britain.

CHAPTER 5.

Science and Religion: A Mid-Victorian Conflict

When reading contemporary discussions of the Darwinian theory, one is struck by the fact that both opponents and supporters often vigorously insisted that the doctrine must be judged on its scientific merits alone. Still, there were few critics indeed who refrained from offering, in addition to the scientific arguments, comments on the religious bearing of the new views — even if only to deny that they possessed any religious significance. This very denial may be taken as proof that they believed that their readers needed such an assurance, which was certainly a correct interpretation of the situation.

We have observed in a previous chapter that there was a clear correlation between attitudes towards the Darwinian theory on one hand, and religious opinions on the other. It would indeed have been surprising if it had been otherwise. The Biblical cosmogony as set forth in Genesis was still the prevalent belief, not only among the masses, but also among the educated.[1] It is true that the discoveries of geology, chiefly in the first half of the nineteenth century, had shaken these beliefs, and necessitated quite an extensive reinterpretation of the Mosaic account of Creation[2] — not to speak of earlier reinterpretations called forth by the Copernican revolution in astronomy. But at each of these advances of natural science the Church had yielded only on the most exposed points: the general accuracy of the Biblical story of the early history of the world, and especially of our race, was not allowed to be called in question.

It is true that many thoughtful Churchmen warned their less liberal brethren that "the Bible was not intended to teach scientific truth." But the Biblical version, even of these historical and scientific matters,

---

[1] See Willis B. Glover, *Evangelical Nonconformists and Higher Criticism in the 19th Century*, p. 49; and H. G. Wood, *Belief and Unbelief since 1850*, p. 50.

[2] See C. C. Gillispie, *Genesis and Geology*, especially Chapters 5 and 6.

had the sanction of universal tradition within the Christian world. Moreover, there was not much that history or science could put in its stead. When Darwin published, Prehistoric Archaeology was in its infancy: such terms as Stone Age, Bronze Age, Iron Age date from the middle of the nineteenth century. Anthropology and ethnology were in an equally rudimentary condition. Geology was barely half a century old as a science. When Voltaire had wished to ridicule the contention that the sea shells deposited in the various rock strata in the Alps were proof of the Deluge, he suggested that they had been dropped by pilgrims or crusaders on their way to the Holy Land, or that they were formed by chemical action.[3]

There was in fact hardly any reason why people should not think that, even if the world as such might date back untold aeons of time, as geologists now claimed, then at any rate the history of mankind stretched only some 6,000 years back, which was the date Archbishop Ussher had arrived at from Biblical data.[4] The Bible and the annals of antiquity took one back, literally, to the dawn of time, to the pristine freshness of mankind's youth.

This view was so deeply rooted and so universal as to seem to most people as nothing but plain common sense. It was not at all necessary to invoke the supernatural sanction of the Divine inspiration of the Bible in order to make the story credible. On the contrary, the fact that the Biblical account seemed so fully in accordance with common sense may even have appeared as an additional reason to believe in the supernatural origin of the Bible. As a writer in the Evangelical *Record* expressed it, "The initial statement of the Bible comes to us with a force, a clearness and a proof such as no conscientious man can repel, for it is backed up not only by all those evidences which prove the Scriptures to be the Word of God, but also by the testimony of every man's conscience and by the universal traditions of the human race."[5] It was easy to say, after the new geological and historical doctrines had been assimilated, that the religious beliefs which had to be changed on account of the scientific discoveries had no religious

---

[3] Voltaire, *Singularités de la Nature*, Ch. XII—XIV.

[4] "The chronology of James Ussher, archbishop of Armagh, is still the standard adopted in editions of the English Bible." (*Dictionary of National Biography*, s. v. Ussher.)

[5] *Record*, 1866, Aug. 27, p. 2.

significance. But how was the common man to tell which of his beliefs were truly religious, and which were not? To most people the body of knowledge that they obtained from the Bible must have appeared at least to some extent religious. Not for nothing was Biblical history sacred history.

The Copernican conflict between science and religion had concerned the constitution of the universe around us. The recent conflict between Geology and Genesis had touched the history of the earth, and the animals and plants living on it, but not man himself, of whom no trace was found below the most recent strata. Darwin's theory struck much nearer the heart of religious beliefs, for now the history of man himself was brought into the center of the discussion. The blow fell the more heavily as concessions made by theologians to geologists had often been coupled with assertions that what was important in Genesis was not the account of the earth's creation, but the creation of man and his early history. In the face of the inroads of geology, theologians had built up a sort of inner defence line around the Bible's version of man's history. In this sphere even the chronology of the Bible was adhered to with extraordinary tenacity. But in the conflicts over Darwinism much more than chronology was of course involved: much more even than a mere reinterpretation of single passages in Genesis. For man's early history, as told in the Bible, was closely bound up with the important religious concepts of the Fall, Original Sin, Atonement, and Redemption. These ideas were explicated in terms of events in sacred history. If those events were to prove fictitious, the concepts themselves would appear to hang in the air.

No sincere Christian could envisage the disintegration of those fundamental concepts of his religion without the gravest misgivings. A man's religion may be said to be an expression of the scale of values that determines his attitudes and actions in various situations in life. The particular beliefs connected with each religion appear to achieve a quasi-symbolical representation of that scale of values or those attitudes: by their verbal and conceptual form they can be easily communicated, and thus serve to create a certain amount of conformity as regards fundamental attitudes within a community. Though the actual concepts, or quasi-symbols, or dogmas, may be of little importance in themselves — what is important is the scale of values they stand for — the connection between the symbol and the thing symbolized becomes to most people so close that one tends to be equated with the other.

Now it is obvious that to many Mid-Victorians, the Biblical account of man's creation and history was part of a system of religious concepts. Because of the confusion between symbol and thing symbolised, acceptance of the Darwinian theory seemed to them to necessitate a complete spiritual revolution, a total change of outlook and attitude towards life. Such a change can hardly be expected to take place without a potent incentive: and to most people, obviously, the possible truth or falsehood of an abstruse scientific theory, which few were in a position to judge on its merits, was not at all a strong enough incentive. They therefore retained their old-established attitude towards life, and with it the concepts and symbols they felt to be connected with this attitude. They by-passed the question of true or false as regards the Darwinian theory, treating it instead as a question of good and bad. The theory was regarded as a religious question, not as a scientific one.

That this was the case, to a great extent, is apparent to any reader of the periodical press of the time. In the popular papers especially, the Darwinian theory was hardly referred to at all except in its relation to religion. And the amount of space and attention given to the theory in the religious publications was immeasurably greater than was normally afforded to scientific questions in those organs.

In some instances, opposition to the theory was openly and squarely based on its possible effect on religion, and ultimately on the well-being of society. To argue in this way was certainly effective from a propagandistic point of view, especially for a public unable to follow the niceties of a scientific argumentation. The underlying attitude was a natural one to take: namely, that the spiritual and moral welfare of the community was a more important consideration than the freedom of scientists to divulge their theoretical conclusions — a freedom which, for instance, the Catholic Church had never granted. One Roman Catholic organ clearly showed that it viewed these matters in quite the same light as the Church had done in Galileo's case. "The salvation of man", it said, "is a far higher object than the progress of science: and we have no hesitation in maintaining that if in the judgment of the Church the promulgation of any scientific truth was more likely to hinder man's salvation than to promote it, she would not only be justified in her efforts to suppress it, but it would be her bounden duty to do her utmost to suppress it ... The truth ultimately can do no harm, although, temporarily, injury may follow from an unreasonable application of it."[6])

Protestant writers were not as a rule quite so explicit. They would not lightly admit that the views they combated might be true. And as the question of the truth or falsity of Darwin's theory was a complex one, it was always possible to cite authorities who vouched for its scientific illegitimacy. In regard to Darwinism, religious people therefore were never compelled squarely to face the dilemma outlined by the Catholic writer just quoted. The theory could be declared scientifically deficient: its theological obnoxiousness was an additional reason for repudiating it.

The two aspects were combined in the argument that religious considerations could legitimately be taken into account whenever there was any doubt at all about the truth of a scientific theory. This was eminently the case with Darwin's theory. In practice therefore, those who employed such an argument were taking the same attitude as the Catholic writer just quoted. One Protestant reviewer expressed it thus: "A theory which is incompatible with views long entertained, and of slow growth, which tends to subvert existing notions, and, indirectly at least, to raise harassing doubts on sacred subjects, should be clearly supported by facts far outweighing those which can be brought forward against it."[7]) Thus theology was not indeed overtly invoked against the theory, but it was allowed to weight the balance. The Bishop of Oxford endorsed this way of looking at the matter in his *Quarterly* review of the *Origin:* "We cannot, therefore, consent to test the truth of natural science by the Word of Revelation. But this does not make it the less important to point out on scientific grounds scientific errors, when those errors tend to limit God's glory in creation or to gainsay the revealed relations of that creation to Himself."[8]) The Darwinians, on their side, recognized that such a theological weighting was bound to occur. Darwin went out of his way in his book to reconcile the feelings of the religious, and his supporters were often at pains to explain that his views harmonized with Christianity.[9])

---

[6]) *Dublin Review*, 44, 1858, 379—381.

[7]) *Edinburgh New Philosophical Journal*, 12, 1860, 234.

[8]) *Quarterly Review*, 108, 1860, 257.

[9]) E. g. *Macmillan's Magazine*, 4, 1861, 247.

When it was asserted that the establishment of Darwinism would
lead to the destruction of Christianity, it was obvious that no Christian
would hesitate in his choice. Those critics who presented the theory in
such a light must have been aware that they thereby in effect condemned
it on theological grounds. One such writer, vaguely feeling that
his argument might recall the Catholic Church's conflict with Galileo,
tried to forestall the expected criticism by asserting — with perhaps
more heat than justice — that Darwin's case was different. "There
are many cases, indeed, in the history of science, where speculations,
like those of Kepler, have led to great discoveries . . ." he wrote. "It
is otherwise, however, with speculations which trench upon sacred
ground, and which run counter to the universal convictions of mankind,
poisoning the fountains of science, and disturbing the serenity of the
Christian world."[10] A Methodist organ painted Darwinism in even
direr colours. "We regard this theory, which seeks to eliminate from
the universe the immediate, ever-present, all-pervasive action of a
living and personal God, which excludes the possibility of the supernatural and the miraculous . . . as practically destructive of the authority
of divine revelation, and subversive of the foundation of both religion
and morality."[11] Nor were such views by any means confined to the
religious papers. The *Edinburgh Review* placed a warning as to the
religious consequences of Darwin's theories at the very beginning of
its review of the *Descent of Man:* "It is impossible to over-estimate the
magnitude of the issue. If our humanity be merely the natural product
of the modified faculties of the brutes, most earnest-minded men will
be compelled to give up those motives by which they have attempted
to live noble and virtuous lives, as founded on a mistake . . . our moral
sense will turn out to be a mere developed instinct . . . and the revelation of God to us, and the hope of a future life, pleasurable daydreams
invented for the good of society. If these views be true, a revolution
in thought is imminent, which will shake society to its very foundations
by destroying the sanctity of the conscience and the religious sense."[12]
The reviewer saw not only religious evil, but also social evil arising out
of the theory. The connection was a natural one. If religious belief

---

[10] Sir David Brewster as reported in *Good Words*, 3, 1862, 3.

[11] *Methodist Recorder*, 1866, Aug. 31, 308. See also *Recreative Science*, 3,
1861—2, 219—220.

[12] *Edinburgh Review*, 134, 1871, 195—6.

was affected, the social fabric itself would disintegrate. A writer in the low-brow and somewhat goody *Family Herald* made this point quite bluntly: "Only let our scientific friends show the people, who are quick to learn, that there was no Adam ... that nothing certain is known, and then that chaos which set in during the lower Empire of Rome will set in here; we shall have no laws, no worship, and no property, since our human laws are based upon the Divine."[13]) That was written in 1861: ten years later the journal was still of the same opinion: "Society must fall to pieces if Darwinism be true."[14]) That the *Times*, in its review of *Descent*, gave prominence to this sort of argument only confirms how widespread was the attitude which gave rise to it. "A man incurs grave responsibility who, with the authority of a well-earned reputation, advances at such a time the disintegrating speculations of this book. He ought to be capable of supporting them by the most conclusive evidence of facts. To put them forward on such incomplete evidence, such cursory investigation, such hypothetical arguments as we have exposed, is more than unscientific — it is reckless."[15])

It must be stressed, however, that unreasoned outbursts of theological zeal were seldom allowed to play the leading part in the attacks against the Darwinian doctrines. More liberal-minded Christians often expressed regret that their co-religionists should reject Darwin on Scriptural grounds.[16 A]) The theory, one said, should be opposed on scientific grounds alone. As a matter of fact, however, this often meant no more than that the religious considerations entered on a more abstract level, in that the scientific arguments were evaluated in terms of a semi-theological philosophy of science. These matters will be discussed more fully in Chapter 9.

Though the incompatibility of Darwin's theory with the traditional beliefs about man's history was the feature that above all drew the general public's attention to the new doctrines, the most serious conflict with religion occurred on another and much more fundamental point

---

[13]) *Family Herald*, 1861—2, 268; answer to correspondent.

[14]) *Family Herald*, 1871, May 20, 44, answer to correspondent.

[15]) *Times*, 1871, April 8, p. 5. See also *Popular Science Review*, 2, 1862—3, 397, *Friends' Quarterly Examiner*, 1867, 57.

[16 A]) *English Independent*, 1871, Mar. 23, 397, 272—3, *Blackwood's Magazine*, 89, 1861, 166, *Temple Bar*, 1, 1861, 543.

than that of the accuracy of the Bible, or the interpretation of certain
Christian dogmas, — namely, the idea of Divine Providence. To the
Mid- Victorians the conviction that the world was placed under the
watchful guidance of a higher power appeared as a fundamental religious
belief.[16 B]) Without Divine supervision, one held, everything would
disintegrate into chaos, for only chaos could result if the universe were
left to the action of chance and blind, inexorable laws. Design and Purpose, the attributes of a Mind, were needed to create and sustain a
Kosmos. Evidence of such design was found especially in the organic
world. Living beings could not have become so perfectly adjusted to
their environment as they undoubtedly were, if Design and purpose
had not been present at their creation.

Now if Darwin was right, the development of the organic world
could be explained without recourse to Divine design and purpose.
Variations that might be called accidental, and the operation of natural
causes, could lead to evolution and progressive change without assuming
design. This was achieved by the theory of Natural Selection, which
was the scientifically significant part of Darwin's doctrine. But it
was also the feature that made his doctrine so difficult to accept for
the English theologians, and also for all naturalists of a religious turn
of mind, who had been accustomed to look at their study as a parallel
to that of theology. Many took quite literally Bacon's words about
God's works and God's word. It is true that the more liberal religious
schools of thought found little difficulty in assimilating the development
theory as such with their religious beliefs. Evolution might be regarded
as having occurred through the providential care of God. But not even
liberal theologians could accept a history of life, including man, where
every reference to Divine Providence appeared superfluous.

The impact of Darwinism on religion would not have been so strong
if British theologians had not been so strongly attached to the tenets
of Natural Theology. The very success of Natural Theology in Britain
— explicable in view of the powerful empiricist tradition of British
thought — had led to that close interdependence of science and religion
which was going to give rise to serious conflict when science advanced
into fields where formerly theology had held exclusive sway.

There is abundant direct evidence that both theologians and scientists

---

[16 B]) This point is elaborated by Gillispie, *Genesis and Geology*, especially 217—228.

in Mid-Victorian Britain did look upon science and religion as closely connected with each other, and dependent on each other. The dependence was logically necessary as long as both science and religion claimed to offer information on matters of fact. Their domains were, so to speak, dovetailed into each other: the limits of each were determined by the other. It is clear that the Darwinian theory acted as a powerful catalyst in exhibiting the dangers of this dependence, and in arousing the latent conflict.

A very marked change in the attitude of theologians towards science, and of scientists towards religion, was taking place in the 'sixties, concurrently with the spread of Darwinian doctrines. The time when Natural History was almost looked upon as a branch of Natural Theology, when every other Church of England clergyman was an amateur naturalist, was passing away. One religious periodical greeted Darwin's *Origin* with the words: "Its publication is a mistake . . . at this time of day, when science has walked in calm majesty out from mists of prejudice and been accepted as a sister by sound theology,"[17]) and expressed the hope and belief that it would soon be forgotten. Instead, it came to mark a decisive stage in the emancipation of science from theology. A few years later one religious writer had to admit that, "taken on the whole, scientific studies have not a religious but a sceptical tendency,"[18]) and another, in a searching analysis, concluded that physical science must be held "the present great enemy of religion."[19])

It may be that the undeniable advance of science during these years did not chiefly dismay the theologians themselves. They were not defenceless: after all, the Church had weathered many storms in the past, and possessed a respectable ideological arsenal. But ordinary religious folk — including many scientists — were in a more exposed position. An instance of the concern felt in these quarters was a declaration published in the *Athenaeum*,[20]) and later in the *Times*,[21]) signed by thirty Fellows of the Royal Society, and forty MDs, expressing the opinion that a scientist "should not presumptuously affirm that his own conclusions must be right, and the statements of Scripture

[17]) *North British Review*, 32, 1860, 486.
[18]) *Spectator*, 37, 1864, 1412, quoting the Duke of Argyll.
[19]) *Rambler*, 6, 1861—2, 390.
[20]) *Athenaeum*, 1864, Sep. 17, 375. Submitted by Herschel, who expressed dissent.
[21]) *Times*, 1864, Sep. 20, p. 7.

wrong." Many leaders of scientific thought, however, publicly dissociated themselves from this action, which expressed the opinion of a minority of the scientific world only. Unable to rally the support of their fellow-scientists, the religious phalanx later closed their ranks, and in 1865 was formed the Victoria Institute, "to investigate fully and impartially the most important questions of Philosophy and Science, but more especially those that bear upon the great truths revealed in Holy Scripture, with the view of defending these truths against the opposition of Science, falsely so called."[22] Violent anti-Darwinianism was the prevailing characteristic of many prominent members of the Institute.

Another fruit of the acute crisis in the relations between science and religion was the Metaphysical Society, founded in 1869 by a group of earnest liberal-minded Churchmen. Unlike the Victoria Institute, they had, however, been at pains to associate with them several of the leading scientists of the time, including T. H. Huxley. The Society met for discussions regularly for many years, and though the tone was gentlemanly, no real *rapprochement* was achieved. This was not indeed to be expected: the parties differed, and in the end agreed to differ, on fundamental points of the philosophy of science.[23]

Among the general public, the problem of science and religion attracted increasing attention. The 1868 Church Congress at Dublin, and the 1870 Congress at Southampton, made it the central point of their debates.[24] In the press the conflict was a constantly recurring theme, and the hardening climate was noticed. The Nonconformist *British Quarterly* asserted that "the age yearns for religious faith, and is disquieted only because its religious faith is disturbed by the readjustments which the advance of science necessitates."[25] One religious writer complained of the "Scepticism of Science, which has increased so rapidly of late years and still daily progresses,"[26] while the freethinkers in the *Westminster Review* found it quite natural that "the theological opinions of the past should be slowly dying out before the scientific opinions of the present."[27]

---

[22] Victoria Institute, *Journal*.
[23] See A. W. Brown, *The Metaphysical Society*, esp. p. 98.
[24] Reports in the *Guardian*, 1868, Oct. 14, 1870, Oct. 26.
[25] *British Quarterly Review*, 49, 1869, 265 265—6.
[26] *Dublin University Magazine*, 65, 1865, 593. See also *Times*, 1872, Feb. 3, p. 4.
[27] *Westminster Review*, 37, 1870, 547.

One did not fail to notice that the advance of science did not concern details only: it was the scientific attitude as such which was inimical to much that had formerly been looked upon as essentially religious. "Physical sciences, when directed against religion," wrote the Roman Catholic *Rambler* in 1862, "only attack it accidentally — in its points of asserted contact with the world of phenomena. But here they wage a war of extermination; they deny the reality of the contact; they account for the phenomena which religion claims as her own upon merely physical laws, and they thus introduce and encourage the suspicion that the claims of religion are due only to the imagination of the pious, or to the imposture of the cheat."[28] It fell to Darwin's theory to bring out the "deep-seated antagonism"[29] between science and religion, and it did so, not chiefly because the evolution theory upset the Biblical creation, story, but because the theory of Natural Selection implied the extension of purely scientific methods and procedures to the domain of organic life, and therefore, ultimately, also to that of mind and soul. In these spheres the theological concept of Design, and the metaphysical ones of final cause and vital force, had hitherto been accepted as indispensable: but they were irreconcilable with the scientific naturalism of which the Natural Selection theory was a fruit, and to which it gave such a strong impetus. The scientific naturalist claimed for science the whole world of sensory experience, and could tolerate no theological enclaves within that domain. This was one of the fundamental differences between the Darwinian and the earlier conflicts between science and religion. Previous advances of science had indeed diminished the area where theological explanations of natural events were acceptable: the conflict was about the relative size of that area. Now the very existence of any such area at all was disputed.

It was not Darwinism alone that caused the violent crisis in the relations between science and religion in the 1860s. There was at least one other potent factor, namely, the introduction into Britain of more modern methods of Biblical criticism.[30] This "Higher Criticism" had long been known and applied in Germany, but British theology seems to have insulated itself from its Continental counterpart. Strauss' *Life of Jesus*, written in 1835 and translated into English in 1846 (by

[28] *Rambler*, 6, 1861—2, 387.

[29] *Academy*, 2, 1870—71, 13.

[30] W. B. Glover, *Evangelical Nonconformists*, p. 13.

George Eliot), was regarded by theologians and religious laymen as a book to be abhorred and denounced rather than discussed. Therefore, when in 1860 six prominent liberal clergymen published the *Essays and Reviews*,[31]) where some of the ideas of the new school of Biblical study were applied in a mild form, laymen and clerics within the English Church were caught unprepared. In the press the book caused a sensation which, quantitatively at least, was considerably greater than that raised by the *Origin of Species*. Scores of articles and pamphlets written in refutation of the *Essays and Reviews* appeared in 1860 and 1861.[32]) A similar reaction was caused by Bishop Colenso's *Pentateuch*,[33]) where the author showed, among other things, that the chronological statements in the Bible were often incompatible with each other, and therefore could not all be accepted as true.

Both the *Essays and Reviews* and the *Pentateuch* were fruits of an application of the same critical methods to the Bible as to all other historical documents. It was this attitude which was new in British theology, and when one of the Reviewers wrote that he assumed that the Bible should be read like any other book, his statement was treated as a designed impiety.[34]) This was the hub of the controversy: whether the evidence of religion should be subjected to scientific tests.

Thus the advance of science in England in the 1860's touched religion in two ways. First, Darwin's theory threatened to oust theological explanations from its last foothold within the world of natural science. Second, the application of scientific principles to the study of the Bible carried over the conflict to the domain of theology itself. Religious people could not but look at the development with grave misgivings. "The danger which to many observers appears to threaten the Christian cause more seriously than any other arises from the application to it of the methods and results of modern science,"[35]) wrote one Evangelical organ, and in the *Contemporary Review* W. B. Carpenter, that year's

---

[31]) *Essays and Reviews*, London 1860. Seven authors: Frederick Temple, Rowland Williams, Baden Powell, Henry Bristow Wilson, C. W. Goodwin, Mark Pattison, Benjamin Jowett.

[32]) The *Literary Churchman* listed no less than 68 titles up to Nov. 1, 1861.

[33]) J. W. Colenso, *The Pentateuch and the Book of Joshua Critically Examined*, 7 parts, 1862—1879.

[34]) See L. E. Elliott-Binns, *English Thought, 1860—1900. The Theological Aspect*, p. 176.

[35]) *British and Foreign Evangelical Review*, 20, 1871, 3.

President of the British Association, declared in 1872 that "the claims of Science have of late been advanced, not only more strongly, but more aggressively, and some of the positions that have been taken up have been such as apparently to threaten, not the outworks only, but the very citadel, of Religious Faith."[36])

The inroads that the Darwinian theory was making into domains which had previously been regarded as the exclusive preserve of religion led to attempts to effect a separation between what was called the "spheres" of science and religion. But the attempts were hardly successful. It was impossible to agree where the dividing line between the spheres should be drawn. According to the ideology of Natural Theology, in fact, the whole world of natural science was included in the even wider world of religion. One obviously could not say that the natural world proved the truths of religion, and at the same time insist that changes in our knowledge of that world had no religious significance. At the other extreme, some scientific men with positivistic and empiricist leanings claimed for science the whole domain of factual experience, whether physical or mental, leaving to religion nothing but the world of ethics. To accept any of these extreme views was to make the conflict inevitable as long as both science and theology were actively pursued. Scientific men tended to view such a result with equanimity, since it was clear that the victories, in the nineteenth century, were almost constantly on the scientific side. Scientists could declare that they would judge their results according to the standards of science alone, and leave it to theologians to look after their own house. Those scientists who had no special theological bent therefore solved the problem for themselves by simply ignoring the conflict and its results.

Theologians and religious people could not afford to be so complacent. For while scientific knowledge could be built up and organized completely independently of theology, theology, at least in its 19th century British form, could not be pursued without reference to science, or more precisely, the facts which science possessed the most powerful tools for ascertaining. Therefore, while those scientists who wished to extend the domain of science over the whole area of human knowledge tended to exclude theology, and could afford to exclude it, those religious people who claimed this same domain for theology had perforce to find

[36]) *Contemporary Review*, 20, 1872, 739. See also: *Theological Review*, 9, 1872, 561.

some accommodation for science within it. Opposition against attempts
to separate the spheres of each therefore came mainly from religious
quarters. A writer in the Unitarian *National Review*, while recognizing
that separation of the spheres was the line "prevailingly assumed both
by liberal divines and by reverential and cautious men of science", could
not himself recommend that solution. "The *savant* cannot help advancing
his lines of thought into human and moral relations, and esteeming
them amenable to him. The theologian cannot help applying his faith
to the universe, for the supernatural is conceivable only in relation to
the natural, and the transcendency of God involves the subordination of
the world."[37]) That was a logical position to take for a supporter of
Natural Theology. And as Natural Theology was so very highly
esteemed by British theologians, the violence and extent of the conflict
raised by the advance of science is not surprising. It is significant that
one religious organ, in its review of the *Origin*, explained that "it would
not be dealing fairly by our readers, and especially it would be unmindful
of the apologetic value of natural theology, were we to look at this
theory from any other point of view than the twofold one of science
and theology."[38])

Now if Natural Theology was popular among theologians, it was
even more so among a certain class of scientific men. A writer in
the *Popular Science Review* took such naturalists to task for their habit
of "dabbling in Divine matters", as he said. "It appears to them that,
unless they drag the Creator into every second paragraph, their essay
will not possess the necessary religious veneering for the public taste."[39])
But even this writer blurred the issue by continuing, "Now, when allusion
is discriminately and respectfully made to the works of the great First
Cause, no fault can be found." As we shall show below, any reference
to the First Cause was likely to create misunderstandings when the
Darwinian theory was concerned.

An extremely common device used by those religious writers who
refused to admit the need for separating the spheres of science and
religion was to insist that there was no real conflict between the two.
Now it was of course impossible to deny that there was some sort of

---

[37]) *National Review*, 11, 1860, 488.

[38]) *North British Review*, 32, 1860, 256. See also C. C. Gillispie, *Genesis and
Geology*, especially p. 3—40.

[39]) *Popular Science Review*, 5, 1866, 215.

conflict. There were two ways of accounting for this, either to declare that the conflict was apparent only, or to insist that the conflicting sides did not rightly represent science or religion. Either way was question-begging: to assert that the conflict was apparent only was to state an (unverifiable) conviction that all problems would be satisfactorily solved some time in future; and to say that true science did not clash with religion was of no avail, as long as true science could not be objectively defined.

The first of these alternatives — asserting that the conflict was apparent, and that only a true interpretation was needed to resolve it — was common among Catholics and Anglo-Catholics, which is hardly surprising, since these denominations undoubtedly possessed the most developed theology. The *Dublin Review*, writing in 1858, put the Catholic view clearly enough. "We much fear that a sort of general impression prevails in some portions of society, that the progress of science is inimical to the interests of religion. We cannot wonder indeed that such a fear should disturb the minds of the more religious class of Protestants ... [But the Catholic] knows that whatever else may be true or false, his religion is infallibly true ... And if in any point they should seem to clash and contradict each other, the Catholic cleaves to that which is certain — his religion, and leaves it to time and inquiry to clear up the difficulty."[40]) Less explicitly, Dr. Pusey said in 1865 that "the right interpretation of God's word would never be found to contradict the right interpretation of the facts of physical science."[41]) These writers recognized that the problem of interpretation did not concern science only, but theology as well, and could view the advance of science without overmuch fear, since the task of harmonizing it with religion was left to expert theologians, not to the natural lights of the common man.

Protestants, with their more direct dependence on the Bible, were generally less willing to admit any latitude of interpretation on the theological side: they therefore tended to choose the second alternative: to deny that the obnoxious theories were truly scientific. This obviously placed them at a tactical disadvantage, since they had to do battle on the scientists' field, not on their own. Generally, in fact, the argumentation did not pass far beyond bare assertions. "Pseudo-science has

---

[40]) *Dublin Review*, 44, 1858, 375—6.
[41]) *Leisure Hour*, 1865, 784, quoting Dr. Pusey.

assailed the foundations of our faith; we have endeavoured to show that true science is a modest but a firm friend to that faith,"[42]) wrote the *British Quarterly Review*. Such statements were extremely common;[43]) and it seems the phrase "true science" was interpreted in two different ways. One was to equate it with the statements of orthodox scientific authorities: "Some narrow scientific men, with a little knowledge, are full of bigotry and intolerance; not so with the masters of science, who have generally been ardent supporters of revealed religion,"[44]) was a typical assertion. The other way was to attempt a more theoretical definition of true science, and especially to assert that it was only the theories, or hypotheses, or speculations, of science that were hostile to religious beliefs, but never ascertained facts. The nature of this argument will be discussed in Chapter 9.

Those who were convinced that science and religion were in perfect harmony, had no incentive to attempt a separation between the spheres of each. In principle, the domain of either was left indefinite and illimited: in practice, scientists were declared to overstep their bounds, or rather, to leave the path of true science, as soon as their conclusions conflicted with traditional religious beliefs. This was the most common attitude among the general public in these matters. The other and more realistic way of solving the conflict was to assign wholly distinct domains to science and religion. This view had adherents chiefly among liberal-minded religious folk.[45])

The most common demarcation line was probably between the physical world, which was given to science, and the spiritual world, which was reserved for theology. Unfortunately the distinction was difficult to uphold, since it could not be made quite clear what was spiritual, and what was physical. To most religious people, there was a causal relation between the spiritual world and the physical one, and if that was the case, it was obviously impossible to assert that the discoveries of religion had no bearing on those of science, and vice versa. This attitude came

---

[42]) *British Quarterly Review*, 41, 1865, 151.

[43]) E. g. *Patriot*, 1862, Oct. 9, *Christian Observer*, 1861, 344, *Nonconformist*, 1865, Sep. 13, 743, *Daily News*, 1866, Aug. 24, p. 4, *London Society*, 13, 1868, 333, *Daily Telegraph*, 1868, Aug. 21, p. 6, *Manchester Guardian*, 1869, Aug. 20, p. 2.

[44]) *Weekly Review*, 1870, Sep. 24, 934. See also: *Journal of Sacred Literature*, 14, 1861—2, 4.

[45]) *National Review*, 11, 1860, 488.

out clearly in the discussion of one of Professor Tyndall's speeches at the British Association.[46]) Tyndall asserted that "the physical philosopher, as such, must be a pure materialist."[47]) So much, said some religious commentators, might be granted, but it only proved that the scientific sphere was a limited one, and theology, which did not limits its purview in this manner, was by implication a higher and nobler study. "If ... they deny the existence of the spiritual world, they cut off from their view half the field of evidence,"[48]) wrote the *Guardian*. Theologians were inclined to demand that scientists should admit "spiritual evidence" as being on a par with the "physical evidence" which they were primarily concerned with. Religion might be excluded fiom the physical world, but it still had a foothold in the mental one. The facts of psychology were not yet given over to science, nor yet the facts of biology: the spheres of science and religion "intersect in the human mind"[49]), as one theologian put it. Therefore, it was maintained, the scientist's failure to take into account "spiritual facts" was a deficiency in his science. "The facts of consciousness are as much facts of nature as any others, and the natural philosopher has no business to reject them from his premises,"[50]) said Canon Mozley, and in the *Contemporary* we read: "The truths of revelation form one connected body of belief based on the wide range of facts and experience which bear their witness to the spititual world. The assault on them too often rests, not on the assured facts of science, but on the groundless visions of speculation; not on the affirmative proof which is certified by observation, but on the negative suspicion that nothing can exist which the sense-philosophy refuses to recognise."[51]) In this way the separation of the spheres was given up, since both domains had to contribute whenever theories were to be constructed. The nature of the argument is analysed more fully in Chapter 9.

A slightly different way of separating the "spheres" was to adopt the style of the Natural theologians, and to maintain that science dealt with secondary causes only, while religion was concerned with the First

---

[46]) See above, Chapter 4, p. 81.
[47]) *Observer*, 1867, Sep. 8, p. 3.
[48]) *Guardian*, 1867, Oct. 9, 1073.
[49]) *Guardian*, 1870, Oct. 26, 1253 quoting Dr. Hannah.
[50]) *Guardian*, 1868, Oct. 14, 1145, quoting Mozley.
[51]) *Contemporary Review*, 6, 1867, 16.

Cause. The solution was not very satisfactory, since those who propounded it had a tendency to allow explanations in terms of a First
Cause as acceptable alternatives to explanations in terms of secondary
causes: and this possibility clearly made nonsense of the separation.

The most effective solution of the whole problem was to reserve for
religion nothing but the world of morality and ethics. This line had
been advocated, among others, by Coleridge, whose authority was now
and then invoked in its favour.[52] It may be surmised that this distinction was sometimes intended by writers who employed the looser term
of spiritual,[53] but by and large religious people seem to have been reluctant to admit the necessity for such a drastic curtailment of the domain
of theology as it implied. The severe treatment meted out to Baden
Powell for advocating such a separation is significant: The *North British
Review* would have nothing of his main assumption: "His whole theory
rests ultimately on an attempt, not only to draw a *distinction*, but to
effect a *divorce* — to establish an actual *separation* between the *physical*
and *moral* departments of nature."[54] It is also significant that the
solution was often commended by empiricist and positivist philosophers
and scientists.[55] Religious people and the majority of the general
public were not yet prepared to accept that complete divorce of the
world of facts from the world of values, which the advance of science
and of scientific method made more and more imperative.[56] That
ultimate solution, however, was in the air. To scientists, and to Darwinian scientists especially, it became increasingly evident that science
had to claim as its sphere not only the world of physical facts, but all
facts.

The attempts to separate the spheres of science and religion may
be said to be but one aspect of the conflict between the two. The
scientists, as the advancing side, were to find no logical halting place
until the whole domain of the world of experience was recognized as
theirs. The theologians, on the other hand, could not in general admit

---

[52] E. g. *Globe*, 1867, Sep. 10, p. 2.

[53] E. g. *North British Review*, 31, 1859, 378, *Nonconformist*, 1868, Sep. 2, 869.

[54] *North British Review*, 31, 1859, 378.

[55] E. g. G. H. Lewes, in *Blackwoods Magazine*, 90, 1861, 545. Similar views
were expressed in *Macmillan's* 4, *Magazine*, 1861, 240, referring to Macaulay; and
Fortnightly Review, 2, 1865, 683.

[56] See Everett W. Hall, *Modern Science and Human Values*.

that religious truths were completely unconnected with the world of experience, which would make nonsense of the idea of Divine Providence, if not of the idea of the Supernatural althogether. Therefore the theologians, as the retreating side, tried to draw the line of demarcation between the two spheres wherever they thought a successful line of defence could be established. The result was continual conflict, as science continued its advance, and the line was pushed further and further back.[57]

---

[57] On the relation of Darwin's theory to theology, see also Tord Simonsson, *Face to Face with Darwinism*, where nineteenth century Swedish Christian discussion is analysed, and further Windsor Hall Roberts, *The Reaction of American Protestant Churches to the Darwinian Philosophy*.

CHAPTER 6.

The Argument of Design

The argument of Design, or the physico-theological argument, as Kant termed it, consists in inferring the existence and nature of God from the order and harmony exhibited by the universe. It was the foundation stone of that Natural Theology which, as was natural in a country whose inclination towards empiricism has appeared in practically every sphere of human activity, was much favoured by British theologians. It was Natural Theology which furnished empirical grounds for religious beliefs. In order to do this, it brought natural science into its service.

It appears that nineteenth century scientists were by no means unwilling to serve as purveyors of material to the theologians in this manner. Rather the contrary, for they thereby found a means of raising the prestige of their own department of study among the general public. In addition, many scientists were clearly themselves genuinely convinced that it was the supposed theological bearing of their study that was the ultimate justification of their pursuit.

Support for the argument of Design was given by no less a scientist than Newton, who had said: "This most beautiful system of the sun, planets, and comets could only proceed from the counsel and dominion of an Intelligent and Powerful Being."[1] These words were very widely quoted by religious writers, as decisive evidence that true science was so far from conflicting with religion that it even directly supported it. Newton's argument had been further elaborated both by scientists and theologians. In various forms, the Design argument continued to be extremely popular all through the 19th century in Britain. One of the best-known expositions of it was written by William Paley, whose *Natural Theology; or Evidence of the Existence and Attributes of the Deity Collected from the Appearances of Nature* was first published in

---

[1] Sir Isaac Newton, *Philosophiae Naturalis Principia Mathematica*, Scholium generale, end.

1802, and went through edition upon edition all through the 19th century.[2]) Two other books by Paley were set books for the B. A. at the universities;[3]) about his *Evidences of Christianity* Darwin says that he greatly admired it — "I could almost formerly have said it by heart"[4]) he wrote to a friend in 1859. Later, however, when theological opposition against his theories sought support from the expounders of the argument of Design, Darwin referred to them, somewhat irreverently, as "Paley and Co."[5])

Most adherents of the Design argument, and notably Paley, sought their chief illustrations from the organic world. The harmonious motions of the heavenly bodies might satisfy a mathematician like Newton; ordinary people were more struck by such instances as the adaptation of the human eye as an instrument of sight, or the wings of birds as instruments of flight, and in general the adaptation of each species of animal or plant to its place in the economy of nature. It was the complexity of these structures, and their harmonious adjustment to other features in the universe, which appeared as convincing proof that they could not be due to accident or blind chance. Therefore, it was argued, they must be due to Design. Further, as the adaptations were on the whole favourable to the organisms, the Design was clearly Benevolent. In other words, they bore witness to God's Providence. It is true that Hume had revived and elaborated the argument of the ancient Epicureans against this reasoning: assuming only the persistence of motion among the finite number of atoms in the universe, a less than infinite number of transpositions of these atoms might, though perhaps after aeons of time, produce some such order as the present — and that order, as we know, reproduces itself. Therefore, the Epicureans said, it was in no way remarkable that animals were beautifully adapted to their environment. Had they not been so, they would have perished, and their remains been absorbed by more viable forms.[6]) Hume did not insist on this argument: he admitted that though it might serve to explain the appearance of some sort of ordered life, it

---

[2]) The British Museum Catalogue mentions 31 editions, the last from 1879.

[3]) Darwin, *Letters*, I, 47.

[4]) Darwin, *Letters*, II, 219, to Lubbock, 1859.

[5]) Darwin, *More Letters*, I, 154, to Lyell, 1860.

[6]) David Hume, *Dialogues concerning Natural Religion*, ed Kemp Smith 1947, Part VIII, p. 182.

could not account for the very high degree of perfection actually reached. Hume's main argument against the hypothesis of Design was not that the order of the organic world could be explained otherwise; it was that the existence of order did not allow of any valid conclusions as to its cause. Order had to be accepted as an ultimate datum, just as the existence of matter itself was an ultimate datum. The utmost that Hume was prepared to grant in the way of Design was that "the cause or causes of order in the universe probably bear some remote analogy to human intelligence."[7] That proposition, however, he declared to be "somewhat ambiguous, undefined, and incapable of extension, variation, or more particular explication."[8] There was not much left of the argument of Design when Hume had done his work. Nevertheless, the fact that anything at all was left after such an assault could be taken as proof of its strength. Moreover, Hume's artful presentation of his thesis prevented many of his readers from appreciating the force of his criticism.[9]

Now Darwin's theory of Natural Selection was directly relevant to the Epicurean hypothesis,[11] as elaborated by Hume. His critics of course recognized this: "The theory is the theory of Epicurus, with the atheism removed."[10] By declaring that the sorting out of the viable from the non-viable was made on the basis of innumerable minute variations as between parents and offspring, and not on the basis of organisms arising complete out of the chance collocation of elemental atoms, Darwin turned the Epicurean hypothesis from a nearly absurd into an eminently plausible one. Moreover, the deficiencies which Hume had pointed out in the Epicurean hypothesis, namely, that it could not account for the high degree of perfection of many organic structures, disappeared in Darwin's theory. Given a constant supply of random variations between parents and offspring, the adaptation of the organisms to their environment followed logically from the Malthusian struggle for life.

By accepting the order and adaptation of the organic world as an ultimate fact, Hume had allowed at least some rudimentary vestige

---

[7] Hume, *Dialogues* Part XII, 227.

[8] Ibid.

[9] Ibid, p. 58—9, ed. introd.

[10] *Contemporary Review*, 17, 1871, 281 (A. Grant).

[11] Comments on this, e. g. *John Bull*, 1859, Dec. 24, 827, *Month*, 11, 1869, 289.

of the argument of Design. When Darwin showed that it was unnecessary
to regard the existing order and adaptation as ultimate facts, he diminished the scope of the argument still further. After Darwin, the
organic world could count for no more than the inorganic in the matter
of Design. Hence in its consequences for the argument of Design Darwin's contribution became more important than Hume's. Hume had
shown that the inference of a Designing Mind was logically invalid,
but as he could suggest no alternative explanation of the adaptation
that existed, the psychological force of the argument was left nearly
intact. Darwin did provide an alternative. The theory of Natural
Selection reduced the order of the organic world to a consequence of
that prevailing in the inorganic. Thereby the Design to be derived
from the one became as abstract and psychologically empty as the
other.

The Design argument is an argument from analogy. From the fact
that some processes and phenomena in our ordinary experience are
regularly explained as due to human design and forethought, is drawn
the conclusion that similar processes in the history of the world, before
the entry of man on the scene, are due to Divine Design and Providence.
The argument is obviously the stronger, the greater is the resemblance
between the phenomena so compared. Darwin's theory concerned the
Design argument so vitally because it removed from its scope precisely
those instances which had formerly been recognized as the best and
most evident. Paley and his followers had insisted chiefly on the minute
adaptation of means to ends in the organic world. And the whole theory
of the permanence of species had been turned to good account for the
purposes of the argument of Design. In the beginning, one said, God
had so framed each species of animal or plant that it should fit into
its proper place in nature, and He had endowed it with the power of
reproducing its like in order that the perfect adaptation should never
be disturbed. This sort of reasoning had to be completely abandoned
if Darwin was right. It shunned any Development theory, and had in
fact been employed against Darwin's predecessors. "The idea of progressive development and transmutation of species," wrote the *Eclectic
Review* in 1859, "is incompatible with the plan on which the different
groups of animated beings have been constructed . . . in all the long
chain of being, from monads to man, we see no evidence that one link
has ever been other than it now is, or that there has ever existed a

tendency, in a creature fitted for one sphere, to usurp that of another."[12]) The publication of the *Origin* did not at once convince everybody that this argument was doomed. Adam Sedgwick, Darwin's old Cambridge professor of geology, declared in the *Spectator* in 1860: "Every organ of every sentient being has its purpose bound up in the very law of its existence. . . . [Darwin's explanations] . . . do not give us one true natural step towards an explanation of the phenomena — viz., the perfection of structures, and their adaptation to their office. There *is* a light by which a man may see and comprehend facts and truths such as these. But Darwin wilfully shuts it out from our sense . . . This is the grand blemish of his work."[13])

The theory of specific permanence and perfect adaptation had originated at a time when the present organic world was still believed to be the only one that had ever existed. The gradual establishment of a long geological history necessitated some modification of the original theory. The simplest and most obvious one was to assert that just as specific permanence and perfection prevailed at the present time, so in previous periods the species which then existed had been fitted to the conditions that then prevailed, but they had been different species. In most cases this view was combined with the belief that each geological period had ended with world-wide destruction followed by a completely new organic creation. Such "catastrophist" views were losing favour with scientists in Darwin's time, but they were by no means extinct, especially in the religious communities. They were also more easily reconciled with the Bible's cosmogony than was Lyell's "uniformitarianism".[14 A])

If species, whether in the present series, or in the geological one, were in fact perfectly adapted to their environment, the whole idea of a struggle for life appeared absurd. Theologically minded naturalists, steeped in the Design theory, found it impossible to believe that God had so ordained that His created beings would not find subsistence enough in the place where He had put them. One popular journal declared that the Natural Selection theory was "thoroughly nullified by another rule of Nature, or of God, and therefore universal. The rule is, that

---

[12]) *Eclectic Review*, I, 1859, 557.

[13]) *Spectator*, 1860, 335. See also *Zoologist*, 3, 1868, 1345.

[14 A]) See C. C. Gillispie, *Genesis and Geology*, especially 102—4; 121.

wherever any animal exists ... there is the food it loves."[14 B]) Was it really true, another critic asked, that, for instance, the species inhabiting the vast prairies of South America had to compete with one another? If not, where then was the struggle for life, and Natural Selection? "We instinctively feel that each would live and find a suitable place in the circle of creation, as the human family forms in the social scale an infinity of definable grades, each unmolested by its neighbour."[15]) We are not surprised to find that the social order among men was also considered as designed: "God has ordained certain proportions of the social scale as essential to the well-being of a community."[16]) The last quotation indicates one of the points of contact between anti-Darwinianism and conservative political views.

As more and more fossil material was collected, and as the science of geology advanced, it became increasingly difficult to maintain that the destruction of old species, and the appearance of new ones, had occurred wholesale, by sudden starts. Both were in the main continuous processes. Such findings, however, did not seriously affect the Design argument. Any individual species could still be represented as specially adapted to the limited stretch of time and space in which it was destined to flourish. But there were other difficulties. It was natural to ask, for instance, why God, being omnipotent, should have limited himself to a very small number of fundamental types when creating those species which were to be perfectly adapted to their environment. Why should birds and reptiles, whose habits were so different, be so closely alike in structure? And why should whales and fishes, whose habits and environment were so similar, be so different in structure? These questions raised the further problem as to whether existing species were really perfectly adapted to their environment. Were not rudimentary organs, such as some skeletal parts of whales, encumbrances rather than helps?

Questions of this sort were answered by supporters of the Design argument in a clearly anthropomorphic manner. The Creator, like a rational human being, had gone to work according to a simple and intelligent plan, starting from a limited number of "archetypal" forms, on which were gradually introduced an infinite number of variations.

[14 B]) *Family Herald*, 1871, May 20, 45.

[15]) *Zoologist*, 18, 1861, 7600.

[16]) *Zoologist*, 18, 1861, 7611.

Paleontology had revealed that the series from the simplest to the most complex reflected, by and large, the chronological sequence of species in the geological history of the earth, and this fact was interpreted as meaning that paleontological research gave an insight not only into the logical structure of God's plan, but also into the manner of its realisation. Species appeared and perished as God unfolded his plan. Development had nothing to do with Descent, nor with the struggle for life. "Such and such forms now extinct had served their day. They had played their part in the Great Creator's plan."[17] The Bishop of Oxford expounded this theory authoritatively in his *Quarterly* review of the *Origin:* "How can we account for all this? By the simplest and yet the most comprehensive answer. By declaring the stupendous fact that all creation is the transcript in matter of ideas eternally existing in the mind of the Most High — that order in the utmost perfectness of its relation pervades His works, because it exists as in its centre and highest fountain-head in Him the Lord of all."[18]

It is obvious that this theory of the Creator's type-plans did not necessarily come into conflict with any *facts* that a supporter of the Descent theory could hope to advance. The same general development could be expected on both theories. The real issue between Darwin and his opponents, therefore, was not a factual one, but concerned the interpretation of the facts, and in particular, the nature of scientific explanation. Darwin's line in the *Origin* was that the theory of type-plans did not really explain anything: it was only a restatement in reverent language of the fact that organisms can be placed in a limited number of classes. "It is so easy", he said, "to hide our ignorance under such expressions as the "plan of creation", "unity of design," & c., and to think that we give an explanation when we only restate a fact."[19] As the only scientifically valid evidence of God's plan was the actually observed paleontological progression, the plan was an *ad hoc* construction, and as such had no explanatory value.[20]

Religious people, however, could not concede that the reference to a Divine plan was no more than restating the facts. If there was a God,

<hr />

[17] *North British Review*, 32, 1860, 476.
[18] *Quarterly Review*, 108, 1860, 259.
[19] *Origin*, 1 ed. r. 408.
[20] See below, Chapter 9.

it seemed obvious that the creation must have taken place according to a rational plan: moreover, God's purposes were at least partly revealed in the Bible. Therefore the existence of a plan, and some features of its contents, were held to be known independently of the actual facts ascertained by paleontology. What one contended for was that this theological evidence should be regarded as at least on a par with the scientific one. "Analogies between varieties and species might be, as we say, part of the Creator's harmonious plan, or, as Mr. Darwin says, part of the law of identity between varieties and species: they might be either; therefore, to quote them as proof on one side of the question, is against every rule of fair evidence,"[21]) wrote the *London Quarterly Review*, and orthodox naturalists agreed.[22])

Indeed, once the idea of a Divine plan was accepted as a genuine explanation at all, it was difficult not to regard it as better than the Descent theory. Darwin's greatest stumbling-blocks were the missing links. Paleontological research could provide only a minute fraction of all the intermediate specific forms which must have once existed if Darwin was right. To account for the enormous gaps he had to invoke the imperfection of the geological record, and so laid himself open to the charge that his theory was not inductive; not based on observation, but on imagination. On the type-plan theory there were no such difficulties. It was compatible with any actually observed development, whether the gradual one assumed by the Descent theory, or a development by jumps and starts, as the catastrophist view of geology assumed. As the latter development was incompatible with Darwin's theory, it was natural for the supporters of the Design argument to try to persuade themselves that the development had in fact been 'catastrophic'. To do this one needed only assume that paleontology had succeeded in establishing a fairly complete history of the organic world — that, in other words, no forms needed to be assumed in addition to those already found. The Design theory could therefore be represented as factual and down to earth, while the Descent theory was hypothetical and imaginary. In short, the Design theory was scientifically superior. "Surely the teaching of the rocks, in the present state of our knowledge, seems rather to point to something like a theory of representation than to

---

[21]) *London Quarterly Review*, 14, 1860, 290.
[22]) *Zoologist*, 18, 1861, 7593.

progressive development,"[23]) said one writer. Representation was
another name for the ideal type theory. The form found in nature
'represented' the type designed by God: as another writer put it,
"before it became a *fact in nature* it must needs have been a *thought
in God.*"[24])

The main opposition against Darwin on the score of Design, however,
was not occasioned by the Descent theory as such. Representation
was not incompatible with the Descent theory. Whether a new form
had flashed into being out of the dust of the earth, or through a slight
variation from a parent form, was strictly unimportant. In both cases
the new form could be taken as a representation of God's type-plan.
It is true that some anti-Darwinians denied this. The well-known
American naturalist Agassiz declared roundly that "species are based
upon relations and proportions that exclude . . . the idea of a common
descent,"[25]) and that "as the community of characters . . . arises from
the intellectual connexion which shows them to be categories of thought
they cannot be the result of a gradual material differentiation of the
objects themselves."[26]) Agassiz evidently assumed that a gradual
material modification must take place independently of a planning
intellect. And it does seem that the opposition against the Descent
theory was partly due to a more or less vague fear lest any such theory
might become a step in the direction of a more naturalistic or even
materialistic view of the world. This was a plausible conclusion. Darwin's theory was clearly naturalistic, and many of his predecessors,
though not so successful, had aimed in the same direction.

But when Darwin's work, and the progress of science generally,
showed that the arguments against the Descent theory were largely
untenable, the supporters of Design changed their tactics, and began
to insist on the perfect compatibility of Design with Descent. As long
as each modification in the series which led to a specific change could
be looked upon as designed, the Descent theory was theologically safe.
Indeed, such a view of God as constantly intervening to adapt each
species to new and varying conditions was in some respects more satisfactory to the religious than the traditional, static view, according to

---

[23]) *Contemporary Review*, 4, 1867, 48.
[24]) *Contemporary Review*, 2, 1866, 124.
[25]) *Annals and Magazine of Natural History*, 6, 1860, 220.
[26]) Ibid.

which God's intervention in the organic world took place only at the original creation. The simplicity of the Biblical account might have to be sacrificed — indeed, the introduction of several creations had already sacrificed it — but on the other hand the ever-present guidance of Providence could be appreciated the more vividly. This was sometimes pointed out in the religious press. "There are some people who think we must give up the Old and New Testament, and all the Gospel promises, if it is shown that species are not invariable . . . Why, from one point of view, Mr. Darwin's theory is even more orthodox than the other: it bids us believe, not in a soulless world, going on by immutable laws, in which all things continue as at the beginning; it tells us that the Spirit, which at first created all things, still moves and works, even to the bringing forth of new forms after His good pleasure."[27]) Liberal-minded religious people, especially, were inclined to accept the principle that the gradual modifications were "themselves called into being or controlled by an intelligent Creator,"[28]) as a leader-writer in the *Pall Mall Gazette* put it. One could persuade oneself that "if we recognize not only the creative but the cooperative power of God, we shall . . . perceive that the Darwinian theory need not weaken our faith — nay, rather that it may give it strength."[29])

However, in order to establish that the modifications were really designed it was necessary to deny that they could be merely accidental. At this point the supporters of the Design argument came up against the crucial part of Darwin's theory, Natural Selection. For on that theory the variations between parents and offspring were accidental in the sense of indefinite and random. They appeared as adaptively directed only because the large majority of them, which were indifferent or non-adaptive, perished in the struggle for life.

Most of Darwin's critics refused to see the point of his Natural Selection theory. One way of evading the problems it posed was to shut one's eyes on all other variations than the adaptive, effective ones which survived. Only the successful results in the trial and error process were counted: the failures were quietly forgotten, or even denied ever to occur. It appears that Darwin's opponents simply refused to entertain the idea that any trial and error process took place. They believed

[27]) *Eclectic Review*, 6, 1864, 35.
[28]) *Pall Mall Gazette*, 1866, Aug. 25, 605.
[29]) *Inquirer*, 1868, Aug. 29, 549.

this the more firmly as Darwin could not give any instance of abortive modifications. Hence the adaptive series, one said, was the only one to be taken into account: the trial and error series existed only in Darwin's imagination. How unscientific, exclaimed one critic, "to trust in an uncertain chance for existence, [rather] than in a principle of adaptive creation."[30])

The most popular exponent of the argument that Descent was compatible with Design was the Duke of Argyll, whose *Reign of Law* was very favourably received in the press.[31]) Natural Selection, said Argyll, did not properly explain the modification of species, since "it can only pick out and choose among the things which are originated by some other law."[32]) A writer in the *Quarterly* expounded the theological implications of this view. After pointing out that the Descent theory "cannot be worked without a principle of design," he went on to declare that Natural Selection "has nothing to do with the creation of any favourable addition to nature; it is only the removal of those who do not possess the addition."[33]) The observation was strictly correct, only it was not the whole truth, since the creation of favourable additions, according to Darwin's theory, occurred in exactly the same way as the creation of unfavourable additions. The critic thus disregarded the non-adaptive variations, singling out, on purely *ad hoc* grounds, only those modifications which appeared as adaptive. The weakness of the argument was pointed out by a reviewer of Argyll's book in the *Spectator:* "On the side on which [Darwin] professes to account for the existence of so many wonderful perfect organisms *without assuming* that such organisms were more expressly provided for than a host of other very imperfect or even wholly *manqués* organisms in the original manufactory of nature, the Duke of Argyll scarcely meets it fairly . . . Darwin . . . has summoned up a picture much more general, of a purely tentative Nature, trying all experiments bad and good, creating fifty different organisms which are not suited to the world in which they are produced, to every one which is."[34A]) This sort of criticism, however,

---

[30]) *Eclectic Review*, 3, 1860, 224.

[31]) *Christian Remembrancer*, 54, 1867, 147, *Tablet*, 1872, May 25, 648 especially stressed this point.

[32]) The Duke of Argyll, *The Reign of Law*, 18th ed. 1884, 219.

[33]) *Quarterly Review*, 127, 1869, 162 (Canon Mozley).

[34A]) *Spectator*, 1867, Jan. 5, p. 18.

was quite exceptional. Among the general public, who of course knew hardly anything of the Natural Selection theory, since the press said very little of it, Argyll's argument was accepted as valid.

Darwin himself clarified his own position in this matter in a letter to Lyell, whose religious feelings were shocked by Darwin's naturalism. Lyell thought that Darwin "deified secondary causes"[34 B]) and held that Natural Selection should in some way be subordinated to Design. Darwin answered: "If you say that God ordained that at some time and place a dozen slight variations should arise, and that one of them alone should be preserved in the struggle for life, and the other eleven should perish in the first or few first generations, then the saying seems to me mere verbiage. It comes to merely saying that everything that is, is ordained."[34 C]) Some theologians were indeed prepared to accept the last proposition. A Catholic writer said, "The theist, though properly attributing to God what, for want of a better term, he calls 'purpose' and 'design', yet affirms that the limitations of human purposes and motives are by no means applicable to the divine 'purposes'."[34 D]) But this was in reality to abandon the argument of Design. The essence of that argument was to infer God's existence from the fact that the human mind can conceive some events as purposeful. To regard perfectly useless variations, or even monstrosities, as serving some higher, unknown purpose was to abandon the only standard that was relevant if the argument was not to become purely tautological: that provided by man's judgment. In the form some theologians gave the argument, God was taken for granted, and Design inferred from his existence.

Some champions of Design stressed the selection rather than the variation aspect of the Natural Selection theory. Granted, for the sake of the argument, that all sorts of variation occur, was it not evident that intelligence and forethought were needed to select the useful variations? Darwin's own parallell of man's selective breeding of domestic animals was cited against him. "This power on the part of man to bring about changes in species such as those referred to by Mr. Darwin constitutes, as it appears to us, an irrefragable proof that the larger changes produced in nature were executed by and under the

---

[34 B]) *Life, Letters and Journals of Sir Charles Lyell*, Vol. II, 361, to J. D. Hooker, 1863.

[34 C]) Darwin, *More Letters*, I, 194, to Lyell, 1861.

[34 D]) *Month*, 11, 1869, 284. See also *Student*, 3, 1869, 269.

direction of a wise and mighty Being, who adapted fresh forms to new conditions of existence,"[35]) wrote one naturalist organ. A religious journal expressed the same view: "The action which [Darwin] attributes to natural selection is clearly regulated action. Why should natural selection favour the preservation of useful varieties only? Such action cannot be referred to blind force; it can belong to mind alone."[36]) Neither of these writers, evidently, grasped the meaning of the Natural Selection theory, and therefore failed to come to grips with the fundamental problems it raised. Their predicament, however, was shared by many, — and certainly by the majority of the general public, — and the assertion that Darwin needed a "principle of design" therefore continued to be very popular. The favourable reaction towards Sir William Thomson's presidential address to the British Association in 1871[37]) is significant in this respect. The general public was preparing to assimilate evolutionism, but it was a pre-Darwinian evolutionism, where Design figured as if Darwin had never propounded the Natural Selection theory.

A variant of the argument that the individual variations were designed was to claim that there were two sorts of variability. Ordinary variations, one admitted, might be random, i. e., sometimes good and sometimes bad. But they never led to specific change. But there were other variations, of another type, which did produce new species: and these variations, it was claimed, were directed and designed. It was natural to assume that the really operative modifications were bigger and more radical than the others. Now there existed many instances of such big, spectacular mutations. Darwin himself gave several examples of them in the *Origin*. Could not then these mutations, asked one writer in the *Popular Science Review*, "serve as a clue to something higher than 'Darwinism?'" He gave the answer himself: "We have here a clue to the mode in which new species may have been brought into existence when circumstances required it. The *type* is here created in complete adaptation to external circumstances and is then, by external conditions . . . perpetuated as a race."[38]) There was, however, little evidence that the sudden big mutations were in fact any more "adaptive" than

[35]) *Quarterly Journal of Science*, 2, 1865, April, 194.

[36]) *London Quarterly Review*, 36, 1871, 275.

[37]) See above, Chapter 4, p. 88.

[38]) *Popular Science Review*, 2, 1862—3, 398.

the ordinary ones, and though the argument was fairly popular, it could not be sustained for long. It will be discussed more fully below, Chapter 12.

Thus acceptance of the Descent theory as such — excluding Natural Selection — did not necessitate any fundamental modification of the Design argument. All the features that on the direct Creation theory had been interpreted as indicative of foresight and Design could still be so regarded. Psychologically, no doubt, the argument lost some of its force as adaptation was seen to be the result of minute, undramatic changes only. This may go a little way to explain why the religious were reluctant to accept Descent. Adaptive changes, by becoming too frequent, were in danger of being regarded by the common man as 'natural' events, and not as supernatural ones, due to the direct inter-position of the Creator.

The supernatural interpretation was threatened in two ways. One danger was that the event in question could be regarded as purely accidental, and therefore by definition not due to Design. Our discussion so far has dealt with the ways in which the supporters of Design tried to avert this outcome by refusing to take into account any but the adaptive variations that occurred. The other danger to the Design interpretation was that the modification in question might be regarded as simply the necessary result of a natural law, and thus not as "designed" for a particular purpose. To use an astronomical example: the fall of a meteor at a particular moment would not be regarded as "designed", since it would appear accidental; while the occurrence of a lunar eclipse at a particular date would not be regarded as "designed", since the eclipse follows as a matter of course from the regular movement of the earth and the moon. To the religious, the elimination of Design, in the direct and palpable sense, seemed to leave the world a prey either to Blind Chance, or to Immutable Law. The latter was, in many people's eyes, as repellent as the first. It was also by far the more important threat at the time, for while scientists were as emphatic as theologians in rejecting the idea of Chance, they did indeed proceed on the assumption that every happening that natural science was concerned with could in principle be explained by reference to fixed, immutable laws. This idea excluded both the idea of accident — an event not due to a definite cause — and the idea of special supernatural interpositions.

Against this the defenders of Design argued that though everything in the world might go on, as the scientists claimed, after immutable laws, the laws themselves were established by the Creator for beneficial ends. It was in this form that the Design argument had been applied by Newton and his successors to the inorganic world, and scientists had long tried to alleviate the fears of the religious by pointing out that the extension of this view of God's manner of acting to the organic world would be a gain rather than a loss to religion. Darwin himself prefaced his *Origin* by a quotation from William Whewell's *Bridgewater Treatise:* "But with regard to the material world, we can at least go so far as this — we can perceive that events are brought about not by insulated interpositions of Divine power, exerted in each particular case, but by the establishment of general laws."[39]

Many religious people, however, were very reluctant to accept this view of Divine action. It smacked too much of Deism, and made God appear too remote, His Providence too impersonal. "The religious mind," wrote the *London Quarterly Review* in its review of Lyell's *Antiquity of Man*, "would still shrink from a system which, while it makes God an original Designer, makes Him nothing more."[40] The idea of God as a Father taking a direct part in the affairs of this world inevitably grew dimmer, and by the same token the analogy between Divine and human action became more remote. Thereby the force of the argument of Design, which was founded upon that analogy, was considerably weakened.

Moreover, the view of God as only making himself known through the mechanism of natural laws seemed to come uncomfortably close to the materialism of Epicurus and Lucretius. Where was the necessity of assuming that the natural laws had been established by a Divine mind at all? Could they not, as Hume said, be regarded as just one of the properties of matter? This seems to have been what many religious people feared, and therefore they would have nothing of development by natural law: "The ever-bearing, self-development principle is inconsistent with a superintending Providence,"[41] wrote one journal. An orthodox naturalist, writing in the *Leisure Hour* as late as 1872, was even more emphatic: "The doctrine, as carried out to its logical

---

[39] *Origin*, 1 ed. 1859, verso of half-title.
[40] *London Quarterly Review*, 20, 1863, 303.
[41] *John Bull*, 1866, May 19, 341.

consequences, excludes creation and theism. It may, however, be shown that even in its more modified forms, and when held by men who maintain that they are not atheists, it is practically atheistic because excluding the idea of plan and design, and resolving all things into the action of unintelligent forces. It is necessary to observe this, because it is the half-way evolutionism which professes to have the Creator somewhere behind it, that is most popular."[42])

The idea that everything happened by natural law, however, was exerting a strong attraction on the public. The scientists would not give it up, and as science advanced, the religious writers were under constant pressure to come to terms with it. The assimilation was a slow process, and a number of different solutions were tried.

One method was to admit that though there was a large area of the world of experience in which the doctrine of natural law was entirely valid, there still remained certain phenomena where something else was involved. This was the "two-spheres" doctrine we have discussed in the preceding chapter. The production of species, one said, might in large part be explained as due to natural causes, yet not entirely: and it was this unexplained residuum, which served as a vehicle for Divine Design and Providential interference. One theologian expressed his view as follows: "The intellect of the scientific world is engaged in a vigorous attempt to discover the relation between the Creator's sustaining Providence and the accidents of existence, and to define the boundaries between them. It is struggling for light as to the proportion and the cases which are for ever being carried out by the ministerial agents of His power or by His own Omnipotent hand; and those in which He has suffered and is still suffering His original work to be modified by the influences and reactions of its constituent parts."[43]) Such a view was not in general acceptable to scientists, who were not willing to admit God's direct interposition, even if it was limited to a small part of the field. For if such interposition was allowed at all, the foundation of the scientific explanatory structure was shaken. Yet among religious writers the view remained in favour. Towards the end of our period, when the difficulties of the Natural Selection theory were widely advertised, many writers gave adherence to similar opinions. The *Guardian*,

---

[42]) *Leisure Hour*, 1872, 459 (J. W. Dawson).
[43]) *Ecclesiastic*, 22, 1860, 82.

quoting Darwin to the effect that his theory left a "residuum of change" not explicable by Natural Selection, commented: "This amounts, as it seems to us, to a confession that Natural Selection may, after all, play only a secondary part in the differentiation of species, — that there may be behind and above it an 'unknown agency,' giving the first impulses in certain definite directions which natural selection cannot alter ... It is a vast concession, and one which seems to us to cover all that [we] need ask for; since it is obvious that we may assign this 'unknown agency' to any cause ... and if we assign it to the finger of God, it will amount to a perpetual divine superintendence of the physical development of the world."[44]) Clearly such a solution of the natural law problem was crude and unsatisfactory: it had to lead to continual retreat on the part of religion as science extended the area where natural laws were found to be valid.

Another way of reconciling the idea of Design with that of invariable natural laws was popularized above all by the Duke of Argyll's *Reign of Law*. All events, without exception, in the natural world, were indeed due to the working of natural laws. But each event was generally to be explained by reference to several natural laws. The result depended on the interaction of the various laws. Thus by arranging or manipulating the laws for a particular purpose, and not by direct interposition, did Providential Design influence the course of events. "And so it is through the whole of Nature: laws everywhere — laws in themselves invariable, but so worked as to produce effects of inexhaustible variety by being pitched against each other, and made to hold each other in restraint,"[45]) said the Duke, who pictured to himself Design as realised by a "Creative Will giving to Organic Forces a foreseen direction."[46]) Argyll's idea was in general favourably received in the press,[47]) and an attitude similar to his seems to have been fairly common among religious folk. One writer said: "No one doubts the existence of what are called secondary laws, or, in other words, that the Creator chooses to work by certain great laws ... but ... they are not self-acting, and they cannot operate except by the continued superintendance of the Law-

---

[44]) *Guardian*, 44, 1871, 936. See also *Contemporary Review*, 17, 1871, 99.
[45]) Argyll, *The Reign of Law*, 93.
[46]) Argyll, *The Reign of Law* 256.
[47]) *Blackwood's Magazine*, 101, 1867, 679.
[48]) — — —

giver."[49]) And a leader-writer in the Congregationalist *Patriot*, commenting on Grove's British Association address in 1866,[50]) found that "the argument for an all-pervading and superintending Providence will become, if possible, all the stronger when we are prepared to believe that so orderly and beautiful a system of physical and organic being as that which now exists, is the result of a most complicated battle between protean forces and living organisms ready to give up their individuality of type to almost every surrounding influence."[51])

Argyll's argument could hardly commend itself to scientists. Huxley, indeed, criticized it scathingly.[52]) Remarkably enough, however, A. R. Wallace was prepared to consider Argyll's suggestion. In 1869 — when he made public his belief that Natural Selection was incapable of explaining the development of man — he wrote in a *Quarterly* article on that question: "While admitting to the full extent the agency of the same great laws of organic development in the origin of the human race as in the origin of all organized beings, there yet seems to be evidence of a power which has guided the action of those laws in definite directions and for special ends."[53])

In fact, neither the attempts to accommodate immediate Divine interpositions beside the series of events occurring in accordance with natural laws, nor the fanciful theory that natural laws themselves were manipulated by a designing Ruler of the universe was acceptable to the scientifically minded. Gradually, therefore, the supporters of the Design argument were driven to the position which Darwin had indicated from the start, in his quotation from Whewell. God's Providential Design was to be inferred solely from the purposefulness of the natural laws that He had established, and never from the properties of particular events. But even if that much was granted, the Design argument still required that the natural laws so discovered should be indicative not only of a purpose, but also of the intelligence and benevolence associated with the Divine Creator. At this point there was still room for disagreement between the scientists on the Darwinian side, and the supporters of Design. What was to be accepted as a "natural

---

[49]) *Edinburgh New Philosophical Journal*, 16, 1862, 277.
[50]) See above, Chapter 4, p. 78.
[51]) *Patriot*, 1866, Aug. 30.
[52]) See below, Chapter 9, p. 181.
[53]) *Quarterly Review*, 126, 1869, 393.

law?" Religious people here had a tendency to stop short at a lower
level of generality than the scientists, since the more abstract and general
the law, the less did it strike the mind as purposefully designed for a
particular end. Moreover, in the case of biology, the "law" of Natural
Selection appeared as not only cruel, but also as wasteful and indeed
wholly devoid of intelligence, and therefore, as it seemed, directly
antagonistic to the assumption of Design. Canon Mozley gave expression
to these views in his *Times* review of Argyll. "Natural Selection . . .
is adaptation by *chance*, and therefore, not by *design*, for a result obtained
by chance is one emptied of design. If chance means anything, it is
the negation of purpose. Natural Selection is adaptation without
purpose. It is, moreover, a theory of waste . . . and in that it does
violence to nature, of which economy is a fundamental law."[54]

In order to avoid accepting Natural Selection as one of the natural
laws established by God, supporters of the Design argument tried to
find alternatives to it.[55] By far the most common of these alternatives
was to postulate progressive development itself as a fundamental,
irreducible natural law. On this view, God was not considered as
constantly intervening to achieve the perfect adaptations of living
forms to their environment. Instead, he was conceived to have foreseen in advance all the various changes that would occur both in the
organic and inorganic world, and to have provided the first created
life-germ with an internal "law of development," which would carry
its descendants infallibly in the direction beneficial to each. In other
words, instead of spreading out the Divine interventions in time, they
were concentrated into one single, and therefore the more wonderful
event: the original creation of the first living organism, provided with
a capability of development which would carry it further and further,
and ultimately to the height reached by man.

Though Darwin, in the *Origin*, insistently argued against such a
law of necessary development,[56] which would indeed have shattered
the foundations of his Natural Selection theory, his Whewell quotation,
and several other references to "laws impressed on matter by the
Creator"[57] led many readers to persuade themselves that Darwin

---

[54] *Times*, 1867, Jan. 31. See also *British Quarterly Review*, 45, 1867, 573.
[55] See below, Chapter 12, p. 267—79.
[56] E. g. *Origin*, 1 ed. r., 266.
[57] *Origin*, 1 ed. r., 414.

regarded the development as designed in this way. "There is nothing in this view of the method of creation that can be regarded as derogatory to the power and dignity of the Great Creator; for the gradual derivation of species from varieties, under the action of a law imposed on organization, is as great an exhibition of power as the occasional infraction of a law, or the constant recurrence of special acts of creation,"[58]) is a typical comment. Darwin's supporters in the press very often expounded his theory in this light,[59]) whether from a desire to reconcile the religious feelings of the public with it, or from a conviction that this interpretation of Darwinism was the true one. Among the more decided anti-Darwinians there was more inclination to insist on Darwin's failure to give enough recognition to Design in his theory. "We should have liked, indeed, to find our author more distinctly inferring that the variations of structure and habit by Natural Selection are all subordinate to the original act of the Creator,"[60]) said one religious organ. But in a very great many instances critics were content simply to assert, without discussion, that it was impossible that any theory of the origin of the organic world could be worked without a principle of Design: this was in fact one of the most common, and certainly the most seriously urged, of the criticisms directed against Darwin's theory.[61])

Support for these views was sought in alleged inconsistencies in

---

[58]) *Dublin University Magazine*, 55, 1860, 718.

[59]) E. g. *John Bull*, 1859, Dec. 24, 827, *Critic*, 1859, Nov. 26, 530, *London Review*, 1, 1860, 59, *All the Year Round*, 3, 1860, 176, *Westminster Review*, 21, 1862, 568, *Macmillan's Magazine*, 4, 1861, 244, *Chambers's Journal*, 1862: 2, 245, *Popular Science Review*, 2, 1862—3, 394, *Daily News*, 1866, Aug. 24, p. 2, *Daily Telegraph*, 1864, Sep. 15, p. 4. *Daily News*, 1866, Aug. 24, p. 2.

[60]) *Ecclesiatics*, 22, 1860, 87. Reprinted in *Literary Churchman*, 6, 1860, 72. See also *Geologist*, 3, 1860, 471.

[61]) E. g. *Annals and Magazine of Natural History*, 5, 1860, 137—8, 6, 1860, 224; 7, 1861, 399, *Literary Churchman*, 8, 1862, 264, *British and Foreign Evangelical Review*, 12, 1863, 480, *London Society*, 13, 1868, 332, *Quarterly Journal of Science*, 3, 1866, 152, *Guardian*, 1868, Sep. 2, 977, *Contemporary Review*, 7, 1868, 137, *Quarterly Review*, 127, 1869, 136, 144, *Nonconformist*, 1869, Aug. 25, 812, *Quarterly Journal of Science*, 6, 1869, 173, *English Independent*, 1869, Aug. 26, 829 *English Churchman*, 1869, 423, *Freeman*, 1869, Aug. 27, 682, *Friends' Quarterly Examiner*, 1869, 454, *London Society*, 18, 1870, 478, *Guardian*, 1870, Sep. 14, 1097, *Record*, 1870, Sep. 16, p. 2, *Tablet*, 1871, Sep. 9, 333, *Edinburgh Review*, 133, 1871, 169, *Methodist Recorder*, 1871, Aug. 8, 438, *Freeman*, 1871, Apr. 28, 199, *Nonconformist*, 1872, Sep. 11, 940.

Darwin's theory.[62]) But above all it was Darwin's inability or unwillingness to try to explain the appearance of the first living form
that was held as a fundamental weakness in his theory. Darwin had
in fact made matters rather easy for his critics by using such expressions
as "life ... having been originally breathed into a few forms or into
one," which he expanded further in the later editions of his book by
saying, "breathed by the Creator."[63]) Now the argument of his critics
was that if the first form had to be created, then the design involved in
that creation included the whole of the subsequent development. This
argument was expressed very clearly by a writer in the *Quarterly Review*.
Design, he said, "would be no more than a legitimate consequence of
an admission which [Darwin] makes upon the very threshold of his
theory. He admits that the first life-germ was a creation. ... the
universal result must be included in that act."[64]) And again: "There
is the original fact of collocation, and design cleaves to that fact ...
if a systematic production is the result, [we must] infer systematic
forces in the cause."[65]) This reference to the wonderful qualities of
the original form was very popular. The *Edinburgh Review*, in its
review of *Descent*, declared that "Evolution pure and simple does not
touch in the least degree the province of religion. It leaves the origin
of life as great a mystery and wonder as ever, and presents a nobler
view of the Creator, who endowed living forms with such wondrous
capacities ... it cannot explain the phenomena without the will of a
directing Intelligence."[66]) Others spoke of "one original germ which
was vitalised by Creative Power."[67]) Darwin did not think much of
this criticism. "It is mere rubbish, thinking at present of the origin
of life; one might as well think of the origin of matter,"[68]) he
wrote in a letter to his friend Hooker where he expressed regret at
having used expressions in the *Origin* which seemed to imply that he
regarded the origin of life as supernatural. He further pointed to
Newton's saying that "it is philosophy to make out the movements

---

[62]) *Annals and Magazine of Natural History*, 7, 1861, 400.

[63]) *Origin*, 1 ed. r., 415; 6 ed., p. 670.

[64]) *Quarterly Review*, 127, 1869, 175.

[65]) *Quarterly Review*, 127, 1869, 159—160.

[66]) *Edinburgh Review*, 134, 1871, 200.

[67]) *Guardian*, 1871, 752. See also *Guardian*, 1871, 582, *Tablet*, 1872, June 8,
711, and Chapter 12.

[68]) Darwin, *Letters*, III, 18, to Hooker, 1863.

of a clock, though you do not know why the weight descends to the ground."[69])

It is obvious that the designed "law of development", like other attempts to establish the Design argument on the basis of an evolutionary view of life without the principle of Natural Selection, had to disregard or deny the occurrence of unfavourable or abortive developments. According to Darwin, each organism, including the primordial germ, had the power to develop into precisely anything. The direction the development would take was not determined in any way by its own properties, but by the contingencies to which its descendants were to be exposed. If the Natural Selection theory was correct, it would be as appropriate or inappropriate to say that an organism had the power to develop into higher forms, as to say that a block of marble had the power to become a masterpiece of sculpture. The "power" could be taken as an indication of a particular Design only if it was a power to become one thing rather than another. The supporters of the Design argument of course assumed that the power of development was just this. They sometimes likened it to the power of a seed or embryo to become an adult individual of one specific kind and no other.[70]) Some subtle machinery or force would carry it infallibly along a predetermined course. The creation of such a machinery, or the establishment of such a force — the two were hardly distinguished — was considered in itself a direct proof of Design, since the development all the time kept the organism in harmony with external circumstances. "Surely it gives a worthier notion of a Creator to suppose that he foresaw all contingencies, rather than that he should be ever remaking and recreating by the direct interposition of his providence,"[71]) was a typical comment.[72])

The adaptation to external circumstances was therefore itself assumed to be predetermined, either by timing the changes in the organism with those of its environment in the way Leibnitz assumed his monads were synchronized, or by so constructing the organism that a certain change in the external conditions should immediately call forth the appropriate

[69]) Darwin Letters, II, 290, to Lyell, 1860.

[70]) *Month*, 1869, 41. See also Chapter 12; p. 273.

[71]) *Zoologist*, 20, 1862, 8252.

[72]) *Edinburgh New Philosophical Journal*, 16, 1862, 43, *Edinburgh Review*, 116, 1862, 391. (Argyll).

organic response. A contributor to the *Edinburgh New Philosophical Journal* expounded this idea: "It would be an insult to reason to deny the power of the Omnipotent to create at once plants and animals out of inorganic or any other kind of matter: on the other hand, it would be equally irrational to doubt His power to ordain and sustain laws by which *originally created* organisms would be adapted to, and modified by external influences. The two modes may be designated, the first, Autotheogeny; and the second, Genetheogeny."[73])

The idea of a predetermined evolution was by no means confined to a narrow circle of religious writers: on the contrary, it was probably the majority opinion among both scientists and the general public. Professor Owen's views went along these lines. His theory of *Derivation*, definitively launched in 1868,[74]) differed from Darwin's precisely because it assumed that organic beings developed according to laws inherent in them, and because it "recognized a Purpose."[75]) Several other scientists held similar views,[76]) and on the strength of their authority lesser writers felt free to give currency to them as scientific truths.[77])

All the formulations of the Design argument which we have discussed so far fought shy of the Natural Selection theory, either by ignoring it, or by denying its validity. There were some attempts, however, to assimilate even the Natural Selection theory with the argument of Design. The crucial difficulty was to recognize that the Natural Selection theory, unlike the theory of predetermined evolution, contained the idea of trial and error, with errors enormously more numerous than successes. It was the wastefulness of the process which seemed to many to be completely incompatible with intelligence and design. Yet the net result of the process was a harmoniously constituted organic world. Now it was surely possible to imagine an infinity of worlds so constituted that such a result would not follow. Accordingly, it was argued, this

---

[73]) *Edinburgh New Philosophical Journal*, 15, 1862, 253.

[74]) Richard Owen, *Anatomy of the Vertebrates* Vol. 3, 808—9. Quoted in the *Examiner*, 1868, 821.

[75]) *Ibid.*

[76]) See below, Chapter 12, p. 271—4.

[77]) E. g. *Eclectic Review*, 8, 1865, 391, *Theological Review*, 2, 1865, 159, *Freeman*, 1866, Aug. 31, 241, *Zoologist*, 1, 1866, 235, *Intellectual Observer*, 10, 1866—7, 337, *Freeman*, 1868, Aug. 28, 692, *Dublin University Magazine*, 74, 1869, 585, (ref. Argyll), *Contemporary Review*, 14, 1870, 184, *British and Foreign Evangelical Review*, 21, 1872, 7, *Annals and Magazine of Natural History*, 6, 1870, 114.

world of ours bore witness to a benevolent purpose. This view was
expressed clearly by W. B. Carpenter in a review of the *Origin* in the
National Review: "It is a matter for regret and surprise that Mr. Darwin
himself should have set forth his hypothesis as excluding the action
of a higher intelligence . . . All the determining conditions of species
— viz., (1) the possible range of variation, (2) its hereditary preservation,
(3) the extrusion of inferior rivals, — must be conceived as already
contained in the constituted laws of organic life."[78]) Another statement
of the argument ran as follows: "To perfect the design argument when
it is applied to elucidate a system of descent with modifications, struggles
with life-conditions, and survival of the fittest, we have to show reasons
for believing that the changes which occur in the organic world follow
a law or set of laws, indicative of intelligence, and capable of working
with beneficial results."[79]) A writer in *Macmillan's Magazine* put it
picturesquely: "The infinitude of small deviations from the parent
type . . . may be regarded as a labyrinth laid out by the hands of the
Creator, through which he furnishes a clue to a higher state of being,
in the principle which rewards every step in the right direction."[80])
The metaphor of the labyrinth is reminiscent of Asa Gray's illustration,
in which he compared the course of evolution with that of a river.
Evolution follows the law of Natural Selection, the river that of gravity,
yet the actual direction taken by either may be designed.[81]) Gray's
main argument was that the Natural Selection theory was *compatible*
with Design, in other words, that even if Darwin's theory was true,
it was possible to believe that the evolution of the organic world was
designed and supervised by God, and not only in general, but in every
detail.[82]) He did not, and could not, establish that Design was a
necessary assumption on the Darwinian theory, or even that that
theory was unplausible without it. Gray himself fully admitted that
Darwin's theory was also compatible with an "atheistic or pantheistic
conception of the universe."[83])

[78]) *National Review*, 11, 1860, 496.

[79]) *Student*, 3, 1869, 272.

[80]) *Macmillan's Magazine*, 4, 1861, 241.

[81]) Asa Gray, "Natural Selection not inconsistent with Natural Theology; A
Free Examination of. Darwin's Treatise . . .", Articles in the *Atlantic Monthly*
for July, August, and October, 1860, reprinted in England in pamphlet form, p. 38.

[82]) Ibid. p. 38.

[83]) Ibid., p. 44.

Yet it would seem that presicely such a necessity to assume Design was required if the *argument* of Design was to be established. The argument required that Design could be inferred from the facts, and such an inference could not be made if the facts were only *compatible* with Design, but could possibly be accounted for without it. And in fact most supporters of the argument of Design were evidently not satisfied with the assurance that Darwin's theory, if true, did not *disprove* Design. They needed a theory which should positively support the argument, for if the argument could not be established on the basis of the organic world, it was doubtful indeed whether it could be established at all. It is small wonder, therefore, that supporters of the Design argument, as we have seen throughout this chapter, in general fought shy of Natural Selection. Even Carpenter and Gray, who were prepared to go a long way with Darwin, would not admit the Natural Selection theory to its full extent.[84] Gray declared that "at least while the physical cause of variation is utterly unknown and mysterious" it ought to be assumed that "variation has been led along certain beneficial lines."[85] This was a gratuitous addition to the Natural Selection theory which Darwin was unable to accept.[86] In order to justify it, it was necessary to assert that the theory by itself was insufficient — and this was, of course, the line taken by the large majority of Darwin's critics. Their arguments will be examined more fully in Chapter 12.

The chief stumbling block in the Natural Selection theory, from the point of view of Design, was its assumption of trial and error variation. While many anti-Darwinians, as we have seen, avoided facing this difficulty, Asa Gray, among others, recognized it, and tried to meet it by insisting that it was not new: the world had known instances of waste in the economy of nature before Darwin pointed them out.[87] As one writer put it, "Why [Nature] is obviously benevolent in a thousand directions, and apparently harsh in a thousand others, we do not *know*, any more from Darwin than we did from Paley, but we certainly are not left in a denser mist."[88]

It is debatable, however, whether the Natural Selection theory

---

[84] See above, Chapter 4, p. 90—92.
[85] Gray, "Natural Selection", p. 38.
[86] See below, note 93.
[87] Gray, "Natural Selection", p. 43.
[88] *Student*, 3, 1869, 273.

did not in fact imply a greater revolution than appears from these words. Before Darwin, beneficial adjustment of means to ends was seen as a fundamental and all-pervading property of the organic world. There might indeed be instances of maladjustment and waste, which were difficulties on the Design theory, but it was not unreasonable to pass them by, leaving them to be explained later, when a fuller insight had been gained into the constitution of the universe. The general conclusion that the instances of adjustment prevailed, remained unshaken: lack of adjustment was the exception. Darwin's theory completely changed this balance. Thousands and millions were born, out of which only a few were destined to live to a mature age. True, the maladjustment was in general only very small — but so was the adjustment. Above all, the process of development worked no more efficiently than if it had been left altogether to chance. There was no evidence that the adaptive modifications were any more numerous than could be expected in a series of purely random variations. Therefore it could hardly be said that the process was, in the above-quoted writer's words, "indicative of intelligence"[89]) It is not surprising that this writer was forced to conclude that it did not "coincide with anthropomorphic conceptions of a Divine plan."[90]) But this was, as we said above,[91]) to erode the foundation of the Design argument, which was based on that analogy.

Darwin's own attitude to the Design problem is not quite clear. In a letter to Gray he confessed that "I cannot see as plainly as others do, and as I should wish to do, evidence of design and beneficence on all sides of us. There seems to me too much misery in the world . . . On the other hand, I cannot anyhow be contented to view this wonderful universe, and especially the nature of man, and to conclude that everything is the result of brute force. I am inclined to look at everything as resulting from designed laws, with the details, whether good or bad, left to the working out of what we may call chance . . . But the more I think the more bewildered I become."[92]) On the other hand, Darwin flatly repudiated the idea that his theory needed the *assumption* of Design. He would have nothing of Gray's development along beneficial lines.[93])

---

[89]) See note 79.
[90]) See note 79.
[91]) See above, p. 125.
[92]) Darwin, *Letters*, II, 312, to Gray, 1860.
[93]) Darwin, *Animals under Domestication*, 1868, end. See *Letters*, III, 62, to Hooker.

The clearest and ablest elucidation of the relation between Darwinism and Design came from Huxley, who expounded the problem at some length in an article in the *Academy* for 1869.[94] "The Teleology which supposes that the eye ... was made with the precise structure which it exhibits, for the purpose of enabling the animal which possesses it to see, has undoubtedly received its death-blow. Nevertheless it is necessary to remember that there is a wider Teleology, which is not touched by the doctrine of Evolution, but is actually based upon the fundamental proposition of Evolution. That proposition is, that the whole world, living and not living, is the result of the mutual interaction, according to definite laws, of the forces possessed by the molecules of which the primitive nebulosity of the universe was composed. If this be true, it is no less certain that the existing world lay, potentially, in the cosmic vapour; and that a sufficient intelligence could, from a knowledge of the properties of the molecules of that vapour, have predicted, say, the state of the Fauna of Britain in 1869."[95] But Huxley did not leave the question at that point. In the manner of Hume, he showed that nothing really followed from the assertion he had made: "The teleological and the mechanical views of nature are not, necessarily, mutually exclusive. On the contrary, the more purely a mechanist the speculator is, the more firmly does he assume a primordial molecular arrangement, of which all the phenomena of the universe are the consequences; and the more completely is he thereby at the mercy of the teleologist, who can always defy him to disprove that this primordial molecular arrangement was not intended to evolve the phenomena of the universe. On the other hand, if the teleologist assert that this, that, or the other result of the working of any part of the mechanism of the universe is its purpose and final cause, the mechanist can always inquire how he knows that it is more than an unessential incident."[96]

Huxley said of Hume's theism, "nothing is left but the verbal sack in which it was contained."[97] That saying applies with equal justification to his own statement of the teleological argument.

---

[94] *Academy*, 1869, Oct. 9, Nov. 13. Reprinted in T. H. Huxley, *Collected Essays*, II, 107—119.

[95] T. H. Huxley, *Collected Essays*, II, 110.

[96] T. H. Huxley, *Collected Essays*, II, 112.

[97] T. H. Huxley, *Hume*, 146.

CHAPTER 7

Miracles

The question of Design was intimately bound up with the question of miracles. It is true that Design was chiefly inferred from the order and harmony of nature, whereas miracles were precisely such events as infringed this order. But the Design argument consisted in tracing the order and harmony back to a Creator, which could only be done by means of an event which was not itself part of the orderly course of nature. In whatever form the Design argument was propounded, it included at least one miracle as an essential constituent — be it the creation of each specific type, or "primordial germ", or the establishment of "intelligent laws of Nature". Therefore, as the idea of Divine Design and Providence was fundamental in Mid-Victorian religion, so also was idea of special dispensations of Providence, or miracles. The high seriousness with which miraculous explanations of natural events were discussed in the 1860's may appear strange and even ludicrous to a modern reader, but unless it is realized what was the attitude of the majority of the general public in those days, the impact of Darwin on his contemporaries will not be understood. Darwin's theory, by threatening to bring the whole domain of organic life within the region of science and natural law, thereby made deep incursions into a field where miracles had hitherto been accepted as a matter of course by practically everybody.

It is probable that the eagerness with which the question of miracles came to be discussed in the 1860' s and 1870's was to a large extent a result of Darwin's theory. In the Metaphysical Society[1]) Huxley argued with philosophers and theologians on the relation of miracles to scientific explanation. Canon Mozley devoted a whole series of Bampton lectures to Miracles, and his book[2]) was widely reviewed and commented on in

---

[1]) A. W. Brown, *The Metaphysical Society*, 64, 140.
[2]) James Bowling Mozley, *Eight Lectures on Miracles*, London, 1865.

the press.[3]) Mozley's theses were taken up by Huxley's friend Tyndall
in the columns of the *Fortnightly Review*,[4]) Mozley replying again in
the *Contemporary*.[5]) A few years later the latter journal allowed Tyndall
to give publicity to a plan to investigate the power of prayer to heal
the sick — prayers were to be said in all the churches of the country for
the recovery of the inmates of a specially selected hospital ward, whose
recovery rate would then be compared with that of the non-selected
wards.[6]) The suggestion was not carried into effect, though a lively
debate in the pages of the Review[7]) showed that it was taken seriously
indeed by both theologians and general public.

As in the case of Design, so in the case of miracles, the Darwinian
theory brought up the fundamental question of the admissibility of a
theological explanation at all. The theory of Descent would in itself
sweep away a fair number of fairly established miracles: the creation of
each individual species. But more important was the fact that Darwin's whole argument had to be based on the assumption that miracles
did not happen.

The popular idea was that a miracle was an extraordinary event
caused by the direct interposition of God for a specific purpose. It
thus had to satisfy two conditions. In the first place, the event was not
to be explicable in terms of natural laws, for then it would occur in the
ordinary course of nature. Everybody agreed that a valid scientific
explanation excluded a miraculous one of a particular event. In the
second place, the miracle must not appear as accidental and trivial,
but as designed and meaningful.

This popular idea was also, on the whole, the theological one. In his
Bampton lectures on miracles, Mozley provided such definitions as ''a
visible suspension of the order of nature for a providential purpose,''
and "an event with a supernatural cause," and these definitions were
quoted with approval in such a serious organ as the *Contemporary
Review*.[8]) It is true that Mozley made the qualification that he meant
by "order of nature" only "the visible portion of the whole" or "that

---

[3]) E. g. *Times*, 1866, June 5, p. 6, *Contemporary Review*, 2, 1866, 302.
[4]) *Fortnightly Review*, 1867, 657.
[5]) *Contemporary Review*, 7, 1868, 481—496.
[6]) *Contemporary Review*, 20, 1872 July.
[7]) *Contemporary Review*, 20, 1872, 434, 761.
[8]) *Contemporary Review*, 2, 1866, 302.

order of nature of which we have experience," implying that if the invisible — i. e. supernatural — portions were included, the total order and harmony of creation would be maintained. Since natural laws were laws derived from the world of experience, Mozley's view implied, however, that miracles were events for which explanations in terms of natural laws were insufficient in principle.

But which events were in fact inexplicable in terms of natural laws? At this point opinions diverged completely. The scientists tended to maintain that no events were in principle inexplicable, though they were of course willing to admit that many were in fact unexplained, and that some explanations might be false. Theologians and religious people in general, on the other hand, never yielded on the point that there were some events that were inexplicable, and they had a natural tendency to claim that as large a portion as possible of the unexplained phenomena were also inexplicable. Moreover, their standard of proof for such explanations as were offered tended to be high, and in fact impossibly high.[9]

Once granted that explanations in terms of miracles were acceptable at all, it was natural for theologians to maintain that they were in many cases superior to scientific explanations, either because the latter were recondite and far-fetched, or because they came into conflict with established views. Darwin's theory, of course, was condemned on both counts. "Why construct another elaborate theory to exclude the Deity from renewed acts of Creation?"[10] wrote the *Athenaeum* in its review of the *Origin*. "Why not accept that interference rather than evolutions of law, and needlessly indirect or remote action?" The great sin that the Darwinians committed was to shut their eyes even on the possibility of direct action by God. "It is the very essence of the development hypothesis to account for *all* phenomena *without* such special interposition; all must be due to 'secondary causes'"[11] complained the *Eclectic Review*. Another religious organ wrote, "Their whole occupation is to trace every fact to some immediate antecedent cause; and they are so anxious to establish regularity of sequence and uniformity of law, that they cannot bear the idea of the Creator stepping in."[12]

[9] See below, Chapter 9.
[10] *Athenaeum*, 1859: 2, 660.
[11] *Eclectic Review*, 3, 1860, 230.
[12] *Patriot*, 1863, Sep. 10, 594.

The diagnosis was evidently correct: scientific explanation consisted largely in showing that the facts to be explained could be regarded as instances of larger classes of facts for which uniformities and regularities had already been established.[13] Therefore Darwin said that the Descent theory ought to be accepted "because so many phenomena can be thus grouped together and explained."[14] The *Christian Observer* retorted, "[They] can all be explained by admitting them to be the result of the free will of an intelligent Creator."[15] This puts the difference between the scientific and the theological view in a nutshell. Darwin was not criticized because the explanation he gave was in itself wrong or inadequate. Indeed, some religious people even went as far as admitting that "there is no question that [Darwin's theory] explains many facts which cannot otherwise be explained if we set aside the direct interference of the Creator."[16] It was, instead, Darwin's philosophical system that was at fault, his failure to take into account anything but the facts of experience. "Views like his rest, in fact, on no demonstrative foundation, but, as we conceive, on *a priori* considerations, and on what appears to us as a restricted, instead of an enlarged, view of . . . Nature,"[17] wrote one of Darwin's orthodox scientific opponents, who also indicated in what direction he considered the view should be enlarged: "It is at this stage of his scientific researches, where the reasonings and methods of science are not longer applicable, that the candid and earnest investigator of truth will turn to any other source from which he conceives that further knowledge is to derived."[18] That other source, needless to say, was Revelation.

Darwin's failure to consider this other sort of evidence was the graver since his theory, if it did not directly deal with, did at least come into close contact with such matters as mind and soul. In this field, it was held, mere science could do little. "The great difficulty on the whole subject," wrote a popular journal, "is that it carries us out of the whole range of science and out of the sphere of reason. The creation of man is a miracle . . . And a miracle is something beyond our

---

[13] See below, Chapter 9, p. 179.
[14] Darwin, *More Letters*, I, 184, to F. W. Hutton, 1861.
[15] *Christian Observer*, 1860, 570.
[16] *London Quarterly Review*, 20, 1863, 302.
[17] *Fraser's Magazine*, 62, 1860, 87. (Hopkins)
[18] *Fraser's Magazine*, 62, 1860, 89.

power to explain or scientifically investigate."[19]) A religious organ expressed the same view. "When scientific men begin to apply the principles of science to objects of a totally different nature, they are sure to fall into error . . . they think they can measure the miracles of God by scientific principles."[20]) The human soul, with its free will, was a direct refutation of the view that everything in nature could be reduced to strict natural laws. "There is a strong probability . . . that the uniformity of nature shall be occassionally diversified by miracle . . . there is in the universe such a thing as liberty or free power. There is in fact such a thing to a certain extent in man himself. But if in man or anywhere else, then assuredly in the Almighty,"[21]) wrote the *Edinburgh New Philosophical Journal*. Support for this idea of occasional miracles could be obtained from no less an authority than Newton, who had stated that God always worked by natural laws except when it was convenient to act otherwise. A writer in the *Quarterly*, quoting Newton, commented with evident satisfaction: "secundum leges accuratas . . . constanter cooperans, nisi ubi aliter agere bonum est" . . . . This last clause secures all. Nothing else is wanted."[22 A]) If Newton, who dealt with the inorganic world, could admit the possibility of miracles, then Darwin's presumption appeared the greater, when he denied it for the organic world. To the religious this was further proof that Darwin was on the wrong track in his researches: "What Newton did for mechanics has still to be done for biology. But it will not be done by any one who starts with the denial of the supernatural,"[22 B]) wrote the *Tablet* in a review of the sixth edition of the *Origin*.

It may be that nothing else was wanted than an admission by the scientists that miracles were acceptable as explanations. But this was wanted badly, for on that admission depended the possibility of connecting the supernatural with the natural, religion with the world of experience, indeed, with everyday human life. "If science will grant no harmony between itself and the Bible, except on the principle of excluding miracle from the annals of the human race, the issue is

[19]) *Temple Bar*, 5, 1862, 219.
[20]) *Church Review*, 1863, Apr. 11, 353.
[21]) *Edinburgh New Philosophical Journal*, 13, 1861, 124.
[22 A]) *Quarterly Review*, 110, 1861, 369.
[22 B]) *Tablet*, 1872, May 25 648.

obvious,"[23]) wrote one religious journal, and another echoed, "The supernatural . . . lies at the fòundation of all theological discussion. If a man deny the supernatural altogether... if all knowledge be but what we mean, or define, as the induction of the senses, what does it matter that we have obtained some little victory over the unbeliever, while he flies back into his fortress?"[24]) Nor were these views confined to the theological organs. The *Times* subscribed to them in its review of *Descent:* "If Mr. Darwin starts with the preliminary assumption that every fact in Nature is capable of scientific explanation — in other words, that no causes have ever operated except natural causes, he will, of course, reject any other causes. But this assumption is the very point to be proved. To argue from it is to assume the whole doctrine of Evolution."[25])

The *Times* reviewer apparently believed, and probably rightly, that he could rely on his readers' common sense to reject the assumption that miracles never happened. The average reader could hardy be expected to distinguish between an assumption made in order to preserve the logical structure of any scientific argument, and an assumption purporting to state an absolute and ultimate truth. It was plain common sense to admit that the field of the unknown is much greater than the field of the known, and that no man is in a position to state any absolute truths about the universe. Hence the conclusion that the scientists' argument was arrogant and baseless. What most people failed to see was that the scientists' assumption did not concern absolute truths and ultimate reality; they only argued that *if* inquiries were to be successfully carried on and fruitfully discussed, it was necessary wholly to exclude miracles from the argument.[26]) One religious writer, admitting that "Science, as such, has not learnt how to tolerate the miraculous," suggested that, if she could not see her way to fitting miracle into her scheme of the universe, she should at least "be content to bow her head and wait."[27]) But for Science to bow her head and wait would mean giving up inquiry in those direction which theology wished to keep under its sway, thus preventing the confirmation of those scientific theories which were under debate.

---

[23]) *London Quarterly Review*, 25, 1865—6, 267—8.
[24]) *Eclectic Review*, 11, 1866, 161.
[25]) *Times*, 1871, Apr. 8 p. 5.
[26]) See below, Chapter 9, p. 182
[27]) *Guardian*, 1870, Oct. 26, 1253.

One of the chief arguments of the supporters of the supernatural was that it was impossible to get away from some original creative act. "The question of the presence of miracle", wrote the *North British Review* in its review of the *Origin*, "is one which has been, with a strange want of logic, almost universally regarded by eminent men with suspicion, Why? We suppose very few, if any, not even excepting Mr. Darwin, would be willing to deny, that there has been the exercise, at some period of the earth's history, of creative power, — in a word, miracle."[28] Other religious writers agreed,[29] sometimes using the terms of meta-physical jargon: "Phenomena we can observe — their laws we are able to ascertain — existence is beyond our ken."[30] Theologically minded scientists were able to refer to William Whewell's philosophy of science, with its insistence on the necessity of assuming a First Cause initiating all chains of secondary causation.[31] As one naturalist reverently put it, "No trains of reasoning have ever yet brought us ... to the absolute origin of the present order of things and [can] unfold to us (what perhaps it was never intended that we should know) the mysteries of creation."[32]

Now once it had been established that an original miracle was inescapable, the religious at once pushed the argument further. By what reasoning, one asked, could it be established that the miraculous interposition could occur only once? "A Being that *can create* may surely do so as often as He pleases; and we have no right to limit that act,"[33] wrote the naturalist just quoted. Another held that "it would be an insult to reason to deny the power of the Omnipotent to create at once plants and animals out of inorganic or any other kind of matter,"[34] and in the *Quarterly* we read that "He who then interfered may interfere at any other point in the series."[35] All the three organs just quoted appealed to educated readers: two of them were read chiefly by natura-

---

[28] *North British Review*, 32, 1860, 485.

[29] E. g. *Rambler*, 3, 1860, 82, *Literary Churchman*, 6, 1860, 282, *Good Words*, 6, 1865, 384—7.

[30] *North British Review*, 34, 1861, 480.

[31] See below, Chapter 9, p. 177.

[32] *Annals and Magazine of Natural History*, 5, 1860, 132.

[33] *Annals and Magazine of Natural History*, 5, 1860, 142.

[34] *Edinburgh New Philosophical Journal*, 15, 1862, 253.

[35] *Quarterly Review*, 109, 1861, 300.

lists. When such views prevailed in these quarters, the state of opinion among the general public may be imagined.

In the *Origin* Darwin repeatedly asked whether the opponents of the Development theory really believed that "at innumerable periods ... elemental atoms have been commanded suddenly to flash into living tissues?"[36] A writer in the *Contemporary* commented, "Has Mr. Darwin proved that they have not? For this is the real task which, in the face of existing circumstances, he should have undertaken."[37] We may wonder what the proof could have looked like. Indeed, the fundamental impossibility of such an undertaking seems to have given the religious side the comfortable assurance that their position was impregnable. A *Quarterly* reviewer evidently implied as much, when he said that "before we can pronounce that [God has never interfered and will never interfere] we must be able to comprehend all His ways, and to fathom all the secret purposes of His all-wise but often most mysterious will."[38]

In his presidential address to the British Association in 1866, W. R. Grove argued, like Darwin, that it was absurd to imagine that "such a gigantic creature as an elephant"[39] could arise out of the dust of the earth. The *Times* sternly objected. "The beginning of the smallest and most rudimental organism is as difficult to conceive as the beginning of an elephant."[40] The more extreme religious organs naturally took the same line. "If there be a God, it is just as likely *a priori* that He would sometimes act suddenly in nature as that He would always proceed slowly and continuously,"[41] was a typical comment. Furthermore, if two events were *a priori* equally likely, the religious would naturally prefer the one for which they believed they had additional evidence: in other words, the one which was intimated by Revelation, or other religious evidence. This argument seemed especially strong in the case of man: "Plain people will probably content themselves with

---

[36] *Origin*, 1 ed. r. 409.

[37] *Contemporary Review*, 2, 1866, 120. See also *London Quarterly Review*, 36, 1871, 271.

[38] *Quarterly Review*, 109, 1861, 300.

[39] See above, Chapter 4, p. 78.

[40] *Times*, 1866, Aug. 24, p. 8.

[41] *Patriot*, 1866, Aug. 30, 572. See also *Press*, 1866, 539, *Morning Advertiser*, 1866, Aug. 25, p 4.

accepting the statement of Revelation as to the creation of man, feeling assured that even the Darwinian development must involve a miracle, and deeming the nobler miracle the more credible."[42])

The foundation of the religious argument was the supposed necessity for Darwin to admit miraculous interference at least once. It did not matter what miracle was used as basic and indubitable; any miracle, if proved, would allow the conclusion that the Supernatural existed, and given the Supernatural, strictly anything could be derived from it. Some writers, instead of employing the original creation of life as their indubitable and basic miracle, preferred to use the wonder of man's superiority over the brutes. This had the further advantage of bringing into play ordinary people's unwillingness to recognize a lowly descent. "If a special creation is admitted for man with his many points of sympathy with the laws of lower organic life, the probability of special creation lower in the scale of nature is at once admitted,"[43]) wrote the *Zoologist*. In the light of this it is easy to understand with what satisfaction the religious press registered A. R. Wallace's declaration that he considered Natural Selection insufficient to produce Man. The *Guardian* immediately drew "the inference which follows from it of the overruling Providence of God."[44])

The very heavy barrage of specific criticism which was directed against the theory of Natural Selection from the end of the 1860's onwards, especially by Mivart,[45]) also brought much grist to the miracle mill. Failing Natural Selection, Darwin had but to confess his ignorance: and for ignorance the religious opposition substituted unknown causes, and ultimately God. "If we require unknown agencies at all, we may surely dispense with natural selection altogether, and attribute the formation of species to these agencies directly, instead of attributing it to natural selection and referring natural selection to the unknown agencies."[46]) Such sentiments were widespread. The *Quarterly Review*, commenting on Darwin's admission, in the later editions of the *Origin*, that he might have attributed too much to Natural Selection,

---

[42]) *Globe*, 1867, Sep. 10 , p. 2.
[43]) *Zoologist*, 18, 1861, 7593.
[44]) *Guardian*, 1870, Sep. 14.
[45]) See below, especially Chapter 12.
[46]) *British Quarterly Review*, 54, 1871, 469.
[47]) — — —

and underrated other causes of change,[48]) said, "The assignment of
the law of 'natural selection' to a subordinate position is virtually an
abandonment of the Darwinian theory, for the one distinguishing feature
of that theory was the all-sufficiency of 'natural selection'"[49]) There
was truth in this contention. Darwin could not tolerate any teleological
— including miraculous — explanation beside Natural Selection. As
he wrote to Lyell, "If . . . I required such additions to the theory of
natural selection, I would reject it as rubbish."[50]) But he never admitted the necessity of any *such* additions.

The writers we have been dealing with so far took miracles in what
may be called the popular sense of direct and immediate interpositions
of Divine power — however such interpositions may have been imagined.
But in the face of the evident reluctance of scientists to allow such
miracles, there were attempts to establish a concept of miracle which
might be more acceptable to the scientific temper. Many of the more
fundamentalist Christians, of course, would not countenance these
attempts, which commonly aimed at adjusting the idea of the miraculous
to that of natural law, and the Methodist quarterly journal warned
that "this system of law, this determination to look on creation as
nothing but law, allows no space for the personal free agency of man
or God."[51]) The prestige of the idea of natural law, however, led to
an increasing demand for a reconciliation with the scientific view of
nature.

Perhaps the most common line of reasoning was that natural laws,
since they merely stated uniformities and regularities, did not indicate
any real causes. A law, said the *Athenaeum*, was "merely a line of
action, or a measure of creative activity."[52]) The activity itself was
still God's. Physical causation was no real causation, since it did not
reveal the Efficient cause, but only "secondary causes".[53]) It revealed
the order of phenomena, not the power producing them, which was in
every case supplied by God. A miracle, said one religious organ, only
serves to remind us of "the true secret of the order of nature: it is the

---

[48]) *Origin*, 6 ed., 657.
[49]) *Quarterly Review*, 131, 1871, 48.
[50]) Darwin, *Letters*, II, 210, to Lyell, 1859.
[51]) *London Quarterly Review*, 14, 1860, 305.
[52]) *Athenaeum*, 32, 1859, 659.
[53]) See below, Chapter 9, p. 178.

Divine action. No laws are suspended; the Divine will is alike the law and its suspension."[54])

This doctrine may be said to be Hume turned upside down. "There is ... no proof that law has any existence save in the mind that conceives it. When, therefore, we use the common expression that God works by general laws, we only mean, in strictness of speech, that his operations take that form to our minds. He is still the cause, and the law is not even an interposed machinery between Him and the effect."[55]) On this view, each separate application of a natural law — in fact, each observable event, — was "inconceivable except as a separate exertion of Divine power."[56]) When applied to the Darwinian doctrine, the argument meant that the production of, say, a new species out of the dust of the earth should be looked upon as no more, and no less, miraculous than an ordinary birth. "It must be as easy for God to cause one *species* to give birth to another, as to cause one *individual* to give birth to another."[57])

In this way the conflict between miracle and natural law, between the natural and the supernatural, was solved by obliterating the difference between them, making every phenomenon equally supernatural, equally miraculous. The solution, of course, was more verbal than real, since in effect nothing else was done than to extend the meaning of the term supernatural to include the natural as well. There still remained the scientific problem whether a particular event belonged to one sub-class or the other. Besides, the assumption that behind every event there was God as "the direct, the present, the immediate, the sole Causer"[58]) remained unverifiable and empty. Yet there are devils in words, and the argument was fairly popular.[59]) The philosophy of science upon which it was based will be discussed in Chapter 9.

A different way of reconciling miracles with natural laws was the one suggested by the Duke of Argyll in his *Reign of Law*, and which

---

[54]) *Freeman*, 1860, Jan. 18, 46.

[55]) *Christian Remembrancer*, 40, 1860, 255.

[56]) *Nonconformist*, 1863, Mar. 4, 176.

[57]) *Contemporary Review*, 11, 1869, 163.

[58]) *Contemporary Review*, 11, 1869, 252.

[59]) Further e. g. *Edinburgh New Philosophical Journal*, 13, 1861, 130, *Edinburgh Review*, 117, 1863, 569, *British Quarterly Review*, 47, 1868, 269, *Family Herald*, 1868—9, 652, *Leisure Hour*, 1870, Sep. 10, 582—3, *London Quarterly Review*, 36, 1871, 275.

we have already met with in the previous chapter. While the reconcilers whom we have discussed above sought to empty the concept of
natural law of every trace of causation, Argyll took the opposite line,
and held that a natural law involved the idea of an operant force. It
was a machinery at the disposal of a will, either man's will, or God's
will. The miracle consisted in a manipulation of the machinery for a
specific purpose, it was not the removal of the machinery in order to
make room for God's direct interposition. God's finger, so to speak, never
had immediate contact with the world: He always operated higher up in
the chain of causation.[60] Argyll's idea, though crude philosophically,[61]
and therefore hardly acceptable either to scientists or theologians,
seems to have harmonized well with the common sense ideas of the
ordinary man, and it was favourably reviewed in most press organs.[62]

The defence of miracles was not always as specific as in the instances
we have referred to above. Instead of seeking to establish a miraculous
interposition as an explanation of this or that event, writers were often
content with the smaller claim that it was impossible to give a scientific
explanation of it, or that, at any rate, man's finite understanding was
totally incapable of the task. The inadequacy of science was stressed
rather than the adequacy of the religious interpretation.

What the defenders of the supernatural wished to establish was that
some things were *in principle* inexplicable by scientific means: a conclusion which squared well with the "two-spheres" doctrine expounded
above.[63] The matters which were thus declared inexplicable naturally
belonged above all to the sphere of organic life and mind. As a popular
journal put it, "Species is a mystery, life is a great mystery, the conscious
rational soul is a greater mystery still. There are such problems in the
universe as physical science will never be able to solve."[64] Among the
general public this was probably the prevalent doctrine.[65] Many
scientists apparently also subscribed to it: an instance in point is the

---

[60] *Literary Churchman*, 1867, 113 criticizing Argyll.

[61] For Huxley's criticism, see below, Chapter 9, note 28, 29, p. 181.

[62] E. g. *Pall Mall Gazette*, 1866, Dec. 31, 2046, *Fortnightly Review*, 1, 1867, 514,
Guardian, 1867, Apr. 24, 454—5, *Christian Remembrancer*, 54, 1867, 147, *British
Quarterly Review*, 45, 1867, 571—2, *British and Foreign Evangelical Review*, 18,
1869, 590—1.

[63] See above, Chapter 5 p. 107—113.

[64] *Illustrated London News*, 1871, 243.

[65] E. g. *Good Words*, 9, 1868, 250, (Argyll), *Eclectic Review*, 14, 1868, 350,

president of the British Association in 1869, Stokes, from whose address we have quoted some relevant extracts above.[66])

To say that a phenomenon is inexplicable is not to assert that it is due to a miracle. But the tendency of the argument was obviously to leave room for a miraculous explanation. The mystery was almost inevitably hypostatized as some supernatural force, as, for instance, by a writer in the *Contemporary:* "Some known forces, indeed, do operate in this work of nature, but they are combined with others mysterious and unseen, which man has so far in vain attempted to discover, but which in our ignorance we should never fail to recognize."[67]) Ultimately, of course, the mysterious force which only thus partly revealed itself, was a manifestation of God.

Now on this assumption it was natural to conclude that the mystery was itself providentially ordained; that God kept man ignorant on purpose. "It is in all respects consistent with what we know, that some intermediate veil should intervene between mortal gaze and the display of infinite power,"[68]) wrote a popular magazine. Indeed, for man to attempt to lift the "veil of mystery"[69]) which enveloped, for instance, the production of species, was not only futile, it was sacrilegious. There were even scientists who gave countenance to this reasoning. At the British Association meeting in 1861, Professor Owen, when asked to explain the distinction between man and the brutes, was reported to have answered, "He confessed his entire ignorance of the mode in which it had pleased our Creator to establish our species, as it was said, 'out of the dust of the eart.' By what marvellous process all that might be accomplished was not told us, nor need it be."[70]) Religious organs were naturally prone to emphasize the latter part of this assertion. "The method by which [man] became possessed of these attributes is a mystery, and must ever remain so,"[71]) wrote the *English Churchman.*

But if God had the power to keep man in ignorance, He had also the power to lift the veil: though the secret, theologians intimated,

---

355—6, *Contemporary Review*, 11, 1869, 246, *English Independent*, 1869, Feb. 18, 150—1, *Contemporary Review*, 14, 1870, 180.

[66]) See above, Chapter 4, p. 83—4.

[67]) *Contemporary Review*, 5, 1867, 78—9.

[68]) *Temple Bar*, 1, 1861, 543.

[69]) *Eclectic Review*, 14, 1868, 350.

[70]) *Daily News*, 1861, Sep. 7. See above, Chapter 4, p. 70.

would not be revealed to man "in this stage of his existence."[72]) An orthodox naturalist ended his hostile review of Huxley's *Man's Place in Nature* as follows, "*He* who teaches the insect to free itself from the pupa-case, will, *at the proper time*, aid mankind in its efforts in the same direction."[73]) Knowledge would come to man in after life. Meanwhile, said the popular *Family Herald* in a leader on Science, "let us rest humbly in our partial blindness, awaiting the day when we shall see more and know more, waiting the great teacher, Death, for ourselves, and the world teacher, Progress, for our children."[74])

In effect, therefore, scientists were exhorted to give up inquiring into what it was ordained man should not know. Inquiry was useless and irreverent. A Roman Catholic journal suggested, quite logically, that it might also be dangerous. Referring to the mystery of man's creation, it said, "Possibly the much-coveted knowledge might be dangerous for us in some way, — a consideration rarely thought of by the eagerness of modern science, pressing onwards so fast."[75])

The aim of the religious defenders of mystery was clearly to ward off scientific incursions into what they considered to be their own exclusive domain. It was for this reason that they were, as the radical *Examiner* expressed it, "enamoured of mystery."[76]) Overtly, one professed humble ignorance; in effect, one substituted a religious explanation for a possible scientific one. Therefore, whether the supernatural was admitted in the form of a specific miracle, or under the guise of a general recognition of mystery, it was bound to come into conflict with the spirit of scientific inquiry. Orthodox writers complained that "the tendency. of modern philosophy is to deny the inscrutable and to throw the plummet into every depth,"[77]) and that science "refuses to entertain mystery. Any theory is better than to admit this even in the most indirect manner."[78]) That was a correct reading of the situation: and the feelings which it gave rise to are proof that one considered the situation as something new. The new factor, in the 1860's, was Darwinism.

[71]) *English Churchman*, 1862, Oct. 9, 987.
[72]) *Ecclesiastic*, 22, 1860, 91.
[73]) *Popular Science Review*, 2, 1862—3, 522.
[74]) *Family Herald*, 1864—5, 142.
[75]) *Tablet*, 1869, Apr. 10, 780.
[76]) *Examiner*, 1872, 603.
[77]) *Standard*, 1868, Aug. 21, p. 4.
[78]) *Nonconformist*, 1871, Aug. 9, 785.

CHAPTER 8

The Bible

In the theological terminology of the time, the discussions on Design
and on the admissibility of miracles concerned *natural religion*, those
fundamental religious propositions which everybody must assent to as
self-evident and indubitable, irrespective of any revelation. Natural
religion, as we have already pointed out, played an important part
in 18th and 19th century British theology, particularly among the
scientifically educated. Yet to many theologians, and even more to
the majority of the general public, natural religion seemed completely
overshadowed by *revealed religion;* i. e., the religious teaching revealed
by God to men through the Bible. Therefore, though the religious
conflicts arising from the incompatibility of Darwin's theory with
received views on teleology and on supernatural explanations were
more serious and more fundamental, in the sense of conflicting with
any religion, natural or revealed, Christian or heathen, most people,
and above all the uneducated, paid much more attention to the theory's
contradiction of the plain statements of the Scriptures. The ordinary
religious man hardly accepted the existence of any other religion than
the revealed religion of the Bible. Abstruse philosophical arguments
on final causes and supernatural interpositions meant little to him:
the Bible meant the more. Though it might be possible to bring the
evolution theory into harmony with natural religion — and the attempts
to establish a Descent theory without Natural Selection aimed in this
direction, and were highly successful on the more popular level — it
seemed impossible to reconcile it with "that biblical Christianity which
is practially of so much higher importance to man than mere natural
theology."[1]

However, in order to extract a meaningful revelation from the Bible
it was necessary to interpret its text correctly, and on this point there
had always been divergent views, both among theologians and among

---

[1] *Edinburgh New Philosophical Journal*, 19, 1864, 57.

laymen. The most common view among the general public, and certainly among the uneducated[2]) was no doubt to regard the Bible as written, so to speak, on God's dictation. Its words were therefore to be interpreted literally. This was the theory of "verbal inspiration" in its extreme form, with which most of the discussion in the present chapter will be concerned.

A less strict school, still adhering to the verbal inspiration theory, allowed a greater latitude of interpretation. Some statements in the Bible, and notably those touching on natural science, though seemingly factual, should be taken in a symbolic or allegorical sense. The allegorical interpretation of the Biblical text, needless to say, had the support of a very ancient tradition.

A subtle elaboration of this line of reasoning was to maintain that God's words, recorded in the Bible, were designed to suit the understanding of the common man, and therefore terms and arguments had to be used which the common man was used to. The Bible was not intended to teach science; its pronouncements on science were incidental to illustrating other matters, e. g. moral lessons, and these were more readily understood if use was made of the ordinary experiences and ideas of the people.

Among a small minority of chiefly liberal theologians, such as the contributors to the *Essays and Reviews*, the Bible was looked upon as a book written by imperfect men whose experience of God had been exceptionally intense, but who had to use their own language and ideas to convey their experience.[3]) According to this view the Bible was not verbally inspired. It was not God's own words, but men's words about God.

Whatever theory was adopted as to the manner of inspiration of the Bible, the primary and fundamental interpretation of particular passages of the text was naturally the literal, straightforward one. The different schools therefore chiefly differed on their interpretation of those passages of the Biblical text which came into conflict with science or ordinary experience. Here the fundamentalist school simply asserted that the Bible was right and science was wrong. Those who admitted interpretations in terms of allegory or metaphor asserted

---

[2]) See L. E. Elliott-Binns, *English Thought 1860—1900: The Theological Aspect*, p. 178, and W. B. Glover, *Evangelical Nonconformists and Higher Criticism*, p. 49.

[3]) W. B. Glover, *Evangelical Nonconformists*, p. 36.

that both science and the Bible were right, but that there was a problem in finding out in what sense either should be taken. The last, and most liberal, school could afford to recognize that the Biblical writers might sometimes be mistaken, especially as regards such things as were of no religious significance, and even as regards such matters as were of importance to religion it was possible to allow a very wide range of interpretation. In practice the last school avoided insisting on the Bible against science except on those points where a retreat would also have involved natural religion. As those conflicts have been discussed in previous chapters, we shall here chiefly deal with those who took a stricter view of Biblical interpretation, and who were, as we have already pointed out, the large majority both among theologians and among the general religious public.

The incompatibility of Darwin's theory with any literal interpretation of Genesis was undeniable. In the more extreme traditionalist organs one was not slow to point this out quite bluntly. "We cannot see," wrote the Roman Catholic *Rambler* in its first notice of Darwin's theory, "how it can possibly be held consistently with the Christian creed."[4] The *Dublin Review* agreed with this judgment.[5] A Nonconformist organ like the *Eclectic Review*, while acknowledging that the evolutionary theory might give, as Darwin himself said, a noble conception of God's work, insisted that it was "utterly at variance with His revealed word."[6] Religious naturalists took a similar line. Sir David Brewster thought Darwin's theory had a tendency to "render the revelation of [God's] will an incredible superstition,"[7] and another anti-Darwinian naturalist asserted that "Mr. Darwin's hypothesis, if true, must destroy the foundations of the religion of the Bible."[8]

This reaction was not confined to the first years of the Darwinian controversy. Though the evolutionary view was gaining ground steadily, it was in a form which could be, and was, combined with even a literal belief in the Genesis story of at any rate man's creation. The solution was to accept evolution as a fact, but reject the Darwinian explanation of the facts. The Darwinian theory remained "irreconcilable with

[4] *Rambler*, 2 1859, 110.
[5] *Dublin Review*, 48, 1860, 75.
[6] *Eclectic Review*, 3, 1860, 230.
[7] Quoted in *Good Words*, 3, 1862, 3.
[8] *Recreative Science*, 3, 1861—2, 219—20 (C. R. Bree).

Scripture,"[9]) "palpably antagonistic to the word of God,"[10]) "simply antagonistic to Christianity and Revelation,"[11]) and so forth.[12])

Many people viewed the whole Darwinian subject exclusively from the point of view of religion. Scientific inquiry had no meaning for them, and therefore they would not regard the Darwinian theory as anything else than a wilful attack on religion. Scientists appeared to them to "pursue their inquiries with a determination to prove the Bible untrue,"[13]) and to construct their theories "in direct and purposed rejection of Revelation."[14]) One daily newspaper, the *Morning Advertiser*, considered it necessary to warn "the many Christian men who belong to the British Association . . . to have their eyes and ears open, and their tongues at liberty, with reference to the insidious attacks which are constantly making at the meetings of the association on the truth of Divine Revelation."[15]) And even if not everybody went as far as asserting that Darwin and his fellow scientists had set as their aim to "batter down the book of Genesis,"[16]) they were shocked by the mere fact that they pursued their work irrespective of the possibility of such consequences.[17]) "Revelation [is] treated with the utmost irreverence,"[18]) warned the *Eclectic Review* in its notice of the *Essays and Reviews*, where Baden Powell mentioned Darwin's theory without expressing dissent. The impiety of Darwin's doctrines, said the *London Quarterly Review*, "consists, first, in denying His express word, but still more in denying *Him*, the Personal Interposer, the Personal Judge."[19]) The basic attitude of these religious critics was expressed in an orthodox naturalist organ: "Scientific men are not at liberty to ignore the statements of Scripture."[20])

[9]) *English Churchman*, 1863, Sep. 3, 854.

[10]) *British and Foreign Evangelical Review*, 17, 1868, 794.

[11]) *Literary Churchman*, 1871, 248.

[12]) E. g. *Freeman*, 1866, Aug. 31, 241, *Zoologist*, 6, 1871, 2615, *Guardian*, 1871, May 24, 624 (letter to ed.), *Sunday Times*, 1865, Sep. 17.

[13]) *Patriot*, 1863, Sep. 10, 594.

[14]) *Press*, 1865, 896.

[15]) *Morning Advertiser*, 1863, Sep. 4, p. 4-

[16]) *Nonconformist*, 1863, Mar. 25, 235.

[17]) *London Quarterly Review*, 14, 1860, 283.

[18]) *Eclectic Review*, 4, 1860, 126.

[19]) *London Quarterly Review*, 14, 1860, 305. Reprinted in *Journal of Sacred Literature*, 12, 1860, 245—7.

[20]) *Edinburgh New Philosophical Journal*, 13, 1861, 130.

The attitude was certainly understandable. If the Bible was the foundation of Christianity, then Christianity would be in grave danger if the Bible was discredited. And with Christianity would go the cement that held society together.[21])

The conflict between Darwinism and the Bible was at its most acute as regards Man's early history. It is true that the doctrine of a continuous evolution and of an enormous time-span for the age of the organic world were impossible to reconcile with belief in a world-wide Deluge, or with any plausible literal interpretation of the Creation story. But there was nothing new in the conflicts that arose on these points. Darwin here did little more than give a powerful support to the uniformitarian view of geology, with which religious people had to live since at least the early 1830's, when Sir Charles Lyell had published his *Principles*.[22]) But earlier geological theories had never implicated man. In fact, all geological evidence pointed to an extremely recent date for man's appearance on the scene. This fact had been used by the religious as a vindication of the Biblical account of man's early history, and had therefore tended to reconcile them with the less palatable parts of the geologists' doctrines. The free-thinking *Westminster Review* commented on the situation that had thus arisen: "It is remarkable how chary Geologists have until recently been of disturbing the popular notion that the creation of *Man* took place in the year 4004 B. C. It has seemed as if they had purchased their right to speculate freely on the anterior history of the Earth by promising to leave untouched that which the Theologian claims as his proper province, the origin and early history of the Human Race."[23]) The *Westminster's* view is plausible enough. We know from their correspondence how careful both Lyell and Darwin were not to hurt religious susceptibilities, and that they went out of their way to reconcile them when they could. Among the religious themselves concessions to the geologists were certainly often coupled with assertions that what was important in the Biblical Creation story was not the inorganic world, nor even the organic world below man, but man himself. "The writings of Moses do not fix the antiquity of the globe. If they fix anything at all it is

---

[21]) See above, p. 101.

[22]) C. C. Gillispie, *Genesis and Geology*, Chapter V.

[23]) *Westminster Review*, 23, 1863, 518.

[24]) *North British Review*, 27, 1857, 345 quoting Dr. Chalmers.

only the antiquity of the species,"[24]) wrote the *North British Review* in 1857, quoting with approval an apologist from 1804. Another Evangelical commentator, writing before the publication of the *Origin*, also took for granted that whatever might be said about the world as a material object, the chronological accuracy of the Bible's history of mankind could not be seriously questioned. "As to the much-controverted 'time question,' it has, if properly examined, less importance than has been attributed to it. We believe as Christians that 6,000 years ago man was called into being; we believe that in the beginning God created, as the original has it, 'the *substance* of the heavens and the *substance* of the earth;' but all thinking men are now agreed that the interval between the two events is long and undefined."[25])

Similar views were widespread among the public, though they were not always made so explicit. It is hardly surprising, therefore, that Darwin decided to touch as lightly as possible on the subject of his theory's application to mankind, "as so surrounded with prejudices,"[26]) as he wrote to Wallace. He could not, however, prevent the question of man from becoming one of the chief points of discussion in the theory, above all among the uneducated. And it was not only the possibility of man's descent from apes that aroused objections, but also the mere fact that the theory would seem to require a much larger time-scale than one had hitherto been accustomed to for the history of mankind.

To the public at large, Darwin's theory was naturally associated with the researches of anthropologists and geologists, whose discoveries of human fossils and sub-fossils were increasingly difficult to reconcile with the view that mankind was no more than 6,000 years old. Sir Charles Lyell's *Antiquity of Man* brought this subject into the limelight of public discussion. Lyell himself, like most geologists, was extremely reluctant to commit himself to any definite dates: still, it was abundantly clear from his discussion that he envisaged a very considerable extension of the traditional chronology. The reception of his book is interesting: practically every press reviewer avoided mentioning any dates whatever, and most stressed the uncertainty of all geological

---

[25]) *Eclectic Review*, 2, 1859, 448.
[26]) Darwin, *Letters*, II, 109, to A. R. Wallace, Dec. 22, 1857.

dating. The public were thus allowed to remain in the belief that the traditional chronology was still acceptable to scientists. Many religious folk seem to have clung to that belief for a very long time. So much seemed to them to be at stake: not only the traditional view of mankind's history, but the trustworthiness of the Bible itself. "We know that the chronology of the Bible has not escaped errors of transcription ... but this is a kind of error that does not shake our faith in the general historic accuracy of the book of Genesis. Could we, however, suppose that the human race is sixty or eighty thousand years old, and that the six days' creation must go for nothing, it would stamp on the book of Genesis that half-mythical, half-legendary, and wholly untrustworthy character which belongs to the unrevealed records of the origin of all ancient nations ... not without clear and ample proof shall we grant that,"[27]) wrote a Methodist reviewer of Lyell's book. Other Dissenting periodicals were equally positive on this point. "The indefinite extension of duration of man's life on earth, also, would, if proved, throw such discredit upon Scripture narratives as to invalidate its testimony upon other matters to a very great extent."[28])

These were admittedly extreme views. The more liberal Church organs did not look upon the disintegration of the Biblical chronology as a theological catastrophe. The *Guardian* emphatically denied that the traditional chronology was a matter of Revelation. "If we should have to enlarge or abandon it, there is no reason why we should sacrifice in the wreck one iota of our belief in the Creation, the Fall, and the Redemption."[29]) But many other religious people felt that precisely these concepts were in fact threatened by the new theories: if the Bible's history of man was only a myth, those concepts, which were based on that history, would have to be given a completely new justification. "Redemption! All honour to man rather, he requires no redemption — he has never fallen ... His own powers and the accidents of nature are all in all,"[30]) wrote one Dissenting organ, and another put its view even more strongly, "It is nothing less than our hopes for a future life that is at stake, with our faith in the plan of our redemption. For in the theory of a 'pithecoid origin' we can see no room for such an

---

[27]) *London Quarterly Review*, 20, 1863, 272.

[28]) *British Quarterly Review*, 38, 1863, 468.

[29]) *Guardian*, 1863, May 20, 477.

[30]) *Eclectic Review*, 3, 1860, 230.

event as man's fall; *a fortiori*, for his Redemption."[31]) Such views were widespread among the religious.[32])

Concurrently with angry assertions and pious protestations that the Bible was right and the scientists wrong, there were many attempts to work out a reconciliation of the two versions while still adhering to the literal truth of the Biblical words. The method was to abandon the traditional and straightforward interpretation of the text, and read a new meaning into it which would not contradict scientific results. This had been done, for instance, when the Copernican astronomy had been accepted by the Church. It had been done again when the findings of geology had called forth a whole range of schemes of reconciliation. The advent of Darwin increased the urgency of adopting some such scheme, and at the same time raised new problems which proved even more difficult to solve.

The process was certainly a painful one to the religious, both because of the natural attachment to old-established views, and because of doubts about the ultimate success of the new scheme of reconciliation. "Geologists have tortured the first chapter of Genesis to suit their theories of the organic world, and now we have to put that unfortunate chapter again on the rack,"[33]) sighed a Methodist reviewer of the *Origin*.

Before Darwin the chief difficulty raised by geology was the chronological one. Genesis said that God took six days to create the present world; geology showed that it must have taken millions of years. The most common way of solving this contradiction seems to have been to interpret the Mosaic day as a geological period — a solution which, said the popular *London Journal* in 1857, "opened at once a path of marvellous attraction for all who desire to harmonise the testimony of Revelation and of science. The obstacles are as nothing as compared with what we have to surmount on every other hypothesis."[34]) But the final establishment of uniformitarian views in geology, to which result Darwin's theory substantially contributed, made this scheme of reconciliation obsolete: and it had never satisfied everybody. Some objected that the interpretation was too recondite. "From some apparent

[31]) *British Quarterly Review*, 38, 1863, 468.

[32]) E. g. *Christian Observer*, 1866, 9, *Guardian* (letter) 1871, 713, *Freeman*, 1866, Aug. 31, 241.

[33]) *London Quarterly Review*, 14, 1860, 283.

[34]) *London Journal*, 26, 1857—8, 173.

contradictions between Scripture and Science, we may be forced to
question long-accepted interpretations of Bible narratives, and to seek
in new ones some ground of harmony. But in such rare cases the new
readings will have nothing *outré* about them,"[35]) wrote one critic.
Others objected that the scheme was tied up too closely with a particular
geological doctrine — the catastrophic one — and therefore made
religious truth dependent on the fate of the scientific theory. Many
theologians therefore strongly advocated a wait and see attitude.
Following Bacon, who asserted that it was impossible for a man to
"search too far or be too well studied in the book of God's word, or
in the book of God's works," (quoted by Darwin at the beginning of the
Origin,[36])) one might contend, as a High Church organ, "It is impossible
that Science and Revelation should really be antagonistic one to the
other, because both emanate from the same Divine Being. At the same
time it is equally impossible that there should not often be an apparent
antagonism until man is thoroughly versed in the whole circle of Science
and knows intimately the whole counsel of God."[37]) Such somewhat
question-begging resolutions of the difficulties — variations of the
argument that true science and true religion were not in conflict with
one another — seem to have been attractive to many minds.[38]) But
most religious people found it very difficult to maintain such a serene
attitude. The urge was often irresistible either to denounce the impiety
of the scientists, or to hail some new scientific discovery as a proof
that the Bible was right after all. It was against both these tendencies
that Dr. Pusey, at the Norwich Church Congress, raised a warning.
"There are two opposite dangers of which believers must beware in
regard to the teaching of science. The first is, the incautious adoption
of such theories as may seem to coincide with Holy Scripture; and the
second is, a fear lest any legitimate results at which science may arrive
should be adverse. We must beware of binding sacred things to any
theory of physical science, and insisting on our interpretation as being

[35]) *North British Review*, 27, 1857, 329.
[36]) *Origin*, 1 ed. 1859, inside of half-title.
[37]) *Church Review*, 1863, Apr. 25, 407.
[38]) *Morning Post*, 1860, Jan. 10, p. 2 *British and Foreign Evangelical Review*,
10, 1861, 626, *North British Review*, 38, 1862, 74, *Family Herald*, 1862—3, 92,
Leisure Hour, 1865, 784, *Church Review*, 1867, Aug. 31, 840, *Eclectic Review* 12,
1867, 328, *Freeman*, 1869, Aug. 27, 683, *Guardian*, 1871, 937.

true."[39]) It is evident that Pusey had grounds for his criticism. We have already discussed instances of theological fear (and, be it added, indignation): there were also many examples of over-eagerness to interpret scientific results as confirming the Scriptural account.[40])

Even parts of the Darwinian doctrine were sometimes put to such use. The evolution theory could resolve such difficulties as the necessity of assuming several creations, and of explaining how all the millions of species now extant could be housed in Noah's ark. "Mr. Darwin's work has done the cause of religion admirable service," wrote the *Dublin Review* after denouncing Darwin's application of the theory to man. "It accounts for the extinction of so many races of animals and plants and the rise of others in a way that is a positive relief after the assertion of some Christian geologists that we are bound to believe in a fresh creation at the close of each Geological period. It accounts for the dispersal of animals and plants without the necessity of recurrence to many — some have said at least five-and-twenty — distinct 'centres of creation' . . . it makes the universality of the Deluge of far easier credence, and diminishes the number of fellow-voyagers with the patriarch in the ark."[41]) At the opposite theological extreme, the Evangelical *Christian Observer* concurred in this view: "We had long indulged a fancy that the practice of classing into separate species had been carried too far, and that many of the objections brought against the details of the ark of Noah originated in errors of that kind."[42])

Other parts of Darwin's theory were of course less welcome to the theological apologist. In spite of Pusey's and many others' warnings against attempting to establish a definite Biblical interpretation to suit the theory,[43]) the schemes of reconciliation were legion. Indeed, some considered it a decided advantage to have available a wide range of such schemes, even though they might be false and incompatible with each other. "Half a dozen schemes of harmony are better than one,"[44]) wrote a contributor to the *British and Foreign Evangelical*

---

[39]) *Leisure Hour*, 1865, 784.

[40]) *Eclectic Review*, 4, 1860, 105, *Edinburgh New Philosophical Journal*, 17, 1863, 168—9, *Morning Advertiser*, 1870, Sep. 20, p. 4.

[41]) *Dublin Review*, 48, 1860, 78.

[42]) *Christian Observer*, 60, 1860, 561.

[43]) *Eclectic Review*, 3, 1858, 61, *Guardian*, 1868, Dec. 23, 1425, *Spectator*, 1861, 648.

[44]) *British and Foreign Evangelical Review*, 10, 1861, 627.

Review in 1861. As regards the vexed question of man's origin, a fairly common way out was to suggest that the Bible's account of man's early history was indeed absolutely true, but did not give the whole truth. When geologists proved that human fossils occurred in the pre-glacial strata, the fact was, though reluctantly, accepted. But these Pre-Adamites could be dismissed as of no importance. "Even if the existence of a rational pre-Adamite rational being on our planet were clearly made out, it would remain to be proved that they had any closer connection with the descendants of Adam than have the possible inhabitants of the Moon,"[45] declared the Quakers' *Friend*. One Evangelical organ put it, "The 'man' who is without the 'living soul', spoken of in Genesis, is not the man of whom the Bible speaks; he is, or was, if indeed such a creature ever existed, a mere animal, and of no more account than the bats and lizards of those long-past times."[46] The *Record* was equally explicit: "But if it could be proved, and we do not anticipate such a result, that there had been on the earth a race of human beings anterior to Adam, this would not affect the history of Moses, or annul the truth repeated in the New Testament, that Adam was the forefather of our race."[47]

Many of the readings suggested by reconcilers who advocated a literal interpretation of the Bible must have seemed strained and forced to ordinary people. We have already referred to the reading of "geological periods" for "Mosaic days". Another suggestion was to modify the ordinary and naïve meaning of creation. Instead of letting it stand for an instantaneous calling into being of a complete and fully developed form, it was taken to imply only creation of the "*substance* of the heavens and the *substance* of the earth"[48] On this interpretation of creation, the doctrine of evolution, in some form, could be assimilated within the Biblical framework. In a somewhat similar manner, a descent theory for man could be assimilated with the Bible if man's creation was said to occur "when God breathed His spirit into man's animal form."[49]

The chronological difficulty was sometimes evaded by the assertion

---

[45] *Friend*, 1863, 65.

[46] *Morning Advertiser*, 1863, Sep. 4, p. 4.

[47] *Record*, 1863, Aug. 28, p. 2. See also *Eclectic Review*, 15, 1868, 471—9.

[48] *Eclectic Review*, 1, 1859, 448.

[49] *Recreative Science*, 2, 1860—1, 151—160, *Dublin Review*, 16, 1871, 486.

that the dates arrived at by adding numbers in Genesis were "indefinite," and "obviously cannot be taken as binding upon men's faith."[50] Certainly everybody could agree that the Biblical chronology was indefinite in a certain sense, and to a certain extent. But many, as we have seen, felt it impossible to grant that it was compatible, on any natural interpretation of the text, with the time-scale demanded by geologists. It is significant that when Sir William Thomson launched his theory of the earth and the sun as cooling bodies, which yielded a geological time-scale of only some 100 million years, it was greeted with great satisfaction by Biblical apologists.[51]

The scientifically minded were inclined to be somewhat contemptuous of the verbal juggleries involved in any attempt to reconcile the Biblical cosmogony with science. "Mere latitude of interpretation ... will not pilot bishop, priest or deacon through the storms of coming controversies,"[52] wrote the rationalist *Westminster Review* in a notice of the Bishop of London's Address on the Harmony of Revelation and of the Sciences, in 1865. Scientists tended to agree with those theologians who preferred to isolate religion from natural science. "The Bible is dishonoured by the very praises of some of its advocates,"[53] declared the *Westminster*, again with reference to one of the many reconcilers of science and religion. The only solution which could commend itself to the scientific mind was to abandon the literal interpretation in favour of some symbolic or mythological reading.

But the time was not yet ripe for such a drastic change in ordinary people's views of the Bible. The strength of the traditional interpretation is well illustrated by the reception of Hooker's address to the British Association in 1868.[54] When Hooker said that many geologists believed that recent geological discoveries confirmed the Scriptural account of man's existence on the earth, the audience, says the *Daily News* correspondent, loudly applauded. But "when the doctor stated that that was not his own opinion, though he was again applauded, it was

---

[50] *Quarterly Review*, 124, 1868, 438—9. See also *Guardian*, 1863, May 20, 477, *Inquirer*, 1863, Apr. 4, 209, *Nonconformist*, 1865, Aug. 2, 627, *Observer*, 1865, June 25, p. 7, *North British Review*, 50, 1869, 549, *Westminster Review*, 23, 1863, 521.

[51] See below, p. 236.

[52] *Westminster Review*, 27, 1865, 263—7.

[53] *Westminster Review*, 41, 1872, 213—14.

[54] See above, Chapter 4, p. 82.

only by a very small proportion of the assembly."[55]) When such was the feeling of an assembly whose scientific education was surely above the average, opinions among the general public must have been even less advanced.

In fact, the defenders of the literal truth of the Bible spent more energy on supporting their assertion that the Biblical account had to be accepted as truth, than on attempts to find schemes of reconciliation with science. Their defence followed two main lines, one emphasizing the credibility of the Bible, the other the unreliability of science. They were naturally often combined.

The fundamental point in the religious argument was that as the Bible was God's word, it could not possibly be untrue. "Take it now as certain that the Scriptures are from God, in their own sense; then, if so, what are our fears? What is it we apprehend?"[56]) Nobody in the 1860's could publicly express doubts about the proposition that the Bible was from God: it was only possible to discuss in what sense it was from God. That question, however, tended to be left open, traditionalists naturally wishing to apply the literal interpretation as far as possible, since it provided, as one writer put it, the "intelligible announcements of revelation."[57]) One contended that revelation "must be accepted on the authority of a Divine statement."[58]) The truths contained in it were said to "rest upon authority more than human."[59]) The Bible should not be read in the light of scientific knowledge: its meaning was conveyed directly. "If it can be proved to be what it professes to be — the Word of God, then it is our duty to receive its statements, however far they may be beyond the range of natural reason."[60])

The literalists did not altogether lack arguments to support these views. There were, as one writer had it, "general grounds for believing that the Bible is the word of God."[61]) The chief of these grounds seems to have been the internal evidence — especially prophecies — provided

---

[55]) *Daily News*, 1868, Aug. 21. p. 3.

[56]) *North British Review*, 36, 1862, 313.

[57]) *Annals and Magazine of Natural History*, 5, 1860, 133.

[58]) *Christian Observer*, 1861, 333.

[59]) *Month*, 4, 1866, 199—201. See also *Globe*, 1867, Sep. 10, p. 2.

[60]) *Freeman*, 1871, Jan. 27, 43. See also *Zoologist*, 6, 1871, 2667—70.

[61]) *Christian Observer*, 1861, 338.

by the Bible itself. Such a view was clearly implied by Cardinal Wiseman, who in 1864 denounced the Darwinian theory, in its application to man, in the following words, "God help us! that many should have allowed themselves to accept such an origin, while a whole host of proofs assigns to us that of revelation."[62])

Others found grounds for their belief in the very wonderfulness of the Biblical story. "An uninspired writer could never have hit upon a theory which so exactly unravels and explains the origin and position of the human family,"[63]) wrote one theological organ.

A more recondite line of argument was to assert that the Biblical evidence was of another nature than that of ordinary experience, or science. "Our revealed knowledge of creation . . . is the evidence of testimony,"[64]) wrote the *Ecclesiastic* in its review of the *Origin*, contrasting it with the inductive evidence of science. The implication was that both had to be taken into account, or that, as others expressed it, "the facts of religion" should be respected as much as scientific facts.[65]) This is of course yet another form of the 'two spheres' doctrine.

The simplest, and probably most effective of the ways of supporting the verbal inspiration and literal interpretation thesis was to maintain that it was the traditional, received opinion, and that therefore the *onus probandi* lay with him who advanced theories which clashed with the established ones.[66]) This argument had to be eked out by the assertion that the scientists who advanced revolutionary views had shirked their duty in this respect: they advanced theories and hypotheses, but they could not prove them. "Scientific men always loudly protest against any appeal to Scripture on questions of science," wrote one reviewer of the *Origin*, "and we do not mean to interfere with their creed. We only ask for received religious belief the same sort of respect that is paid to received scientific belief, namely, the respect due to established position. It has an *a priori* claim on our reverence, that is all,

---

[62]) *Athenaeum*, 1864, May 28, 744. See also *Record*, 1866, Aug. 27, p. 2, *British and Foreign Evangelical Review*, 17, 1868, 187, *English Independent*, 1868, Aug. 27, 902

[63]) *Church Review*, 1861, 91; 1863, Apr. 25, 408.

[64]) *Ecclesiastic*, 1860, 89.

[65]) *British and Foreign Evangelical Review*, 8, 1859, 311, *Dublin Review*, 17, 1871, 39.

[66]) *Annals and Magazine of Natural History*, 17, 1866, 70, *London Quarterly Review*, 36, 1871, 273, *Quarterly Review*, 131, 1871, 69.

a claim to be honoured as truth until beyond all question it can be proved to be error."[67]) A host of defenders of the Bible found comfort in the assurance that none of the obnoxious theories could be strictly proved: only the hypotheses or theories of science clashed with the Bible.[68]) The nature of this argument will be discussed more fully in the next chapter.

It was not only that hypotheses were declared suspect as such: the fact that scientists continually advanced new theories to supersede old ones was looked upon as yet another reason to distrust them, and to stick to the unchanging text of the Book. Some of Darwin's theories, wrote a Scottish periodical, were "absolutely untenable except by those who are willing to set a changing and flitting science, which geology is now, has ever been, and ever must be, against the simple statements of the only authority that we can have on creation, or on any kindred subject."[69]) "*Uncertainty* is written upon every geological page,"[70]) wrote the *Christian Observer*, and the *Quarterly Review* found that when such uncertain questions as those of species "have received surer solutions than we at present possess or have reason to expect, it will be time enough for considering in what relation they stand to things of higher value."[71]) The *Church Review* drew attention to the conflicts of opinion among the anthropologists appearing at the debates at the British Association. "It is refreshing to find that the vacuum which such a disposal of Revelation leaves behind it does not appear to have been filled up to the satisfaction of anybody. Each speaker's theory was opposed to his neighbour's, and remained peculiarly his own."[72])

Some defenders of traditional Biblical views were less assertive. Instead of maintaining the Biblical account in direct opposition to the scientific one, they claimed that the testimony of the Bible should

---

[67]) *London Quarterly Review*, 14, 1860, 283.

[68]) *Dublin Review*, 48, 1860, 79. *Recreative Science*, 3, 1861—2, 225, *Record*, 1862, Oct. 8, p. 2 *London Quarterly Review*, 20, 1863, 272, *Nonconformist*, 1863, Mar. 25, 235, *Church Review*, 1863, Apr. 11, 353, *Patriot*, 1863, Sep. 10, 594, *Dublin University Magazine*, 65, 1865, 594, *Nonconformist*, 1866, Aug. 29, 705, *Press*, 1871, 262, *Tinsley's Magazine*, 8, 1871, 397, *Punch*, 1868, Dec. 19, 264 (comic verse).

[69]) *Tait's Edinburgh Magazine*, 26, 1859, 557—8.

[70]) *Christian Observer*, 1863, 362.

[71]) *Quarterly Review*, 114, 1863, 415. See also *Patriot*, 1863, Sep. 3, 576, *Morning Advertiser*, 1864, Sep. 15, p. 4 *Methodist Recorder*, 1864, Sep. 23 364.

[72]) *Church Review*, 1864, Oct. 15, 1015. See also *Guardian*, 1863, Sep. 9, 841, *Nonconformist*, 1864, Sep. 21 767.

at least be accepted at those points where science could provide no guidance at all. In practice, of course, the difference between the two attitudes was small, since no agreement could be reached as to the extent of the domain where scientific methods could yield valid results. As so often in these discussions, therefore, the difficulty was transferred to a more abstract level. The religious wished to reserve as much as possible for the field of revelation; the scientists were not content to abandon any part of the field of actual experience. In particular, the contested area concerned man's history and man's mental constitution. It is here, said a reviewer of the *Origin* in *Fraser's Magazine*, "where the reasonings and methods of science are no longer applicable, that the candid and earnest investigator of truth will turn to any other source from which he conceives that further knowledge is to be derived."[73] The implication could not be missed; others stated it explicitly. "Here Revelation steps in."[74]

It is evident that belief in the literal truth of the Bible was a position which could not lightly be given up by the religious. But as the theory of evolution gained ground among the scientifically informed, an increasing number of religious people found it convenient to leave the most exposed parts of their front in order to put up a more effective defence of the Bible from new positions. But they did not wholly abandon the literal interpretation of the Bible. As the exposed parts were solely those that concerned science, it was only here that an evacuation was carried out. The literal interpretation was given up only for the statements that touched science. As a rule, the question as to how, in that case, those statements were to be interpreted, was in most cases left open. One might be content with a general assertion, such as, "Holy Scripture is not a work of natural history,"[75] or more fully, as a pre-Darwinian apologist, "Since Scripture is not designed to teach us Natural Philosophy, it is altogether beside the mark to attempt to make a cosmogony from its statements, which are not only too brief for the purpose, but are expressed in language not fitted or intended to convey such information."[76]

---

[73] *Fraser's Magazine*, 62, 1860, 89.
[74] *Edinburgh New Philosophical Journal*, 16, 1862, 285. See also *Nonconformist*, 1867, Sep. 18, 767, *Tablet*, 1871, Apr. 15, 456.
[75] *Athenaeum*, 1862: 2, 54.
[76] *North British Review*, 28, 1858, 554—5.

Others preferred to express their opinion by an affirmative rather than a negative statement, explaining what the Bible did convey rather than what it did not convey. "Only the religious element of the Bible is to us of undoubted truth and value, let all the rest be criticised and judged by the ordinary principles of historical proof and scientific fact,"[77]) wrote the Unitarian *Inquirer*, but did not specify what the religious element was, except for the obvious implication that it did not include the details of the Creation story. Instead of the word *religious*, we often meet with *moral* or *spiritual*. "Let us beware, to use the fine image of Macaulay, how we bring down the ark of the covenant into the battle; let us tremble to link our trust in God with any but moral truth,"[78]) said a pro-Darwinian contributor to *Macmillan's Magazine*. Likewise, a *Fortnightly* reviewer, acknowledging that knowledge of Divine matters could not be derived wholly from the Bible, without reference to "God's natural Revelation," held that "it can . . . be demonstratively shown that the morality of Scripture is preceptively right."[79])

Some also found it possible to harmonize the non-literal interpretation of the scientific statements of the Bible with the verbal inspiration theory. Since the language of the Bible was directed to the ancient Jews, the argument ran, it had necessarily to take into account the limitations of their scientific knowledge, especially as scientific facts were not as a rule mentioned for their own sake, but in order to convey some other message from God to man. "Scripture language is and necessarily must be popular . . . poetic or vividly pictorial,"[80]) wrote one expounder of "Scripture and Geology." Others declared, with Cardinal Manning, "Holy Scripture does not contain a revelation of what are called physical sciences; and when they are spoken of, the language is that of sense, not of science, and of popular, not of technical usage."[81]) From this position the transition to a symbolic, or mythological view of the Biblical narrative was easy indeed.

---

[77]) *Inquirer*, 1860, July 7, 566.

[78]) *Macmillan's Magazine*, 4, 1861, 240.

[79]) *Fortnightly Review*, 2, 1865, 682. See also *Morning Post*, 1860, Jan. 10, p. 2, *Nonconformist*, 1868, May 27, 538.

[80]) *British and Foreign Evangelical Review*, 10, 1861, 626.

[81]) *Contemporary Review*, 6, 1867, 9. See also *Journal of Sacred Literature*, 4, 1863—4, 9, *Guardian*, 1870, Oct. 26, 1253 *Friends' Quarterly Examiner* ,1871, 591.

Traditionalists, however, were not in general inclined to allow such a transition. Those who took the view that the story of creation is only a well-composed myth, wrote one orthodox naturalist, forget that Moses "inserted in the selfsame book which contains this 'fiction' the Ten Commandments."[82] In fact, the more fundamentalist Christians tended to regard all retreats on the part of theology with profound alarm. "If we propose to examine the statements of Scripture, we are instantly met with the cry that the Bible was not given to teach science! We freely admit that there is a sense in which the Bible was not designed to teach physical science. But it would be easy to show that in precisely the same sense the Bible was not designed to teach the science of morals or even the science of theology,"[83] wrote a Methodist reviewer of the *Descent of Man*. His fears are understandable enough. But they proved exaggerated — at least as far as the 19th century was concerned. After the big battles of the 1860's and early 70's, when an evolution theory which satisfied the religious had been elaborated and widely accepted among the educated, a non-literal interpretation of the Biblical cosmogony was quietly conceded by practically all theologians, not only, as before, primarily by liberal ones. As at the same time increased attention was given to the moral element of the Bible, the lessening of its authority in the scientific sector was effectively counterbalanced.

On the other hand, it is obvious to everybody[84] that the Bible, and Christianity itself, has considerably lessened its hold on the British people during the century which has passed since Darwin's publication of the *Origin of Species*. As a long-term prophecy, the view of the 19th century traditionalists has at least not been refuted. The decline of Biblical Christianity, and the spread of scientific knowledge about the world's and man's early history coincide in time; it is hardly far-fetched to see a causal relation between the two. The development has taken longer than 19th century traditionalists feared, and than positivistic scientists and philosophers hoped, and its result is not quite what either party expected. But it would be difficult to deny that by and large, for better or for worse, the prophecy has been ful-

---

[82] *Annals and Magazine of Natural History*, 6, 1860, 205.

[83] *London Quarterly Review*, 36, 1871, 271. See also *Eclectic Review*, 9, 1865, 391, *North British Review*, 35, 1861, 185.

[84] See W. B. Glover, *Evangelical Nonconformists, English Thought*, 1860—1900, Elliott Binns, and H. G. Wood, *Belief and Unbelief*.

filled. The religious community certainly felt that the Darwinian doctrine touched them to the quick. While its repercussions on Biblical interpretations in all probability affected popular views of dogmatic Christianity, its even more fundamental repercussions on the argument of Design, and on attitudes towards scientific as against supernatural explanations, influenced the opinions of the educated sections of the people. The establishment of Darwin's theory has been the decisive step towards that complete separation of science and religion, of facts — of whatever kind — and values, which is the distinguishing mark of the intellectual climate of our own age.[85])

---

[85]) See Everett W. Hall, *Modern Science and Human Values.*

CHAPTER 9

Mid-Victorian Philosophy of Science

One of the criticisms most commonly made of the Darwinian theory in the years after the publication of the *Origin* was that it was not *inductive:* it was based on assumptions instead of facts. Darwin, it was said, had deserted the true British scientific tradition, inaugurated by Bacon and brought to fruition by Newton.

There was some truth in this criticism. Darwin's Descent theory was essentially hypothetical. He did not produce any experimental evidence to prove that any animal of one species had, through a series of generations, given rise to an animal of a distinctly different species. Indeed, no such direct evidence was to be obtained during Darwin's lifetime, and this deficiency of direct proof was held by T. H. Huxley as the chief weak spot in the Darwinian armour. As long as all the available evidence was still consistent with the traditional view that species only varied within narrow limits, and that boundaries between species were always clearly marked, Darwin's opponents threw the burden of proof on him, and extremists considered it justifiable to characterize his theory as a baseless speculation.

The Natural Selection theory was equally hypothetical. Darwin did not *prove* that variations in nature were "indefinite". Now if they occurred in certain directions only, or if they stopped after reaching a certain point, there would not always be material for selection to work on. The survival of the fittest could lead to progressive evolution only if at least some of the offspring were more fit than the parent forms.

What Darwin did was to show that a great number of well-known facts about the resemblances of animals and their distribution in time and space could be explained *on the assumption* that they were related to each other by descent, more or less remotely. He further showed that their divergence could be explained *on the assumption* that the offspring of the same parents are not exactly alike, but vary in all directions around an average, generally coinciding with the average parent type, and that among these variations there are generally some that give their possessors a better chance to survive in the struggle for life.

Darwin's critics did not always, nor even commonly, discuss these assumptions on their merits or demerits. They did not assert that they were improbable. Instead, many critics attacked the assumptions as such, insisting that a scientist had no right to make assumptions at all, but should only base his theories on ascertained facts. Others recognized that hypothetical concepts, which had to be assumed or postulated, but could not be directly proved, were indispensable in science. These people criticized instead the way Darwin derived and employed his assumptions. Either view inevitably led to considerations of scientific method and the nature of scientific explanation. The Darwinian controversy thus brought into view some fundamental problems of the philosophy of science. What was a scientific explanation of an event? What was a natural law? What was a cause? What was induction, and inductive proof? Such questions as these were implicit, and sometimes explicit, in much of the discussion around the Darwinian doctrines.

Broadly speaking these questions were answered in two different ways. One was an empiricist answer, the other an idealist one, and in the Darwinian controversy Darwin's supporters quite consistently sided with the empiricists, while his opponents almost equally consistently took the idealist line.[1]) The Darwinians found philosophical support in the writings of J. S. Mill,[2]) and the long British empiricist tradition, while the anti-Darwinians found theirs in the idealistic philosophical tradition from Plato onwards, and in the writings of the foremost philosopher of science of the age, William Whewell.[3])

If the periodical press can be taken as evidence, the idealistic, anti-Darwinian view of scientific philosophy was favoured by the majority of the educated public in the 1860's. The explanation is probably not

---

[1]) See also my article on "The Darwinian Theory and 19th century philosophies of science", *Journal of the History of Ideas*, 18, 1957, 362—393, on which part of the text in this chapter is based.

[2]) John Stuart Mill, *A System of Logic*, 1 ed., 1843. I quote from the 9th ed., 1875 [Mill, I, II].

[3]) William Whewell, *Philosophy of the Inductive Sciences*, 1 ed., 1840. [Whewell, I, II].

far to seek. Idealistic philosophy was traditionally the philosophy of the Church, which had always been suspicious of thoroughgoing empiricism.

The conflict between idealism and empiricism as regards the philosophy of science turned on the concept of causation. To the empiricist a cause was nothing but the name given to any event or phenomenon that invariably preceded another under certain specified circumstances.[4] Now it was sometimes possible to explain the occurrence of particular sequences of phenomena by showing that they were special instances of more general sequences. This, said the empiricists, was fundamentally what was meant by scientific explanation. But it is obvious that such an explanation never touches the question why there are any uniformities *at all:* this has to be accepted as an ultimate datum. J. S. Mill put the matter as follows: "An individual fact is said to be explained, by pointing out its cause, that is, by stating the law or laws of causation, of which its production is an instance ... And in a similar manner, a law or uniformity in nature is said to be explained, when another law or laws are pointed out of which that law itself is but a case, and from which it could be deduced ... What is called explaining one law of nature by another, is but substituting one mystery for another; and does nothing to render the general course of nature other than mysterious: we can no more assign a *why* for the more extensive laws than for the partial ones ... Every such operation brings us a step nearer towards answering the question ... comprehending the whole problem of the investigation of nature, viz.: What are the fewest assumptions, which being granted, the order of nature as it exists would be the result? What are the fewest general propositions from which all the uniformities existing in nature could be deduced?"[5]

On the idealistic view, a cause was something else than merely an event or phenomenon. If one class of events was invariably, under specified conditions, followed by events of another class, then, idealists argued, this proved that the first events contained within themselves some unobservable entity, some force or power which produced the result.

The contemporary idealistic view was expressed by Whewell as follows: "By Cause we mean some quality, power, or efficacy, by which

---

4) Mill, I, 376.
5) Mill, I, 540, 549.

a state of things produces a succeeding state ... None of our senses
or powers of external observation can detect ... the power or quality
which we call Cause. Cause is that which connects one event with
another, but no sense or perception discloses to us, or can disclose,
any connexion among the events which we observe ... cause is to be
conceived as some abstract quality, power, or efficacy, by which change
is produced; a quality not identical with the events, but disclosed by
means of them."[6])

Whewell particularly stressed that the cause should be regarded
as a substantial entity with real existence. For instance, he made a
sharp distinction between what he called the induction of "laws of
phenomena" and "the induction of causes" The latter only, he said,
"appear to conduct us somewhat further into a knowledge of the essential nature and real connexions of things."[7]) Such a distinction was
not acceptable to empiricists. Mill found it "not philosophically sustain-
able, inasmuch as the ascertainment of causes, such causes as human
faculties can ascertain, namely causes which are themselves phenomena
is ... merely the ascertainment of other and more universal Laws of
Phenomena."[8]) This was pure nominalistic milk.

At the root of the difference between idealists and empiricists as
regards causation lay the empiricists' insistence that we know nothing
except through the experience of our senses, while idealists equally
insistently urged that we have additional sources of knowledge, be it
intuition, or conscience, or Divine Revelation. This was again clearly
expressed by Whewell. We must lay down the principle, he said, that
every event has a cause, and from this principle "we infer that the
world itself must have a cause; that the chain of events connected by
common causation, must have a First Cause of a nature different from
the events themselves. This we are entitled to do, if our Idea of Cause
be independent of, and superior to, experience: but if we have no Idea
of Cause except such as we gather from experience, this reasoning is
altogether baseless and unmeaning."[9]) Indeed, Whewell directly
criticized the arch-empiricist Hume for adopting a view of causation
which excluded such an important inference. Hume had observed

---

[6]) Whewell, I, 159, 160, 170.
[7]) Whewell, II, 571.
[8]) Mill, II, 394.
[9]) Whewell, I, 160.

that as no experience could reveal anything else in causation than simple recurrence, and since all knowledge is derived from experience, therefore all we can know of causation was contained in that idea of recurrence, or invariable sequence. Whewell commented: "Our inference from Hume's observation is, not the truth of his conclusion, but the falsehood of his premisses: — not that, therefore, we can know nothing of natural connexion, but that, therefore we have some other source of knowledge than experience."[10]

Whewell accepted, apparently, only such premisses as allowed him to draw the conclusions he believed desirable, namely, that it is possible for us to know, albeit very imperfectly, the ultimate cause of all that exists; in other words, God, or the First Cause.

We have already met with this notion of God as the First Cause in our discussion of the argument of Design. In that discussion, "first" was taken in the sense of anterior in time. Observable phenomena were causally connected with each other, forming chains of causation where each effect was in turn the cause of another effect. At the beginning of the chain, or chains, was the Cause which was not itself the effect of something else. In contradistinction to the First Cause, the *causa causans*, those causes which were themselves caused were called *secondary causes*.

But the term First Cause was not only conceived as anterior in time, but also as higher in order of generality. A temporal chain of causation leading to the movement of, say, a billiard ball — Whewell's example[11]) — might include such things as another ball, the queue, the arm of the player, the contraction of a muscle, etc. But instead of tracing this temporal chain of causation, one could establish the order of generality of the causes. The movement of the billiard ball was a particular instance of the movement of bodies in space, and in a sense caused by gravity. And from here one could go on. As Whewell said, "In contemplating the series of Causes which are themselves the effects of other causes, we are necessarily led to assume a Supreme Cause in the Order of Causation, as we assume a First Cause in Order of Succession."[12]) Exemplifying this aphorism, he said, "Gravity is the cause of the motions

---

[10]) Whewell, I, 72.
[11]) Whewell, I, 170.
[12]) Whewell, I, xlvii, Aphorism LVI.

of the planets, but what is the cause of gravity?"[13]) It is evident that both the Supreme Cause and the First Cause were taken by Whewell as names for God.

On both these scores, the idealist-religious view clashed with the empiricist one. As regards the temporal chain of causes — the Order of Succession — empiricists insisted that only the observable links actually occurring in that chain were properly to be called causes: nothing else could be known. And as regards the explanation of one cause in terms of another with a higher degree of generality — the Order of Causation — it was, as Mill said, "but substituting one mystery for another." Empiricists regarded it as a misuse of language to say that gravity was the cause of the movements of the planets. Gravity was, like the word *force* itself, a "logical fiction,"[14 A]) systematizing a large number of observable facts. Gravity was not a cause distinct from the moving bodies, or residing in them: it was the same cause or causes expressed in different language. In the same manner, the only meaning an empiricist could give to the phrase, "the cause of gravity," was that of a concept comprehending a still larger number of facts than that of gravity — yet another logical fiction.

Empiricists and idealists therefore held different views regarding the ontological status of hypothetical, non-experiential concepts. Empiricists denied their substantial existence, idealists affirmed it. And this difference had important practical consequences, notably in the Darwinian controversy. For from the postulate that any temporal chain of causation originated in the First Cause it followed that at whatever point scientists were unable to explain a sequence of events in terms of secondary causation — whenever there was, as it was expressed, a break in the chain — then they had to recognize the possibility that this was a point where a chain originated, where the cause antecedent to the ultimate ascertained secondary cause was the First Cause. This was the doctrine of miraculous intervention in idealistic philosophical jargon.[14 B])

Furthermore, the conceptual realism of the idealistic school also had the important consequence of leading them to regard such concepts as Vital Force and Final Cause as substantial entities or agencies,

---

[13]) Whewell, II, 580.

[14 A]) Mill, I, 382.

[14 B]) See above, Chapter 4, p. 83.

instead of as logical constructions. These concepts were of supreme importance in pre-Darwinian biological discussion. Again we may cite Whewell for an authoritative statement of the idealistic view. "In order to obtain a Science of Biology," he said, "we must analyse the Idea of Life. It has been proved by the biological speculations of past time, that organic life cannot rightly be resolved into mechanical or chemical forces, or the operation of a vital fluid, or of a soul. Life is a System of Vital Forces; and the conception of such Forces involves a peculiar Fundamental Idea."[17] Vital Forces, Whewell admitted, like any other force, were known to us only through their effects,[18] nevertheless they were truly causes, and as such revealed the "real connexions of things."[19]

The idea of Final Cause was equally important in Whewell's and the idealists' system. Though final causes must not be employed in physics, they were indispensable in biology. "In mere Physics, Final Causes, as Bacon has observed, are not to be admitted as a principle of reasoning. But in the organical sciences, the assumption of design and purpose in every part of every whole, that is, the pervading idea of Final Cause, is the basis of sound reasoning and the source of true doctrine."[20] It is obvious that the admissibility of Final Cause was a fundamental prerequisite for establishing the argument of Design. Vital Force, in the sense employed by the conceptual realists, was useful chiefly by serving as a vehicle for the establishment of final causes.

On all these points the idealistic view clashed directly with the Darwinian doctrine, and it is indeed not surprising that Darwin and Huxley appear as wholehearted supporters of empiricism. Huxley several times acknowledged his debt to Mill's *System of Logic*,[21] and his book on Hume[22] is clearly the product of a sympathetic agreement on fundamental points. Darwin was much less articulate than Huxley on philosophical questions. He admitted that he had not "a metaphysical head."[23] But he often expressed his high appreciation of Huxley's

---

[17]) Whewell, I, xxxiv.
[18]) Whewell, II, 442.
[19]) Whewell, II, 571.
[20]) Whewell, II, 86.
[21]) E. g. *Lectures and Essays*, 54, note; *Lay Sermons*, 95.
[22]) T. H. Huxley, *Hume*, London, 1881.
[23]) Darwin, *Letters*, II, 290, to Lyell, 1860.

views.[24]) The most explicit statement from Darwin's own pen is probably a passage in the 6th edition of the *Origin*, where he clearly repudiates conceptual realism. "It has been said that I speak of natural selection as an active power or Deity; but who objects to an author speaking of the attraction of gravity as ruling the movements of the planets? Every one knows what is meant and is implied by such metaphorical expressions; and they are almost necessary for brevity. So again it is difficult to avoid personifying the word Nature; but I mean by Nature, only the aggregate action and product of many natural laws, and by laws the sequence of events as ascertained by us."[25])

Darwin did not flatly reject the current metaphysics of the First Cause, though he avoided the phrase. He contented himself with declaring that any discussion of such matters would take the whole case "out of the range of science."[26]) Nor did he mean by this phrase to subscribe to the current "two spheres" view: he made it sufficiently clear that he would not entertain a reference to the First Cause as a valid alternative to a scientific explanation. To accept it as such, he said, would be "to enter into the realms of miracle, and to leave those of Science."[27]) The main thing, for Darwin as for any empiricist, was to put the observable phenomena in definite relations to each other. A theory involving a miracle, if it could be conceived, would infringe this empiricist rule.

Huxley made explicit much that was implicit in Darwin. With respect to conceptual realism he said, for instance, "Nobody knows anything about the existence of a 'force' of gravitation apart from the fact ... all we know about the 'force' of gravitation, or any other so-called 'force' is that it is a name for the hypothetical cause of an observed order of facts."[28]) And again, "The tenacity of the wonderful fallacy that the laws of nature are agents, instead of being, as they really are, a mere record of experience, upon which we base our interpretations of that which does happen, and our anticipation of that which will happen, is an interesting psychological fact."[29]) As to final

---

[24]) Darwin, *Letters*, III, 3, to Hooker, 1863.

[25]) *Origin*, 6 ed., 99. See also *Origin*, 1 ed. r., 408.

[26]) Darwin, *More Letters*, I, 191.

[27]) *Origin*, 6 ed., 318. See also *Origin*, 1 ed. r., 299.

[28]) Huxley, *Essays on Controversial Subjects*, 287.

[29]) *Ibid.*, 253.

causes, Huxley emphatically endorsed Bacon's dictum about them as "barren virgins"[30]), and refused to make any exception, as Whewell had done, for the science of biology. The idea of vital force received short shrift. In his famous essay on "Protoplasm", Huxley asked, "What justification is there, then, for the assumption of the existence in the living matter of a something which has no representative, or correlative, in the not living matter which gave rise to it? What better philosophical status has 'vitality' than 'aquosity?' ... If the phaenomena exhibited by water are its properties, so are those presented by protoplasm, living or dead, its properties."[31])

The idea of a First Cause and its correlative of secondary causation was as devoid of meaning to Huxley as to Hume or Mill, and he consistently avoided the phraseology. Suppose, he said, for the sake of the argument, that we admit that an event is produced by a supernatural cause. What then? "How much wiser are we; what does the explanation explain? Is it any more than a grandiloquent way of announcing the fact, that we really know nothing about the matter? A Phenomenon is explained when it is shown to be a case of some general law of nature; but the supernatural interposition of the Creator can, by the nature of the case, exemplify no law."[32]) The measure of Huxley's agreement with Mill and Hume on these matters appears from the following words, written in 1871: "In ultimate analysis everything is incomprehensible, and the whole object of science is simply to reduce the fundamental incomprehensibilities to the smallest possible number."[33])

The word *incomprehensible* was sometimes taken up by mystery-mongers as proof of the inadequacy of mere science. But Huxley vigorously rejected the idea that the inability of science to solve that problem was any justification for bringing the supernatural into the picture. In the first place, the supernatural explanation was spurious: it did not really explain. In the second place, it worked "not only negative, but positive ill, by discouraging inquiry, and so depriving man of the usufruct of one of the most fertile fields of his great patrimony, Nature."[34]) Thirdly, it also damaged science by bringing logical con-

---

[30]) Huxley, *Lectures and Essays*, 129.
[31]) Huxley, *Lay Sermons*, 151.
[32]) Huxley, *Lectures and Essays*, 164.
[33]) *Contemporary Review*, 18, 1871, 466.
[34]) Huxley, *Lectures and Essays*, 166.

fusion to its theories. "The fundamental axiom of scientific thought," said Huxley, "is that there is not, never has been, and never will be, any disorder in nature. The admission of the occurrence of any event which was not the logical consequence of the immediately antecedent events, according to these definite, ascertained, or unascertained rules which we call the 'laws of nature', would be an act of self-destruction on the part of science."[35]) Or again, "It is wholly impossible to prove that any phenomenon whatsoever is not produced by the interposition of some unknown cause. But philosophy has prospered exactly as it has disregarded such possibilities."[36]) Huxley realized that the fundamental difference between empiricist and idealistic views of causation could not be resolved by logic or observation. It could not be *proved* that one was right and the other wrong. He justified his choice of the empiricist view by pointing out that of the two it was the only one that made scientific inquiry and scientific progress possible.

On one point empiricist and idealistic philosophers of science were agreed: they rejected the popular view that scientists should only describe facts and avoid hypotheses. Both Whewell and Mill insisted that no scientific inquiry could proceed without hypotheses, and that Newton's much-quoted *Hypotheses non fingo* should not be taken at face value.

But the two philosophical schools did not agree about the contents of the hypotheses that they regarded as allowable. When the hypothesis concerned a substantial cause, empiricists must inevitably require that "the supposed cause should . . . be a real phenomenon, something actually existing in nature,"[37]) as Mill said: for only such phenomena were recognized by empiricists as causes. The hypothetical cause might as yet be unobserved, but "its existence should be capable of being detected, and its connection with the effect ascribed to it should be capable of being proved, by independent evidence."[38]) Therefore Mill declared it to be "a condition of the most genuinely scientific hypothesis, that it be not destined always to remain an hypothesis, but be of such a nature as to be either proved or disproved by comparison with observed facts."[39])

---

[35]) Huxley, *Essays on Controversial Subjects*, 247.
[36]) Huxley, *Life*, I, 174.
[37]) Mill, II, 15, 16.
[38]) Ibid.
[39]) Mill, II, 14.

The main function of hypotheses, on the empiricist view, was to serve as guides to scientific enquiry. Laws of causation were hypothetically postulated, their consequences deductively calculated, and the deductions in their turn confronted with the observed facts. No other restriction was placed on the *framing* of hypotheses than that they should be capable of being tested by experience in this way. But strict conditions had to be satisfied before any hypothesis could be accepted as proved.

Whewell was as emphatic as Mill on the necessity of employing hypotheses. But he placed restrictions on the framing or construction of hypotheses: the criterion to be applied, on his view, was not whether their consequences were testable by experience: but that they should themselves, at the outset, be "close to the facts, and not connected with them by other arbitrary and untried facts."[40] Whewell explained that he meant by this that they should be "clear and appropriate", terms which he discusses at some length.[41] The first of these requirements put a premium on mathematical concepts — such as gravity —, the second, which Whewell himself recognized was difficult to apply,[42] was in reality a conservative criterion, serving to exclude such hypotheses as clashed with established view. As instances of inappropriate conceptions he cited mechanical and chemical hypotheses to explain vital powers.

The result of applying both these requirements to the Darwinian hypotheses was obvious; and Darwin's opponents did not fail to employ them against the Darwinian doctrines as Whewell expressly said they should be employed, "as a caution and prohibition."[43]

It is significant that Whewell used the word *hypothesis* very sparingly, and tended to equate it with, and indeed to replace it by, "conception superinduced upon the facts" which loomed so large in his theory of induction: which was, he said, "the peculiar import of the term Induction."[44] Mill called attention to this terminological peculiarity when he remarked that "Dr. Whewell's doctrine of conceptions might be fully expressed by the more familiar term Hypothesis."[45] Whewell's

---

[40] Whewell, II, 438.
[41] Whewell, II, 183—7.
[42] Whewell, II, 186.
[43] Whewell, II, 187.
[44] Whewell, II, 215.
[45] Mill, II, 193.

choice of terms, however, was not accidental. At the time when he wrote, hypotheses were regarded by the general public as something suspect. It was a negatively loaded term. Induction, on the other hand, was positively loaded: it gave certain and firm knowledge. By making the hypothetical "conception" a constituent of the inductive, and not of the deductive part of scientific inquiry, it was natural for him to place on them the restrictions which he did: but it also enabled him to confer on those hypotheses which he did allow the reliability which everybody recognized as belonging to inductive knowledge.

Therefore, though Whewell repudiated the popular view that scientists should only concern themselves with facts and desist from framing any hypotheses, in relation to such theories as Darwin's his attitude coincided with the popular one: biological science had not arrived at the stage where "clear", mathematically formulated hypotheses could be framed, and the objectionable Evolution hypothesis could easily be represented as "inappropriate"[46]). In the Darwinian discussion the general public could therefore find support from the leading scientific theorist of the time when they branded Darwin's theory as baselessly hypothetical, and not founded on induction. This was a grave charge. The inductive method was looked upon as a peculiarly British contribution to scientific progress, and to discard it in favour of a deductive or speculative system appeared to the general public as not only scientifically unwise, but as morally reprehensible. It was a sort of betrayal of the national cause.

The general public, who naturally knew little or nothing of Whewell's theory of induction, did not use the term in any very precise sense. By and large, it was taken to imply that scientists should be concerned with ascertaining facts only, and abstain from framing hypotheses. The hypothetical element of traditional scientific — or religious — view was overlooked. Bacon and Newton came to stand as guarantors of safe, conservative opinions.

To condemn Darwin for having neglected induction in favour of a deductive and hypothetical speculation was, as we said, an extremely common way of arguing against his doctrine. One of the very first published comments on it, a brief article in the *Zoologist* on Darwin's and Wallace's papers to the Linnaean Society *Journal* for 1858, was on these

---

[46]) Whewell rejected the Darwinian theory; see I. Todhunter, *William Whewell*, p. 433, letter to Rev. Dr. D. Brown.

lines. The commentator asked whether the views presented in those papers were not "founded upon the imaginary probable, rather than established by induction from ascertained facts", which latter process he did not "hesitate to pronounce the only solid and satisfactory basis of a new opinion."[47] When the *Origin* appeared this sort of criticism took formidable proportions. Sedgwick, Darwin's old teacher at Cambridge, expressed his "detestation of the theory . . . because it has deserted the inductive track, the only track that leads to physical truth."[48] The Bishop of Oxford, in the *Quarterly Review*, declared that "when subjected to the stern Baconian law of observation of facts, the theory breaks down utterly."[49] Professor Owen, in the *Edinburgh Review*, laid it down that if there was a "preordained law or secondary cause" bringing about specific change, such a law could "only be acquired on the prescribed terms,"[50] that is to say, by induction. One writer called Darwin an "apostate . . . from the grand induction of the special creation of organized beings."[51] Thus *induction* became one of the key words in the early criticism of Darwinism.[52]

Darwin was himself sensitive on this point. It must be remembered that the *Origin* was not he sort of book that the had originally planned. Chiefly due to Wallace's article, he decided to publish his conclusions without the full factual documentation which he recognized had to be supplied.[53] Describing his forthcoming book in a letter to Asa Gray, he said, "What you hint at generally is very, very true: that my work will be grievously hypothetical, and large parts by no means worthy of being called induction, my commonest error being probably induction from too few facts."[54] On the other hand, of course, he was not willing to admit that his method was scientifically unsound: His arguments were the common empiricist ones: see below, p. 194.

---

[47] *Zoologist*, 1859, Mar., 6474.

[48] *Spectator*, 1860, 285.

[49] *Quarterly Review*, 108, 1860, 239.

[50] *Edinburgh Review*, 111, 1860, 532.

[51] *Recreative Science*, 3, 1861—2, 219—26.

[52] See also: *British Quarterly Review*, 38, 1863, 495, *Watchman*, 1861, Sep.18, 309, *Athenaeum*, 1866: 2, 120, *Watchman*, 1868, Aug. 26, 281, *Globe*, 1868, Aug. 20, p. 2 *John Bull*, 1871, Apr. 6, 234, *Edinburgh Review*, 128, 1868, 414.

[53] *Origin*, 1 ed. r., 2.

[54] Darwin, *More Letters*, I, 126, to Asa Gray, 1859.

Huxley, considerably more pugnacious than Darwin, was not inclined to let the opponents have it all their own way in this matter. He did not mince words: "I do protest that, of the vast number of cants in this world, there are none, to my mind, so contemptible as the pseudo-scientific cant which is talked about the 'Baconian philosophy.'[55]) Or again, "Critics exclusively trained in classics or in mathematics, who have never determined a scientific fact in their lives by induction from experiment or observation, prate learnedly about Mr. Darwin's method, which is not inductive enough, not Baconian enough, forsooth, for them. But even if practical acquaintance with the process of scientific investigation is denied to them, they may learn, by the perusal of Mr. Mill's admirable chapter 'On the Deductive Method', that there are multitudes of scientific inquiries in which the method of pure induction helps the investigator but a very little way."[56]) This was the obvious Darwinian defense: to show that scientific inquiry was not, and could not be, inductive only; that hypotheses were necessary, and that in reasoning upon hypotheses one was bound, as Mill said, not by the laws of induction, but by those of deduction. One of the fullest statements of this view was put forth by Henry Fawcett in *Macmillan's Magazine*,[58]) and later at the British Association in 1861. It is interesting to note that Fawcett, like Huxley, supported his views by quoting Mill.[59])

As the term induction, to the general public, had come to mean little more than the collection and accurate description of facts, the chief ground for asserting that Darwin's theory was not inductive was usually that he did not bring forward any new facts. Professor Owen in the *Edinburgh* was the leading exponent of this line of argument. In his review of the *Origin*, after pointing out some four or five instances of factual description from the book, Owen commented, "Having now cited the chief, if not the whole, of the original observations adduced by its author . . . our disappointment may be conceived."[60]) He then examined Darwin's illustrative examples, for

---

[55]) Huxley, *Lectures and Essays*, 46.

[56]) *Ibid.*, 172.

[57]) — — —

[58]) *Macmillan's Magazine*, 3, 1860—61, Dec. 1860, 81—92.

[59]) See also above, Chapter 4, on Huxley at the British Association meetings in 1862 and 1863. Also *Examiner*, 1867, Nov. 30, 757, and 1871, May 13, 492—3.

[60]) *Edinburgh Review*, 111, 1860, 495.

instance the little passage on the connection between the number of cats in a village, and the occurrence of *trifolium pratense* in the neighbouring fields, with biting sarcasm, "the more sober, or perhaps duller naturalist would no doubt appreciate more highly a dry statement of investigations."[61]) Another naturalist reviewer wrote, "That a book having the name of Charles Darwin on its title-page would be extensively read, is a matter of course, but that, without containing the smallest tittle of new evidence on the subject of the evolution of one species from another, it should have been regarded as establishing a theory, may well excite our surprise."[62])

The criticism of the naturalists was taken up by the less scientific opponents. "Since the publication of Mr. Darwin's *Origin of Species*", wrote the *Guardian*, "it seems more than ever necessary to urge the Baconian principle of facts before theories."[63]) The *John Bull* found that "we may fairly refuse to accept an hypothesis in natural history which relies on observation not yet made, upon fact not yet revealed."[64])

Some critics, in an effort to be fair to Darwin, took the line of distinguishing between the facts and the theory. The facts were valuable in themselves, though they did not bear out the theory which Darwin had unfortunately built on them. The *Daily News* ended its hostile review by commending Darwin as a "painstaking and most accomplished naturalist", and was convinced that the completed work which Darwin promised would be "a valuable assemblage of facts."[65]) A speaker at the British Association in 1861 was applauded for declaring that "the facts brought forward in support of the hypothesis had a very different value indeed from that of the hypothesis."[66]) Such views were especially common after the publication of Darwin's *Orchids*, of which the *Guardian* said that "It is not necessary for the enjoyment of this exquisite and fascinating book to be a believer in the Darwinian theory of the mutability of species."[67])

---

[61]) Ibid. The Darwinian passage in the *Origin*, 1 ed. r., 64.

[62]) *Annals and Magazine of Natural History*, 7, 1861, 399.

[63]) *Guardian*, 1861, Feb. 6, 134.

[64]) *John Bull*, 1863, Feb. 21, 124. See also: *Press*, 1859, 1243, *Quarterly Review*, 108, 1860, 238, *Daily News*, 1859, Dec. 26, p. 2, *Record*, 1862, Oct. 8, p. 2,*Blackwood's Magazine*, 92, 1862, 90.   [65]) *Daily News*, 1859, Dec. 26.

[66]) *Manchester Guardian*, 1861, Sep. 9, reporting speech by Dr. Lankester.

[67]) *Guardian*, 1862, Nov. 5, 1054. See also above, Chapter 4, p. 72.

Thus whether the facts adduced were regarded as few or many, the implication was always that they were insufficient to support the theory. But even more serious than Darwin's failure to provide such supporting facts, one alleged, was his way of constructing his theory on the basis of probable, possible, or hypothetical facts. He replaced observation by imagination. This line of criticism was vigorously pursued in the most influential reviews of the *Origin*, Owen's in the *Edinburgh*, and the Bishop of Oxford's in the *Quarterly* and at the British Association in 1860. A correspondent at the British Association meeting found that the prevailing attitude among those present was that "a series of plausible hypotheses had been skilfully manipulated into solid facts."[68] A host of press voices accused Darwin of mingling together "facts and assumptions, probabilities and speculations . . . in most illogical confusion",[69] and of using his imagination to supply the lack of factual observation.[70] A commentator on Hooker's address to the British Association meeting in 1868 therefore found it natural to sneer, "Once involved in a dissertation on Darwinism, there was scarcely a fact or a speculation that could not be introduced with more or less appropriateness."[71]

In fact, like the term *induction*, the word *hypothesis* became a key word in the Darwinian controversy. Induction was the good, positive ideal which Darwin had deserted for its bad, negative counterpart, variously named hypothesis, theory, speculation, assumption, imagination, fancy, guess, and the like. Again and again we meet with such expressions as "mere conjecture"[72] "crude theories,"[73] "reckless speculation,"[74] "rash speculation,"[75] "bold theorizing."[76] A science worthy of the name, one said, should content itself with patiently collecting facts. That was what Bacon and Newton had done: it was also

---

[68] *John Bull*, 1860, July 7, 422.

[69] *Record*, 1862, Oct. 8.

[70] E. g. *Christian Observer*, 60, 1860, 565.

[71] *Standard*, 1868, Aug. 21, p. 4. See also *Church Review*, 1862, Oct. 11, 636, *Christian Observer*, 1871, 889, *Edinburgh Review*, 135, 1872, 114, *John Bull*, 1872, Nov. 30, 824, *Times*, 1872, Dec. 13, p. 4, 1872, Feb. 12, p. 2.

[72] *Guardian*, 1863, May 20, 476.

[73] *Nonconformist*, 1867, Sep. 18, 767.

[74] *Contemporary Review*, 7, 1868, 138.

[75] *Manchester Guardian*, 1872, Aug. 16, p. 5.

[76] *Morning Post*, 1872, Aug. 15, p. 4.

what was done in the most advanced branches of science. Commenting
on the British Association discussions of 1867, a correspondent noted
that "in sections where facts are scanty, a speculative spirit was most
rife."[77] The criticism was not that the hypotheses were unplausible
or disproved — though this, needless to say, was usually taken for
granted — but that they occurred at all. The Darwinians were the
"hypothetical school"[78], and that was enough to condemn them.[79]

The full force of the popular prejudice against hypotheses was brought
out in the discussion that followed upon Tyndall's British Association
lecture "On the Scientific Use of Imagination."[80] The *Times*, as we
have seen, was severe. The *Record* gravely remarked that "most Natural
Philosophers since the time of Bacon, and notably Newton, have warned
us of the danger of imagination."[81] Another religious organ said, "We
are content with the facts, the *savans* may have all the speculation to
themselves."[82] The argument was so popular that it even found its
way into the pages of *Punch:*

> "Hypotheses non fingo,"
> Sir Isaac Newton said
> And that was true, by Jingo!
> As proof demonstrat*ed*
> But Darwin's speculation
> Is of another sort
> 'Tis one which demonstration
> In nowise doth support."[83]

Though the majority of those who criticised the hypothetical nature
of Darwin's theory chose to condemn hypotheses wholesale, there were

---

[77] *Nonconformist*, 1867, Sep. 18, 767.

[78] *Nonconformist*, 1866, Aug. 29, 705.

[79] See also: *Church Review*, 1861, 28, 1863, Apr. 11, 353, *Public Opinino*, 1865,
July 8, 40, *Contemporary Review*, 4, 1867, 49, *Athenaeum*, 1868, Feb. 15, 243,
Morning Advertiser, 1868, Aug. 21, p. 4, *Globe*, 1868, Aug. 20, p. 2, *Weekly Review*,
1869, 820, *Record*, 1870, Sep. 16, p. 2, 1871, Aug. 9, p. 2, *Literary Churchman*,
1872, 158.

[80] See above, Chapter 4, p. 87.

[81] *Record*, 1870, Sep. 23, p. 2.

[82] *English Independent*, 1870, Sep. 22 922. See also *Methodist Recorder*, 1870,
Sep. 23, 538, *John Bull*, 1871, Apr. 6, 234, *Times*, 1871, Apr. 7, p. 3, *Contemporary Review*, 6, 1867, 16, *Examiner*, 1861, Sep. 7, 568, *Record*, 1862, Oct. 8, p. 2,
Contemporary Review, 2, 1866, 122, *John Bull*, 1867, Mar. 30, 231.

[83] *Punch*, 69, 1871, Apr. 8, 145. See also June 10, 234, and 55, 1868, Dec. 19, 264.

some who concentrated their attack on a narrower front. The *Dublin Review*, for instance, admitted that men of science should not be deterred "from framing the best theories they can." But the journal indicated one of the most potent reasons for the conservative and religious distrust of hypotheses, when it added that scientists should be wary of putting them forward "especially when they appear to clash with portions of truth already firmly established."[84] The same recommendation was apparently implied by the reviewer who wrote that the Darwinians should "confine their speculations within the limits of common sense."[85]

Others again fully admitted that hypotheses were necessary tools. But hypotheses had to be proved: and it was very common to insist that as long as Darwin had not proved his theory, the traditional view should stand. "The weight of proving the truth of a new theory lies with him who puts it forward,"[86] wrote the *Guardian*. As long as it remained unproved, traditional views should apparently not be disturbed. "Let us frame hypotheses if we will but keep them in their proper place, as stimulating inquiry,"[87] as Stokes put it at the British Association in 1869.

Others agains sought a reason for rejecting Darwin's hypotheses, not in their startling consequences, but in the nature of the hypotheses themselves. At this point it was possible to make use of Whewell's stipulation that hypotheses should be "clear" and "appropriate." Hopkins, who reviewed the *Origin* in *Fraser's Magazine*, took this line. "We are not denying our author's right to reason upon his hypothesis ... He has an unquestionable right, according to the laws of all philosophical reasoning, to do so,"[88] he wrote. But he had made a distinction between "geometrical laws" and "physical causes" clearly reminiscent of Whewell, and concluded that in biology conceptions were not yet clear enough to allow inferences as to causes. "We cannot but smile at the thought that we should at present hope to arrive at even the most remote approximation to the real solution of the problem of vital

[84] *Dublin Review*, 1, 1863, 523.

[85] *Church Review*, 1861, 28.

[86] *Guardian*, 1860, July 4, 606. See also: *Fraser's Magazine*, 62, 1860, 80, *Popular Science Review*, 2, 1862—3,3 392, *Contemporary Review*, 2, 1866, 120, *London Quarterly Review*, 36, 1871, 282—3.

[87] See above, Chapter 4, p. 83. *Daily News*, 1869, Aug. 19, p. 4.

[88] *Fraser's Magazine*, 62, 1860, 77.

phenomena."[89]) Almost the same argument was put forward by the
critic who declared that "until physiology and kindred sciences have
entered into their deductive stage, it is not probable that the question
of the Origin of Species will be definitely solved."[90]) This sort of criticism was fairly common, though in general critics contented themselves
with the quite general charge that the theory was "hastily or prematurely" constructed,[91]) that it "had not a sufficient basis of induction to
rest on"[92]), that it contained "hypotheses which are not legitimate
deductions from facts,"[93]) and similar phrases.[94])

It is evident that many opponents by "insufficient basis of induction"
really meant the empiricist basis of Darwin's theory. Darwin did not
take into account such things as First Cause and Final Causes. He
considered only natural causes. Against this no objections were raised
as long as the investigation remained on the purely descriptive level.
But Darwin and the Darwinians went further. In order to connect
the observed natural facts, they constructed theories where again nothing
but natural facts occurred, or were assumed. It was here that their
opponents set in their attack. It was here, they said, that facts and
hypotheses, observation and imagination, induction and speculation,
were mingled together. The world of observation might be studied
and described in empiricist terms. But when empiricist principles were
carried over to the unobserved world as well, idealists had perforce to
react.

The attitude of the theological idealists was expressed quite clearly
by a Methodist reviewer of Lyell's *Antiquity of Man*. The empirical
scientists, he said, "come into the field unembarrassed by belief, not
asking and not caring what received truths their opinions may support
or upset. If, of two classes of facts, one be stronger than the other,
— if, of two theories, one have less difficulties than the other, — they
can be satisfied to accept the better evidence and the easier theory.

---

[89]) *Fraser's Magazine*, 61, 1860, 742—4, 62, 1860, 84.

[90]) *Intellectual Observer*, I, 1862, 333. See also Sir William Thomson's address
to the British Association in 1871, *Report*, Ciii.

[91]) *Edinburgh Review*, III, 1860, 499.

[92]) *Watchman*, 1861, Sep. 18, 309.

[93]) *Patriot*, 1863, Sep. 10, 594.

[94]) See also *Daily News*, 1859, Dec. 26, p. 2, *London Quarterly Review*, 20, 1863,
272, *Dublin University Magazine*, 65, 1865, 594, *Nonconformist*, 1868, May 27, 538,
London Quarterly Review, 36, 1871, 268, *Times*, 1871, Apr. 8, p. 5.

But it is otherwise with those who begin an investigation under the influence of settled convictions . . . If one party is open to the accusation that previous conviction blinds them to the force of facts, the other is subject to the reproach that the want of such conviction makes them injure the cause of truth by hasty conclusions, and generalisations founded on insufficient data."[95]) Another religious writer accused scientists of framing hypotheses based on the "negative suspicion that nothing can exist which the sense-philosophy refused to recognize."[96]) A *Saturday* reviewer was more specific. Referring to examples of parallel developments in genetically distinct branches of the animal kingdom, he asked, "Do they not give distinct intimation of the presence of another law of structure, another *vera causa* . . . besides . . . natural selection . . . Is it not more philosophical to search patiently for the true nature of the cause that lies behind, recognized, though not seen, when we use the word 'analogy' as applicable to the relations of all organized beings, than to disregard the obvious indications of its presence and fill up the broad chasm of our ignorance by assumptions that may be admitted to be within the range of bare possibility, but which are certainly not sustained by the available evidence?"[97])

The fundamental objection against Darwin's hypotheses was, accordingly, that they disregarded the supernatural. As one popular writer put it, "All the mischievous and absurd theories of our savans have been founded on inferences, and . . . generally collapse of their own inherent fallacy . . . the discoverer or his disciples *invent* a hypothesis which shall serve to explain the phenomenon . . . Why then is it invented? Simply to gratify the impatient desire to solve some mystery."[98]) Religious people had no need of hypotheses, since they could always explain some phenomenon by means of a supernatural interference. The idealistic school of scientific philosophy desired to leave this possibility open, and therefore it is hardly surprising to find idealists such as Whewell adopting a restrictive attitude against hypotheses. But to empiricists a hypothetical argument was absolutely essential. Denouncing hypotheses was therefore one of the ways of attacking the empiricist philosophy. If that philosophy could be thoroughly discredit-

---

[95]) *London Quarterly Review*, 20, 1863, 272—3.
[96]) *Contemporary Review*, 6, 1867, 16.
[97]) *Saturday Review*, 1860, May 5, 573. [Mivart?]
[98]) *Globe*, 1868, Aug. 20, p. 2.

ed, the defence of religion against science would clearly become much easier.

Unfortunately ordinary readers were not willing to pursue the discussion very far on this rarefied level. As, moreover, explanations in empiricist terms often commended themselves to common sense, idealists found themselves in a position of what they took to be unfair disadvantage, something which was well brought out by a reviewer of an anti-Darwinian work in the *Morning Post*: "The triumph of the materialist comes from this — that in treating upon tangible facts alone, and ignoring every power which is not palpable to the sense, he appeals successfully to practical minds, averse to ascending into the more elevated contemplations of a sounder philosophy. Great moral courage is indispensable in the opponent of views easily advanced, of a popular nature, and susceptible of refutation only by the employment of the most severe form of abstract reasoning."[99] This sort of argument might at least establish that such questions as the Darwinian one were too complicated for ordinary minds to grasp: the problems were, as one writer put it, "for the logician, not merely for the naturalist."[100] Darwinian views ought to be kept within the somewhat narrow limits of the educated public.

The Darwinian defence against the charge of using hypotheses was the simple empiricist one. Commenting on the maxim that scientists should "observe and not theorize," Darwin said, "About thirty years ago there was much talk that geologists ought only to observe and not theorise; and I well remember some one saying that at this rate a man might as well go into a gravel-pit and count the pebbles and describe the colours. How odd it is that anyone should not see that all observation must be for or against some view if it is to be of any service!"[101] Again, on the criteria for judging hypotheses he agreed with Mill. His hypothesis took into account only natural phenomena, and he wished it to be judged solely on its power of accounting for as many facts as possible. "I am actually weary of telling people that I do not pretend to adduce direct evidence of one species changing into another, but that I believe that this view is in the main correct, because so many phenomena can be thus grouped together and explained ...

---

[99] *Morning Post*, 1861, Sep. 6. p. 2.

[100] *Illustrated London News*, 58, 1871, Mar. 11, 243.

[101] Darwin, *More Letters*, I, 195, to Henry Fawcett, 1861.

I generally throw in their teeth the universally admitted theory of the undulations of light . . . admitted because the view explains so much."[102]) Huxley also emphatically declared that "All science starts with hypotheses"[103]), and that "there cannot be a doubt that the method of inquiry which Mr. Darwin has adopted is not only rigorously in accordance with the canons of scientific logic, but that it is the only adequate method . . . what Mr. Darwin has attempted to do is in exact accordance with the rule laid down by Mr. Mill; he has endeavoured to determine great facts inductively, by observation and experiment; he has then reasoned from the data thus furnished; and lastly, he has tested the validity of his ratiocination by comparing his deductions with the observed facts of nature."[104]) On this basis Huxley could not but conclude, as he did, that Darwin's hypothesis was the only one that had any scientific existence.[105]) Lesser Darwinians in the press had little else to do than to reaffirm these empiricist views.[106])

When illustrating the argumentation of the general public on these philosophical matters, we have in this chapter been concerned chiefly with those arguments which were advanced against the alleged hypothetical, non-inductive nature of Darwin's theory. The popular argumentation on the question of causation will not be gone into here. It has been illustrated in the chapters on Design and Miracles; it will be further discussed in connection with the various criticisms of the Natural Selection theory, below, Chapter 12.

It is quite obvious that the public discussion of the philosophical and methodological foundations of the Darwinian theory was largely motivated by its religious implications. It was impossible to accept the theory without effecting changes in a whole system of religious and metaphysical beliefs sanctioned by tradition, or, conversely, to preserve that body of beliefs intact without rejecting the theory. To the religious, the theory was an incubus which had to be cast out, or at any rate isolated and neutralized. To attack the theoretical

---

[102]) Darwin, *More Letters*, I 184, to F. W. Hutton, 1861.

[103]) Huxley, *Hume*, 55.

[104]) *Westminster Review*, 17, 1860, 566—7.

[105]) Huxley, *Lectures and Essays*, 287.

[106]) *National Review*, 10, 1860, 214, *Temple Bar*, 1, 1860—1, 542, *Examiner*, 1867, Nov. 30, 757, *Popular Science Review*, 9, 1870, 73—4. *Athenaeum*, 1870, Feb. 26, 286.

foundations of the theory was one of the ways, and an important one, of achieving this result: thereby the theory could be, if not directly refuted, at any rate represented as no more than a loose speculation, scientifically unjustifiable, and without any foundation in fact.

The Darwinians, and first among the Darwin himself and Huxley, justified the theory on frankly empiricist grounds. They recognized only observable phenomena as evidence, they refused to regard hypothetical concepts as anything more than abstractions serving to comprehend observable facts, and they judged their value solely according to their power of thus comprehending a large number of observable facts. In order to accept an explanation as a causal one, they required that one set of observable phenomena — the effect — should be connected with another set of independently observable phenomena — the cause. This ruled out explanations in terms of hypostatized tendencies and forces, which were, in conformity with empiricist views, no more than abstractions from the actual facts to be explained. It also ruled out final causes and miracles.

It is natural that the Darwinians should wish to lay down these empiricist methodological principles. Indeed, once they were granted, the Darwinian theory could hardly be resisted, as their opponents were ready to admit. If the theory was to be resisted, it was therefore practically necessary to appeal to non-empiricist principles. On the lower levels this was of course not recognized. Indeed, when the doctrine was attacked for not being inductive, the critics considered themselves more empirical than Darwin, because they advocated factual observation rather than theorizing. One did not admit that the traditional explanatory superstructure which was raised on these facts, with such constituents as Divine interpositions and Design, was itself, from the scientific point of view, altogether hypothetical.

The better informed critics, who admitted the right of scientists to argue hypothetically, adopted another procedure. They rejected the empiricist view that a hypothesis should be judged solely according to its power of comprehending a large number of facts: the further requirement was added that it should contain, in Whewell's phrase, a "clear and appropriate" conception. It was never quite clear what that phrase meant exactly, and it was therefore easy to declare arbitrarily that such hypotheses as one wished to reject were inappropriate or lacking in clarity.

The Mid-Victorian discussion between empiricists and idealists as regards the philosophy of science may be said to have been only superficially concerned with theoretical problems. Its motives lay deeper. The parties disagreed about the fundamental scale of values, symbolized by their different religious beliefs and disbeliefs. One of the results of the appearance of Darwin's theory was to bring that underlying disagreement into full view.

CHAPTER 10

The Immutable Essence of Species

The main assumption of the traditional biological doctrine which Darwin's theory was to supersede was that there was a fundamental difference between species on one hand, and varieties, genera, orders, classes, etc., on the other. *Classis et Ordo est sapientiae, Species naturae opus,* Linnaeus had declared. Class and order, in other words, were arbitrary terms invented for convenience of classification, and the same was true of variety. Naturalists might differ on the application of these terms in particular instances, while still agreeing on every factual question concerning the particular forms under discussion. But in the case of species, one held, things were different. Whether two forms were specifically distinct or not was a factual question, not a question of classification only. It is true that it was difficult to explain precisely what was the essence of species. The difference between varietal and specific distinctions was not as clearcut as one could wish. But one remained convinced that it was real and fundamental. Varietal distinctions were accidental and fluctuating, while specific ones were essential and permanent. Thus the immutability doctrine was, as it were, guaranteed by the current definition of species, and this was one of the first obstacles that Darwin had to remove in order to prepare the way for his transmutation theory. His task was greatly complicated by the fact that the traditional definition of species had been neatly fitted into a closely-knit framework of idealistic metaphysics. At this point, therefore, as at so many others, the scope of the disagreement between Darwinians and traditionalists was widened so as to become a conflict between empiricist and idealistic philosophies of science.

The Definition of Species

The traditional, orthodox definition of species was simple and straightforward. A species was a group of animals or plants which had all descended from the same original pair. Differences between species were looked upon as instituted by God at the original Creation. Varieties, on the other hand, had not been directly formed by God. Varietal differences had arisen through the influence of various external conditions, such as food, climate, or artificial selection, on the originally created specific form. Accordingly, varietal differences were at least in principle reversible, whereas specific differences were irreducible.

Now it is obvious that a definition of species, in which identical ancestry was the only criterion employed, could not be practically applied except at best under conditions of domestication, where it was sometimes possible to follow the line of descent of two forms up to their common parents. But forms found in nature did not carry their genealogical tables about them. Other criteria than that of actual common descent had to be employed.

The most common and most effective of these other tests of species was that of sterility. If the two forms could produce no offspring, or infertile offspring only, they were held to be specifically distinct. But if the forms produced fully fertile offspring, they should be considered as varieties of the same species.

Another criterion, less strict than the previous one, was that of reversion. If animals of a domestic breed were left to fend for themselves, they would lose several characteristics and revert to the type of their supposed wild ancestors, showing thereby that they were a variety only, not a distinct species.

Such criteria as these, however, could not be applied by naturalists except in a small proportion of cases. For instance, it was often necessary to work from dead specimens. Further, experiments in cross-breeding were laborious and costly, and results were often inconclusive, since it was difficult to decide whether any observed infertility was due to a general incompatibility, or to individual peculiarities, or to experimental conditions. Above all, criteria of this type were wholly irrelevant for the whole paleontological material, which was naturally supremely important in any investigation into the question of the origin of species.

In the overwhelming majority of cases naturalists therefore had to use other than experimental criteria to determine whether differences between forms were specific or not. The only facts that could be taken into account were the actual similarities and differences between the forms as found. In other words, one had to rely on systematic, taxonomic criteria, and interpret them in the light of evidence drawn from material

which could be subjected to experiments. Since observation showed
that the offspring of the same parents were never exactly alike, a
certain range of variations had necessarily to be allowed. The question
was, how wide a range? When did a certain number of points of divergence amount to a specific difference?

It is obvious that a decision on that question had to be to some
extent arbitrary. It depended on the observed range of variation in
undoubted species. It further depended on the number of intermediate
forms that were known. If two forms occupied the end-points of a
continuous series, they might be assigned to the same species, even if
they differed more widely from each other than two other forms between
which no connecting links had been found. Further, the geographical
distribution had to be taken into account. Evidently naturalists differed
on the weight to be given to the various considerations, and borderline
cases might be classed by some as distinct species, by others as varieties
only. Darwin had little difficulty in finding numerous examples of
just this, thus proving that the actual systematic criteria employed
to determine species were in fact to some extent arbitrary. He also
adduced a great mass of evidence to show that the commonly employed
experimental criteria of species were not absolutely infallible. Reversion
did not always occur, or occurred with different degrees of completeness.
And the phenomena of hybridism and infertility were in reality much
more complicated than many seemed to suppose.[1] To some extent
it must be a matter of arbitrary choice precisely what degree of infertility
should be accepted as a criterion of specific difference.

On the strength of the evidence he presented, Darwin concluded that
though species was an indispensable term of classification, it was
nothing more: "I look at the term species as one arbitrarily given for
the sake of convenience to a set of individuals closely resembling each
other . . . it does not essentially differ from the term variety, which
is given to less distinct and more fluctuating forms. The term variety,
again, in comparison with mere individual differences, is also applied
arbitrarily, and for mere convenience' sake."[2]

Naturalists before Darwin had of course been aware that what they
actually described as a specific type was an abstraction from what they

---

[1] See below p. 206 ff.
[2] *Origin*, 1 ed. r., 45.

knew of individual instances. But many of them had taken it for granted that these abstractions were at least an approximation to something real: the original prototype created by God. This was a natural attitude to take as long as the traditional theory of the fixity of species and their separate creation was adhered to. Darwin, after he had ceased to believe in the traditional theory, called attention to the fact that none of the criteria employed by naturalists was capable of unequivocally distinguishing species from varieties. He thereby cleared the way for the suggestion that the search for an absolute and fundamental criterion of species might be futile. The assumption that the specific type existed otherwise than an abstraction could be regarded as a piece of conscious or unconscious conceptual realism. The prototype, or the abstract specific type, had been assimilated with an idea in the mind of God, or with the first created form. On the empiricist view, which was Darwin's, the prototype was nothing but an hypothetical construction.

The metaphysical assumptions underlying the orthodox view were clearly expressed by the well-known American naturalist J. D. Dana: "One system of philosophy ... argues ... that species are but an imaginary product of logic ... Another system infers, on the contrary, that species are realities, and that the general or type-idea has, in some sense, a real existence. A third admits that species are essentially realities in nature, but claims that the general idea exists only as a result of logical induction.[3]" The first of these is the empiricist view, the second and third varieties of the idealistic one. Dana himself adopted the third: "Species are realities in the system of nature, while manifest to us only as individuals; that is, they are so far real, that the idea for each is definite, even of mathematical strictness (although not thus precise in our limited view) ... They are units fixed in the plan of creation, and individuals are the material expressions of those ideal units."[4] Dana unhesitatingly identified the specific type with the form created by God. "When individuals multiply from generation to generation, it is but a repetition of the primordial type-idea; and the true notion of the species is not in the resulting group, but in the idea or potential element which is at the basis of every individual of the

[3] *Annals and Magazine of Natural History*, 20, 1857, 495.
[4] *Ibid.*

group."[5]) Species are, he said, "divine appointments which cannot be obliterated."[6])

The article from which the above quotations are taken was published in 1857, but the journal in which they appeared[7]) clearly still adhered to similar views when it reviewed the *Origin*. "It is quite evident that there is an *idea* involved by naturalists in the term 'species' which is altogether distinct from the fact (important though it is) of mere outward resemblance . . . and therefore it is no sign of metaphysical clearness when our author . . . refuses to acknowledge any kind of difference between "genera", "species," and "varieties," except one of *degree*."[8]) Darwin's lack of metaphysical clearness was of course the result of his adoption of an empiricist philosophy of science.

Another exponent of the idealistic theory of species was the famous Swiss-American naturalist Agassiz. His view was even more purely idealistic than Dana's. "While individuals alone have a material existence, species, genera, families, orders, classes, and branches of the animal kingdom exist only as categories of thought in the Supreme Intelligence, but as such, have as truly an independent existence and are as unvarying as thought itself after it has once been expressed."[9]) From this postulate Agassiz concluded that "as the community of characters [of individuals belonging to one species] . . . arises from the intellectual connexion which shows them to be categories of thought, they cannot be the result of a gradual material differentiation of the objects themselves."[10])

Agassiz' brand of idealism was not considered sufficiently orthodox by everybody. Most traditionalists wished to assimilate the specific type not only with God's thought, but also with the material form created by God. "The species," wrote Professor J. W. Dawson, "is not merely an ideal unit, it is a unit in the work of creation. No one indicates better than Agassiz does the doctrine of the creation of animals, but to what is it that creation refers? Not to genera and

---

[5]) *Ibid.*, 487.

[6]) *Ibid.*, 491.

[7]) The *Annals and Magazine of Natural History* reprinted his article from the American *Silliman's Journal*.

[8]) *Annals and Magazine of Natural History*, 5, 1860, 133.

[9]) *Annals and Magazine of Natural History*, 6, 1860, 229.

[10]) *Ibid.*, 220.

higher groups: *they express only the relations of things created;* — not to individuals as now existing: *they are the results of the laws of invariability and increase of species;* — but to certain original individuals, proto-plasts, formed after their kinds or species and representing the powers and limits of variation inherent in the species, — the 'potentialities of their existence,' as Dana well expresses it. The species, therefore, with all its powers and capacities for reproduction, is that which the Creator made; — His unit in the work, as well as ours in the study."[11]) We have here a direct synthesis of idealistic philosophy and Biblical orthodoxy, which was evidently very widespread among the early opponents of Darwin's theory, though most of them did not express themselves so clearly.[12])

Other critics, whose theological bias was not so marked, also drew attention to the thoroughly empiricist basis of Darwin's definition of species. Hopkins declared in *Fraser's* that "all theories like those of Lamarck and Mr. Darwin — which assert the derivation of all classes of animals from one origin, do, in fact, deny the existence of natural species at all,"[13]) and Professor Owen, in the *Edinburgh*, sought support from Linnaeus against Darwin and concluded that "the evidence we have does not favour the theory that 'the species, like every other group, is a mere creature of the brain'.[14])" But though Hopkins and Owen were clearly critical of Darwin's position, they did not wish to maintain that the traditional theory should be accepted as correct on *a priori* grounds. They recognized that both the theory of specific permanence, and that of transmutation, were hypothetical. However, when judging the hypotheses, they applied Whewell's criteria rather than Mills'.[15]) What they considered important was the degree to which the hypothesis as such could be directly verified — i. e., to what extent it could really be shown that species did change indefinitely. "There is no more reason *a priori*," wrote Hopkins, "for believing that [variation] has the power of raising an organic being from a lower to a higher stage of organic life, than there is for supposing its power

---

[11]) *Ibid.*, 207. quoting Dawson's *Archaia*.

[12]) E. g. *British and Foreign Evangelical Review*, 9, 1860, 423, *North British Review*, 32, 1860, 463, *Recreative Science*, 2, 1860—1, 275.

[13]) *Fraser's Magazine*, 61, 1860, 747.

[14]) *Edinburgh Review*, 111, 1860, 532.

[15]) See above, Chapter 9, p. 183—5; 196.

restricted to those limits which the distinctive characters of natural species would impose on it. The onus of proving the existence of such power must rest with him who asserts it."[16]) Owen declared, in a similar vein, "Cuvier admits the tendency to hereditary transmission of characters of variation. Neither he nor any other physiologist has demonstrated the organic condition or principle that would operate so as absolutely to prevent the progress of modification of form and structure correlatively with the operation of modifying influences, in successive generations. But those who hastily or prematurely assume an indefinite capacity to deviate from a specific form are as likely to obstruct as to promote the solution of the question."[17]) This way of looking at the question was quite common: one argued that as long as Darwin could not positively verify his hypothesis, the traditional and orthodox one should be accepted.[18])

It is obvious that Darwin's theory could have no standing-ground whatever as long as biologists postulated as a fundamental truth that the specific type existed as an independent and absolutely fixed entity behind the variety of forms under which it manifested itself under different external conditions. One assumed, in other words, that there was a limited range of variation around the basic fixed type, and whenever transmutationists succeeded in proving a greater range of variation than had hitherto been admitted, it was possible blandly to assert that such a discovery had no bearing on the question of the fixity of species, since it would only prove, as one writer put it, that "we may have too much limited our idea of 'species.'[19]) When Darwin showed that the widely different kinds of domestic pigeons were all probably descended from one primitive type, traditionalists urged that, precisely on this ground, "we must characterize them not as species, but as permanent varieties."[20]) This way of arguing placed the question outside the range of empirical tests. A transmutationist could only hope to prove his case by showing that two forms which on one criterion had to be reckoned as distinct species, could, on another criterion, be shown not

---

[16]) *Fraser's Magazine*, 62, 1860, 76.

[17]) *Edinburgh Review*, 111, 1860, 499.

[18]) See above, Chapter 8, p. 169.

[19]) *Leisure Hour*, 1863, 192.

[20]) *North British Review*, 32, 1860, 470. See also: *Recreative Science*, 1, 1860—1, 272—3.

to be distinct. The traditionalists countered by admitting one of the criteria only as valid. The *impasse* which this led to was discussed by Grove in his presidential address to the British Association in 1866:

"The great difficulty which is met with at the threshold of inquiry into the origin of species, is the definition of species; in fact, species can hardly be defined without begging the question in dispute.

Thus if species be said to be a perseverance of type incapable of blending with other types, or which comes nearly to the same thing, incapable of producing by union with other types offspring of an intermediate character which can again reproduce, we arrive at this result, that whenever the advocate of continuity shows a blending of what had been hitherto deemed separate species, the answer is, they were considered separate species by mistake, they do not now come under the definition of species, because they interbreed.

The line of demarcation is thus *ex hypothesi* removed a step further, so that, unless the advocate of continuity can, on his side, prove the whole question in dispute, by showing that all can directly or by intermediate varieties reproduce, he is defeated by the definition itself of species."[21]

Pro-Darwinian writers had all along agreed with Darwin on the empiricist principle that "the *thing* Species does not exist, the term expresses an *abstraction*"[22] as G. H. Lewes expressed it. Others insisted that we search in vain for "the undiscovered and undiscoverable essence of the term species."[23] Gradually, as the Descent theory gained ground, even those press organs which were generally opposed to Darwin's views came to accept the empiricist view of species. A writer in the *Popular Science Review*, which was on the whole a traditionalist organ, acknowledged in 1863 that "the name of a new species had, until recently, something almost magical in its attractions ... Thanks mainly to Mr. Darwin, this ... is at an end."[24] The next year a reviewer in the same journal declared flatly that "species is a purely abstract term."[25]

[21] British Association *Report*, 1866, lxxi. Grove had been prompted by Hooker. See Chapter 4, p. 78—9

[22] *Cornhill*, 1, 1860, 443.

[23] *Westminster Review*, 17, 1860, 299. See also: *Gardener's Chronicle*, 1860, Jan. 21, 45—6, *All the Year Round*, 3, 1860, 175, *John Bull*, 1859, Dec. 24, 827, *Times*, 1859, Dec. 26, p. 8—9.

[24] *Popular Science Review*, 2, 1862—3, 252.

[25] *Popular Science Review*, 3, 1863—4, 501.

An *Edinburgh* reviewer expressed the same view, though he gave it an anti-Darwinian slant: "When . . . Mr. Darwin contends that variety is an incipient species, and that species, genera, and orders are merely certain stages of descent from some one remote ancestor, he is merely attacking general terms that are the result of an induction formed for our own convenience, and not anything that can be considered sacred and above criticism."[26]) Pure empiricism was advocated by the *Pall Mall* writer who said, "Does there exist such a thing as Species? If it should turn out that no such thing exists, there would be obvious futility in arguing whether *it* were fixed or otherwise. And we believe that the denial of objective existence is as imperative in the case of Species as in the case of all other Abstract Names which for many years usurped, in philosophy, the place of realities."[27]) The extent of the change of public opinion on this subject may be inferred from the fact that even the *Annals and Magazine*, the old stronghold of idealism in these matters, admitted a contributor who directly denounced what he called "the old Platonic school" of natural history.[28])

The assumption that the specific type was something really existing had, however, other support than the conceptual realism of the idealistic school. There were above all two classes of facts which helped to give substance to the idea: they concerned the processes of reproduction and variation, and may be classed under the heads of sterility and reversion. The sterility of hybrids was seen as a sign of the irreducible distinctness of the specific type; reversion as an indication of its essential permanence in spite of the fluctuations due to individual variability.

The Sterility of Hybrids.

Sterility was widely recognized as one of the most useful tests of species. What we are now concerned with, however, is the use of the facts of sterility as a direct refutation of the transmutation theory. In order to achieve this, it was necessary to establish not only that those species which had come under observation were in fact infertile *inter se*, but also that the constitution of any species was such that sterility necessarily resulted from crossing with any other species. On that basis the falsity of the Darwinian theory followed as a logical consequence.

---

[26]) *Edinburgh Review*, 128, 1868, 418.
[27]) *Pall Mall Gazette*, 1868, Feb. 15, 636.
[28]) *Annals and Magazine of Natural History*, 6, 1870, 34 (E. Ray Lankester).

No two specific forms could have a common origin, for even if it was granted, for the sake of the argument, that an individual might so diverge from its parents as to become specifically distinct from it, such an individual would not be capable of propagating itself, since it would — by definition — be sterile with the rest of the race. It was hardly necessary to consider seriously the possibility of two individuals so modified being born, meeting, mating, and producing offspring: such a contingency would obviously enter into the realm of the miraculous.[29])

The implications of assuming sterility to be a fundamental characteristic of interspecific crosses were expounded in a pre-Darwinian article in the *Quarterly Review* as follows: "A species is a kind of animal or plant, so distinct from all others that the continuation of the species is only possible between a pair of individuals belonging to that species. Offspring cannot be produced by a pair of individuals of different species except when those species are very nearly allied, and then the progeny is barren either in the first, or, at the farthest, in the second generation. It seems to result from this that all the individuals of species are the descendants of a single pair."[30])

It was not easy, however, actually to prove that the sterility of hybrids was inherent in the very constitution of species. One did not know what caused sterility: one only knew that most species were in fact infertile when crossed with other species. This was not enough to establish the conclusion that all interspecific crosses were *fundamentally* incapable of reproduction. Conceptual realism therefore was again called in aid. The observed fact of sterility was generalised into a principle of sterility, and this in turn was converted into the desired cause of the observed facts. The idea was expressed in various ways. "The tendency of peculiarities to perpetuate themselves is so much weaker than the immutability which causes sterility or reversion,"[31]) wrote the *Athenaeum*. Ultimately, of course, sterility was a Divine provision. "God forbids hybrids to breed",[32]) wrote a popular journal, and a religious Review declared that "the sterility of true hybrids affords another evidence of the jealousy with which the Creator regards all

---

[29]) See below, Chapter 12, p. 245, and 14, p. 304—5.

[30]) *Quarterly Review*, 106, 1859, 148.

[31]) *Athenaeum*, 1868, Feb. 15, 243.

[32]) *Family Herald*, 1867—8, 188.

attempts to introduce confusion into His perfect plan."[33]) In less theological language, it was said to be "a clear proclamation of nature against the confusion which the variety necessary to Mr. Darwin's theory implies."[34])

Thus the sterility of hybrids was looked upon as a law of nature, designed and instituted by God.[35]) It was in this way that readers of the time interpreted such expressions as "the law of sterility affixed to hybridism,"[36]) "the law of mutual unfruitfulness,"[37]) "the barriers established in nature between species by the laws of their reproduction,"[38]) and so on.[39]) By means of this reasoning, what may have originally been a generalisation from the facts was changed into a deduction from the metaphysical assumptions of religion.

Darwin and the Darwinians, true to their empiricist philosophy of science, looked upon the sterility of hybrids as a question of facts only. Was it, or was it not, the case that hybrids of what naturalists considered as distinct species were invariably infertile? And when sterility occurred, with what facts or phenomena was it associated? Darwin discussed these problems at length in the *Origin*, and came to the conclusion that though hybrids were generally sterile, there were cases of hybrids of undoubted species which were fully fertile, and even occasionally more fertile than the parent forms. Moreover, it was possible to arrange the forms according to the fertility of the offspring in a graduated series. This was entirely consistent with the Darwinian theory, according to which it was natural to regard sterility as an incidental result of the modifications which had gradually made the forms more and more unlike each other.

---

[33]) *North British Review*, 32, 1860, 484.

[34]) *News of the World*, 1860, Jan. 8 p. 6.

[35]) See above, Chapter 6, p. 128 ff.

[36]) *Quarterly Review*, 108, 1860, 245.

[37]) *Watchman*, 1861, Sep. 18, 309.

[38]) *Leisure Hour*, 1872, 344.

[39]) See e. g. *Daily News*, 1859, Dec. 26, p. 2 *Examiner*, 1859, Dec. 3, 772—3, *British Quarterly Review*, 31, 1860, 418—19, *Eclectic Review*, 3, 1860, 234, *Christian Remembrancer*, 40, 1861, 261, *British Quarterly Review*, 38, 1863, 493, *Popular Science Review*, 2, 1862—3, 392, *Quarterly Journal of Science*, 3, 1866, 168, *Zoologist*, 3, 1868, 1385, *Lancet*, 1868:1, 622, *Christian Observer*, 1870, 38, *Family Herald*, 1871, May 20, 44, *London Quarterly Review*, 39, 1872—3, 107—11.

[40]) See e. g. *Times*, 1859, Dec. 26, p. 8, *Cornhill*, 1, 1860, 604, *Westminster Review*, 17, 1860, 555, *Student*, 1, 1868, 184, *London Review*, 16, 1868, 179.

The pro-Darwinian writers naturally adopted Darwin's view of this question,[40]) but even the more sober anti-Darwinians sometimes agreed that he had fought the sterility difficulty with "marked success"[41]) Most of the opposition on this point, in fact, came from the more extreme opposition. Towards the end of our period, moreover, when the Descent theory was gaining ground, the importance of the sterility argument diminished, as critics were concentrating their efforts on the Natural Selection theory.

Reversion and Limited Variation

While the phenomena of hybridism could be taken as proof of the distinctness of species, the phenomena of reversion were adduced as evidence of their immutability.

Reversion had primarily been observed with domestic animals or plants: they often showed a tendency to "revert", to reassume the characteristics of their wild ancestors. Such reversion could be explained several ways. In the first place, it could be taken as due to the direct influence of natural conditions on the artificially produced domestic species. In the second place, it could be seen as a consequence of the intercrossing of varieties which under domestication had to be kept separated by means of artificial selection. In the third place — and this is the explanation with which we are here chiefly concerned — it could be taken as proof that the specific character of varieties was not fundamentally changed, though their outward form might under certain conditions remain modified for any number of generations. The fact that the original type reappeared proved that it had never been really destroyed, but had all the time been latent, as a potential force within the successive generations of modified individuals.

The tendency to revert was very often interpreted, like the sterility of hybrids, as a providentially ordained law, established to preserve the purity of the created specific type. It may be surmised that this was the implication conveyed whenever critics rejected the transmutation theory by referring to what they called the "law of reversion", "an overpowering tendency in the variety to return to the original type."[42])

---

[41]) *Dublin Review*, 48, 1860, 71.

[42]) *Contemporary Review*, 4, 1867, 48. See also *Daily News*, 1859, Dec. 26, p. 2, *Quarterly Review*, 108, 1860, 236, *Annals and Magazine of Natural History*, 5, 1860, 348, *London Quarterly Review*, 14, 1860, 288, *North British Review*, 32, 1860,

On this view the facts of reversion were only an extreme case of the influence of the inherent power moulding the organism to approximate to the ideal specific type: "The law of variation, combined with the law of reversion, seems to point to the conclusion that variation is limited, and that whenever the limits are approached, the tendency is not to further variation but to a return towards the original type."[43]

A good deal of attention was given to the determination of the laws of variation. Whereas Darwin asserted that "our ignorance of the laws of variation is profound,"[44] and therefore considered himself at liberty to postulate the hypothesis that variation was in principle unlimited, his opponents, Whewellian fashion, disregarding the power of that hypothesis to explain the wider questions of biology and paleontology, asserted that the hypothesis was unwarranted by the facts. The chief objection against Darwin on this score, therefore, was that he had gone beyond the factual evidence. He had not established his hypothesis by induction, for the plain fact was, said one early and not very logically minded critic of the Darwinian theory, that there was not "in the whole range of natural science, a single instance of indefinite progress."[45]

Traditionalists admitted that variations occurred, but maintained that it was only proper to assume that the variations constantly kept within the specific frontiers. It might be impossible to indicate exactly what the limits were, yet the existence of the limits could not be put in doubt. Dana expressed these ideas in a pre-Darwinian article on Species: "There is a central or intrinsic law which prevents a species from being drawn off to its destruction by any external agency, while subject to greater or less variations under extrinsic forces ... The limits of variation it may be difficult to define among species that have close relations. But being sure that there are limits, — that science, in looking for law and order written out in legible characters, is not in fruitless search, we

---

466, *Recreative Science*, 2, 1860—1, 275, *Good Words*, 3, 1862, 4, *John Bull*, 1863, 123, *North British Review*, 46, 1867, 282, *Morning Post*, 1869, Aug. 23, p. 2 (rep. McCann) *Tablet*, 1869, Apr. 10, 780, *Month*, 11, 1869, 41—2, *Tablet*, 1871, Apr. 15, 456, *Illustrated London News*, 58, 1871, 243, *London Quarterly Review*, 39, 1872 —3, 107—11.

[43]) *Rambler*, 1860, 367.
[44]) *Origin*, 1 ed. r., 144.
[45]) *Zoologist*, 1859, 6359.
[46]) *Annals and Magazine of Natural History*, 20, 1857, 492.
[47]) *Annals and Magazine of Natural History*, 5, 1860, 139.

need not despair of discovering them."[46]) The British journal in which Dana's article was published applied similar views in its review of the *Origin*. "Whilst 'individual' variation (in each species) is literally endless, it is at the same time strictly prescribed within its proper morphotic limits ... *even though we may be totally unable to define their bounds.*"[47]) What is especially remarkable about these statements — and more could be instanced[48]) — is the insistence on the necessity of assuming limits in spite of the clear recognition that they could not be determined by observation. The basis of this belief in the existence of the limits was naturally the conviction that the species were designed by God, who would not allow any confusion in His perfect plan. Doubt about the specific limits appeared as almost morally reprehensible: the naturalist made things too easy for himself by assuming, as Darwin did, that something which he could not find did not exist.

Less conscientious writers did not stress, or perhaps were not aware of the gap between observation and theory. They simply asserted as a fact that "the limits of variation differ for every species ... [but] in all [they] may be ascertained by patient inquiry,"[49]) or that "observation teaches the limits to which variation extends,"[50]) or that man "cannot force any being across certain fixed barriers of being,"[51]) and the like.[52]) Darwin's hypothesis of unlimited variation was at any rate totally unacceptable in the light of the positive evidence: "The law of Natural Development may, for all we know, be a law of indefinite variation and approximation within certain limits, and nothing can justify us in assuming the contrary except the ascertained fact that such limits have really been passed,"[53]) wrote the *Times* in its review of *Descent*.

Some writers, however, advanced more elaborate experimental evidence of the existence of a fixed limit to variation. It consisted in pointing out that variations were becoming progressively smaller as

---

[48]) See e. g. *Edinburgh New Philosophical Journal*, 12, 1860, 236, *Dublin Review*, 48, 1860, 78.

[49]) *Annals and Magazine of Natural History*, 6, 1860, 207.

[50]) *Athenaeum*, 1863:1, 587.

[51]) *Eclectic Review*, 14, 1868, 346.

[52]) See e. g. *Annals and Magazine of Natural History*, 7, 1861, 401, *Guardian*, 1869, Sep. 1, 985 (rep. Freeman) *Month*, 11, 1869, 41—2, *Popular Science Review*, 10, 1871, 188, *Lancet*, 1872:2, 297.

[53]) *Times*, 1871, Apr. 8, p. 5.

they carried the species further away from the original type. Professional breeders were said to have found that they could indeed change any character fairly quickly, but that "each step in advance, in whatever direction it may take place, becomes more and more difficult."[54]) It might be possible to increase the length of the beak of a pigeon by an eighth of an inch in a few generations, but a further sixteenth of an inch presented extreme difficulties. Variations in the same direction became smaller and smaller, indicating that too great a divergence from the normal specific type could not be achieved: "The rate of variation in a given direction is not constant, is not erratic; it is a constantly diminishing rate, tending therefore to a limit."[55])

A special and interesting form of the argument was to deny that any variations at all were *improvements*. Losses might occur, but not gains. This was a natural consequence of conceiving the specific type as an ideal and perfect form of which the individuals within the species were only imperfect copies. Since this specific type created by God was perfect, any change, any deviation from its organisation must inevitably be unfavourable. "The perfection of any organized being consists in its unity as a whole — in its adaptation to all the phases of its being; and a special modification in one direction, so far from rendering it more suited for different circumstances, would in all probability produce an opposite result,"[56]) wrote an early critic. The perfection of the type was guaranteed by Divine Design. "When God saw that every living creature which he made was good, we cannot doubt that the type of each of them was perfect. The struggle for life, therefore, is to *prevent*, and not to promote a change in the original form,"[57]) declared one religious writer. Another critic objected that he had been taught to consider the "law of divergences" as "a backward, and not a progressive tendency in creation."[58])

Nor did the anti-Darwinians feel that they had in this instance to rely on general, *a priori* arguments only. "What does observation say as to the occurrence of a single instance of such favourable variation?" asked the Bishop of Oxford in the *Quarterly*. "Has any one such instance

---

[54]) *Fraser's Magazine*, 62, 1860, 76.

[55]) *North British Review*, 46, 1867, 282.

[56]) *Zoologist*, 1859, Jan., 6358.

[57]) *Good Words*, 3, 1862, 5.

[58]) *Critic*, 25, 1862—3, 132.

ever been discovered? We fearlessly assert not one."[59]) However, if this appeal to experience had been carried out in detail, it would soon have become clear that such assertions were difficult to uphold. It could naturally not be denied that some variations were better adapted to particular conditions than others. Now the traditionalists' point was that no variation was ever an improvement on the ideal type created by God. But how was that ideal type to be described in terms of observational data? As Darwin insisted, "better adapted" had no definite and absolute meaning: the adaptation had to be to particular conditions of life. A form might be well adapted to conditions in one part of its range, less well adapted in other parts. In order to sustain the assertion that the ideal type was perfectly adapted, traditionalists evidently had to define it in such a way as to beg the question at issue.

In fact, if an acceptable definition of "favourable variation" was to be retained, it was difficult to uphold the thesis that no favourable variations ever occurred. A variant of the argument was therefore developed, according to which improvements were indeed recognized as occurring, but were assumed to be counterbalanced by deterioration in other characters in the organism. How these different changes were to be weighed against each other was never made clear: it is evident, once again, that the idea was due to *a priori* considerations, though it was ostensibly supported by factual evidence. The rule might seem plausible in the case of domestic animals: "The varieties . . . that have been introduced, however beneficial to man, exhibit no improvement in the typical character of the animal."[60]) But it was extended to cover the whole organic world. There was, one asserted, a "law of balance"[61]), or of "correlation", which kept the species within its prescribed limits: "Correlation is so certainly the law of all animal existence that man can only develop one part by the sacrifice of another,"[62]) wrote Bishop Wilberforce.

On their side, Darwinians were able to assert that though observation had not indeed confirmed — how could it? — that variation could proceed

---

[59]) *Quarterly Review*, 108, 1860, 238. See also: *Eclectic Review*, 3, 1860, 234, *News of the World*, Jan. 8, 1860, p. 6 *British Quarterly Review*, 38, 1863, 492.

[60]) *John Bull*, 1863, 124.

[61]) *Intellectual Observer*, 3, 1863, 85.

[62]) *Quarterly Review*, 108, 1860, 238.

indefinitely, or even across the specific frontier, neither could it confirm that there was in fact a limit. Hence there was no reason to reject the hypothesis that variation was indefinite, if that hypothesis could be supported on other grounds. The metaphysical assumption that the species as originally created by God must have been perfect was naturally rejected along with all mixing of theology with science. "There are no such things known as finest kinds,"[63] a wrote one reviewer. In regard to the theological question, it was even possible to take a leaf out of the opponents' book: "All agree that change, *within certain limits*, is universal. ... Why or how shall man, when change is admitted, not knowing and being incapable of knowing the whole, presume to define those limits? ... prescribing limits by ... arbitrary and convenient definitions to the Almighty Power."[64]

Moreover, one denied that the factual evidence supported the limited variation theory. "A race, once produced, is no more a fixed and immutable entity than the stock whence it sprang; variations arise among its members, and as these variations are transmitted like any others, new races may be developed out of the pre-existing ones *ad infinitum*,"[65] wrote Huxley in one of his reviews. On the other hand, it does not seem that the Darwinians ever came to grips with the difficulty presented by the diminishing rate of variability in the same direction. Wallace's solution was hardly satisfactory; according to him, "the real question ... is, not whether indefinite and unlimited change in any or all directions is possible, but whether such differences as do occur in nature could have been produced by the accumulation of variations by selection."[66] Darwin hardly took up the challenge. He defended his own assumption on general grounds: "Seeing that individual differences of the same kind perpetually recur, this can hardly be considered as an unwarrantable assumption. But whether it is true, we can judge only by seeing how far the hypothesis accords with and explains the general phenomena of nature." But he fairly acknowledged that "the ordinary belief that the amount of possible variation is a strictly limited quantity is likewise a simple assumption."[67] This was the best he could do at

---

[63]) *Gardener's Chronicle*, 1860, Jan. 21, 46.
[64]) *London Review*, 1, 1860, 625—6.
[65]) Huxley in *Westminster Review*, 17, 1860, 551. Quot. from *Lay Sermons*, 296.
[66]) *Quarterly Journal of Science*, 4, 1867, 486.
[67]) Origin, 6 ed. 103.

the time: the puzzle could not be solved until more had been discovered of the laws of variation, and until the science of genetics had been able to distinguish between the fund of mutant genes that exist in a population, and the new mutations which supervene.[68]) As long as the breeder can draw on the extant fund, he makes rapid progress, but when he has to wait for new mutations to occur, the rate of change becomes very slow.

As regards the reversion of domestic races turned wild, Darwinians had a simple answer. It could be explained by natural selection: if a domestic race was left to fend for itself in natural conditions, it would inevitably develop those characters which were most useful in such conditions, and they must be expected to approximate to the wild type rather than to the domestic one.[69]) Moreover, it was a mistake to think that reversion always occurred when domestic races were exposed to wild conditions: and this also was in accordance with the Darwinian theory. Reversion was not an absolute and irreducible law, as the idealistic opposition claimed, but depended on the circumstances. Many domestic races would simply die if exposed to natural conditions, others would revert, some would retain their acquired characteristics.[70])

[68]) Theodosius Dobzhansky, *Genetics and the Origin of Species*, 3 ed., 1951, esp. ch. 4 and R. A. Fisher, *The Genetical Theory of Natural Selection*, Oxford, 1930.

[69]) See e. i. *Lancet*, 1868:1, 622, *National Review*, 10, 1860, 207, *Contemporary Review*, 8, 1868, 149, *Fortnightly Review*, 3, 1868, 367.

[70]) See *Edinburgh Review*, 128, 1868, 440.

CHAPTER 11

Missing Links

By far the most common factual — as distinct from religious or ideological — objection against Darwin's theory was that no direct evidence had been produced of the innumerable intermediate forms which, if the theory was true, must have once existed, and ought continually to arise. This argument of "missing links" was presented under several different forms. If Darwin was right, one said, the whole series of organic beings in the world should even at the present time contain an infinite gradation of forms. Yet this was patently not the case. In general there was little difficulty in assigning each individual organism to a distinct species. Connecting links, in the sense of forms which could equally well be put in one species as in another, were rare or missing. Secondly, if the theory was true, experiments ought to give evidence of the appearance of successive variations, leading by a series of small steps from one species to another. No such evidence had been produced. Thirdly, the historical record ought to show up specific changes in the animals that had been associated with man since ancient times. And if that time was too short for a specific change to take place, it ought at any rate to be possible to estimate from it the rate of change in the organic forms, whereby the data of paleontology could be correlated with the geological time-scale. Fourthly, the paleontological record itself was of supreme importance. Darwin's theory assumed that the existing species had passed through innumerable transitional forms by almost insensible steps. The geological strata ought therefore to contain a record of the changes that had taken place. Yet in the paleontological record the forms seemed to be specifically distinct, just as in the present-day organic world. Everywhere links were missing which, it was held, ought to be there on Darwin's theory.

The Present Organic Series

In the *Origin* Darwin himself asked the question which naturally arose on his theory from the absence of connecting links among the living species: "Why, if species have descended from other species by insensibly fine gradations, do we not everywhere see innumerable transitional forms? Why is not all nature in confusion instead of the species being, as we see them, well defined?"[1]) His own answer was that "extinction and natural selection will . . . go hand in hand. Hence, if we look at each species as descended from some other unknown form, both the parent and all the transitional varieties will generally have been exterminated by the very process of formation and perfection of the new form."[2])

Darwin argued that many distinct species which now inhabit a contiguous area may have arisen at a time when that area was broken up into isolated regions. Secondly, he pointed out that not even in contiguous areas must we expect species to form a continuous series, since the adaptation of forms is less to climate, which changes gradually, than to other species which are generally distinct, from whatever reason. Thirdly, Darwin argued that when intermediate forms between two species did exists, experience showed that the intermediate form was usually less numerous. Therefore it would become more liable to extinction in times of difficulty. Fourthly, a less numerous stock would be less capable of turning out a sufficient number of favourable variations, and would thus be liable to extinction because of competition with more adaptable forms.

Darwin's opponents seldom took up a discussion with him on these points — indeed, the argument of the discreteness of the present organic series was a rather unimportant variant of the missing link argument. The most common objection was that Darwin's explanation was altogether hypothetical. For instance, in what cases could Darwin prove by geological facts that a species had been isolated at the period of its formation? His suggestions were baseless speculations, unworthy of a true inductive scientist.[3])

Other writers simply disregarded Darwin's explanation, interpreting his theory as implying the denial of all specific distinctions. According to the reasoning of the idealistic school of naturalists, the independent existence of species as categories of thought was the sole guarantee of order and symmetry in the organic world. If species were not immutable, they could not be categories of thought, and if they were not that,

---

[1]) *Origin*, 1 ed. r., 147.
[2]) *Ibid.*, 148.
[3]) See e. g. *Critic*, 1859, Nov. 26, 528—530, *Patriot*, 1860, Jan. 19, 45.

everything would necessarily become confused and disordered. This view had been stated quite definitely by Dana. If God had not established distinct limits between species, he said, then "the very beauties that might charm the soul would tend to engender hopeless despair in the thoughtful mind, instead of supplying its aspirations with eternal and ever-expanding truth. It would be to [make] the temple of nature fused over its whole surface, without a line the mind could measure or comprehend."[4]) Agassiz expressed similar ideas in his criticism of the *Origin*,[5]) and the argument remained quite common in the more fundamentalist religious organs all through the 1860's.[6])

On the Darwinian side the argument seldom went beyond Darwin's own contribution. Some writers, however, chose to argue instead that the existing dividing lines between species were not in fact as sharp and distinct as was sometimes said.[7]) When new evidence on this point was produced — for instance, when H. W. Bates declared that he had found a whole series of transitional forms linking distinct species of tropical butterflies with each other, pro-Darwinian writers tended to acclaim the discovery as a confirmation of the transmutation theory,[8]) while anti-Darwinians tended to play it down. Were the species distinct in the first place? one asked: "Who shall say that we know enough of the habits of [this butterfly] to prove respecting it the fact on which its specific distinctness depends?"[9]) wrote one traditionalist, thus providing an instance of that shifting of ground as regards the definition of species which we have discussed in the previous chapter.

Experimental Evidence

In a country where empirical methods were as highly esteemed as in Britain, it was natural that considerable stress should be laid on

[4]) *Annals and Magazine of Natural History*, 20, 1857, 488—9.
[5]) *Annals and Magazine of Natural History*, 6, 1860, 231.
[6]) *Zoologist*, 18, 1861, 7591. *Eclectic Review*, 3, 1860, 236, *Nonconformist*, 20, 1860, 1034—5, *British Quarterly Review*, 38, 1863, 493, *Nonconformist*, 1866, Aug. 29, 705, *Record*, 1870, Sep. 23, p. 2, *Family Herald*, 1861—2, 268, *Fortnightly Review*, 2, 1865, 681, *Month*, 1869, 41—2, *Methodist Recorder*, 1866, Aug. 31, 308 *Family Herald*, 1867—8, 188; 1871, May 20, 44, *Christian Observer*, 1870, 38, *John Bull*, 1872, Nov. 30, 824.
[7]) E. g. Asa Gray in the *Athenaeum*, 1860: 2, 161.
[8]) *Spectator*, 1863, 2009—10, *Guardian*, 1863, Nov. 4, 1036.
[9]) *London Quarterly Review*, 22, 1864, 60.

Darwin's inability to produce experimental evidence of specific changes. It was also a considerably more effective form of the missing link argument than the one we have just discussed. Darwin could indeed easily give both simple and plausible explanations of the absence of connecting links in the present series by pointing to the myriads of forms which, as paleontology showed, had once lived, but had been extinguished. But the fact that, in spite of long-continued efforts, no single instance could be adduced of the production of an indubitably new species by selective breeding, remained an awkward difficulty.

Huxley considered the deficiency of experimental proof of specific change as the chief weakness of the Darwinian theory. "In my earliest criticisms of the "Origin", he wrote in 1888, "I ventured to point out that its logical foundation was insecure so long as experiments in selective breeding had not produced varieties which were more or less infertile; and that insecurity remains up to the present time."[11]) Yet in spite of his emphasis on the deficiency of proof, Huxley maintained that the Darwinian theory was the only one that had any "scientific existence."[12]) Darwinians in general were inclined to hold, like Darwin, that the theory was to be accepted, at least provisionally, since it "explained so much."[13]) Asa Gray, for instance, adressing a traditionalist public, quoted the Swiss naturalist De Candolle as saying that the Descent theory was "the most natural hypothesis."[14]) The Darwinian argument was that though the basic hypothetical proposition could not be directly confirmed, it had not been disproved either. To conclude, as the most violent anti-Darwinians seemed to do, from the lack of positive confirmation of specific change, that species were in fact immutable, was patently a conclusion from negative evidence. As one correspondent in the *Spectator* said, the evidence was "not enough to warrant so positive a conclusion."[15]) The Darwinian theory was *consistent* with all the known facts. Its value, according to empiricist principles, depended on the total number of facts and classes of facts which it

---

[10]) — — —

[11]) Article by Huxley in Darwin, *Letters*, II, 195. See also report of Huxley's lecture, *Illustrated London News*, 1860, Feb. 18, 163.

[12]) T. H. Huxley, *Lectures and Essays*, 287.

[13]) Darwin, *More Letters*, I, 184, to F. W. Hutton, 1861.

[14]) *Annals and Magazine of Natural History*, 12, 1863, 86.

[15]) *Spectator*, 1859, 1210.

could account for, and there was no doubt that in this respect it was immensely superior to any other hypothesis.

But anti-Darwinians, strengthened by Whewell's philosophy of science, refused to extend their purview beyond the basic hypothetical proposition. Within that limited field it was not difficult to declare Darwin's hypothesis to be less "close to the facts" than the traditional view. And among the general public, whose prejudice against hypotheses we have already discussed, the lack of experimental evidence was naturally given much prominence. Owen declared in the *Edinburgh:* "We have searched in vain, from Demaillet to Darwin, for the evidence, or the proof, that it is only necessary for the individual to vary, be it ever so little, in order to the conclusion that the variability is unlimited, so as, in the course of generations, to change the species, the genus, the order, the class. We have no objection to this result of 'natural selection' in the abstract; but we desire to have reason for our faith."[16] The Bishop of Oxford, in the *Quarterly,* was heavily sarcastic: "We plead guilty to Mr. Darwin's imputation that 'the chief cause of our natural unwillingness to admit that one species has given birth to another and distinct species is that we are always slow in admitting any great change of which we do not see the intermediate steps' . . . In this tardiness to admit great changes suggested by the imagination, but the steps of which we cannot see, is the true spirit of philosophy."[17] Hopkins, in *Fraser's,* also took his stand on the lack of experimental proof. Until such proof had been furnished, he asid, "no theory like Mr. Darwin's can have a real foundation to rest upon, or can ever secure the general convictions of philosophers in its favour."[18]

A host of lesser writers testified to the prevalence of these anti-hypothetical views. They spoke of the "total absence of direct evidence"[19] which made Darwin's theory an "unconfirmed speculation,"[20] and so forth.[21] Indeed, one maintained that Darwin's failure to obtain any

---

[16] *Edinburgh Review,* iii, 1860, 521.

[17] *Quarterly Review,* 108, 1860, 250 quot. *Origin.* See i ed. r. 407—8.

[18] *Fraser's Magazine,* 62, 1860, 80.

[19] *Eclectic Review,* 3, 1860, 234.

[20] *Nonconformist,* 1863, Mar. 4, 176.

[21] *British and Foreign Evangelical Review,* 17, 1868, 185, *Times,* 1870, Feb. 12, p. 4, *Daily News,* 1859, Dec. 26, p. 2, *Recreative Science,* 2, 1860—1, 273, *British Quarterly Review,* 35, 1862, 255, *Record,* 1862, Oct. 8, p. 2, *Popular Science Review,* 2, 1862— 3, 392, *Fortnightly Review,* 2, 1865, 681, *Morning Advertiser,* 1866, Aug.

direct proof of transmutation really only strengthened the case against him. "The fact that . . . no species has been known practically to overpass these [specific] limits, should logically bring us to the opposite conclusion, viz., that the laborious investigations of Mr. Darwin have more than ever established the fixity of species, though they have shown reason to believe that many so-called species are mere varieties."[22]

The argument that the experimental evidence failed to confirm the theory of transmutation hinged on the infertility criterion. On almost any other criterion the races produced by breeders might qualify as species. About his pigeons, for instance, Darwin said, "Altogether at least a score of pigeons might be chosen, which if shown to an ornithologist, and he were told that they were wild birds, would certainly, I think, be ranked by him as well-defined species."[23] It was therefore natural for Darwin to try to limit the scope of the sterility test. His general criticism of that test has been discussed above.[24] But he also specifically criticised the use of the sterility test in the case of the domestic breeds. Under domestication, he argued, there is reason to believe that animals become considerably more fertile than in the wild state. Therefore the sterility test ought not to be applied to domestic breeds. If the pigeons had not grown up under artificial conditions, it was quite possible that they would in fact have been sterile *inter se*, in which case they would have qualified as species on any test.[25]

This explanation, however, did not commend itself to the public. It was too hypothetical. Besides, anti-Darwinians were able to turn the difference between domestic and wild races against Darwin. It might be, one argued, that domestication increased fertility. But then, on the other hand, was it not patently true that it also increased variability? In that case, conclusions drawn from the widely divergent domestic races to the variations in a state of nature were invalid. "We deny that any parallel can be drawn from [domestic animals] on a

---

25, p. 2, *Quarterly Journal of Science*, 3, 1866, 154, *Morning Post*, 1869, Aug. 23, p. 2, *John Bull*, 1871, Apr. 6, 234, *Tablet*, 1871, Feb. 25, 232, *Popular Science Review*, 10, 1871, 188, *Times*, 1871, Apr. 8, p. 5, *John Bull*, 1872, Nov. 30, 824, *London Quarterly Review*, 39, 1872—3, 107—11, *Contemporary Review*, 4, 1867, 48.

[22] *Edinburgh New Philosophical Journal*, 19, 1864, 56.

[23] *Origin*, 1 ed. r., 19.

[24] See Chapter 10, p. 206—9.

[25] See e. g. *London Review*, 16, 1868, 179.

general scale, in the feral world; for everything tends to prove that the
whole system of certain species . . . when under domestication, tends
to become plastic,"[26]) wrote one orthodox naturalist, and several others
agreed: "A kind of artificially produced plasticity of character of both
animals and plants, that have been domesticated, must be admitted,
but whether this, which is certainly proportionate to the artificial
circumstances, has any relation to 'variation under nature,' seems
scarcely implied."[27])

Sometimes the fact that domestic breeds remained fertile with the
parent forms was taken as proof that variability was of a different
nature under domestication. That fact, said an *Edinburgh* reviewer,
"destroys the validity of the argument that because the one is the
result of small variations selected by man, the other is the result of
small variations selected by nature."[28]) In this way, however, any
experimental evidence on the matter was *ipso facto* declared inad-
missible. The argument therefore simply revealed the fundamental
difference between empiricist Darwinians and idealistic anti-Darwinians on scientific explanation. Darwin's opponents, following
Whewell, required that the hypothesis itself should be "close to the
facts." i. e., in this instance, that the actual causes of the individual
variations, whether under nature or under domestication, should be
laid bare. "Mr. Darwin . . . seems to us not to have proved that the
cause of variability is the same . . . and until the cause in each case
is shown to be identical, one class of Mr. Darwin's objectors will not
be refuted,"[29]) wrote the *Spectator* in its review of Darwin's *Animals
under Domestication*.

The belief that variability differed under nature and under domestication was not only an inference from the observed facts. Like so many
other inferences in the domain of organic nature it had also been drawn
into the orbit of the Design argument and had thereby come to derive
support from the general belief in Design. The variability of domestic

[26]) *Annals and Magazine of Natural History*, 5, 1860, 140.

[27]) *Zoologist*, 18, 1861, 7590. See also: *Examiner*, 1859, 772—3, *London Quarterly
Review*, 14, 1860, 287, *Spectator*, 1868, 318, *Month*, 11, 1869, 41—2, *Illustrated
London News*, 58, 1871, 243.

[28]) *Edinburgh Review*, 134, 1871, 201.

[29]) *Spectator*, 1868, 319. See also Agassiz in *Annals and Magazine of Natural
History*, 6, 1860, 225.

animals was providentially ordained by God for the benefit of man. "It will not be denied, we presume," wrote one traditionalist organ, "that animals were created for the use of mankind ... the provision for becoming serviceable to man, breeding in confinement or restraint, and accommodating themselves to circumstances, whether of climate or country, is very marked."[30]) Domestic animals, said a religious writer, were "influenced by a principle of domesticability, or a power of adaptation purposely given to them, as *intended* to be man's companions."[31]) That such ideas as these were not confined to the extreme religious press is shown by the fact that Professor Owen maintained that the horse had been "predestined and prepared for Man"[32]). Less theologically minded naturalists, of course, held aloof from such arguments.[33])

Altogether, though Darwin's attempt to explain away the fertility of the domestic races was hardly successful, his opponents were no more successful in establishing a fundamental difference between domestic and natural variation. Darwinians had to recognize that the modifications effected by man's selection did not yet reach beyond specific limits. Anti-Darwinians, for their part, had to admit that as considerable modifications had in fact been achieved experimentally, it was at least possible that similar variations could also appear in nature under favourable circumstances. The Darwinian theory was not indeed proved, but neither could unprejudiced observers regard it as disproved, or even as unplausible.

Historical Evidence

A minor variant of the missing link argument brought into view the evidence of recorded history. Changes ought to have occurred in the animal and plant world during the several thousand years that had elapsed since the dawn of history. Yet no new species had been produced. They were all, as far as one could judge, specifically the same in ancient times as now.

This argument must have had great force with those — the majority of the scientifically innocent — who believed that the age of the world

---

[30]) *Edinburgh New Philosophical Journal*, 11, 1860, 283.
[31]) *Christian Observer*, 1860, 572.
[32]) Richard Owen, *Anatomy of the Vertebrates*, III, 1868, 796.
[33]) *Gardener's Chronicle*, 1868, 1343: "borders on the ludicrous".

was no more than six thousand years. And even if geologists and the more informed public had given up that chronology, the habits of thought established by the traditional view were slow to change. Moreover, we have already seen that one of the ways of reconciling the Biblical chronology with geology was to assume that the present creation was no older than six thousand years, whatever the age of the world as such.

A frequently recurring observation was that the species portrayed in the ancient Egyptian pyramids were the same as those we meet at the present day. The pyramids were some four thousand years old, which was, on the traditional view, a very substantial proportion of the time granted for the present organic world as a whole. Now as there was no reason to suppose that the rate of change — if there had been any change — had been slowing down, the conclusion was that no change could have occurred since the Creation. The Egyptian representation of the ibis, for instance, was exactly like the present bird. Therefore, one argued, "if the permanence of a species can be proved for such a length of time as 3,000 years . . . we are . . . furnished with an answer to Mr. Darwin's theory."[34] Nor was the ibis the only instance. "From the early Egyptian habit of embalming, we know that for 4,000 years at least the species of our own domestic animals, the cat, the dog, and others, has remained absolutely unaltered,"[35] wrote the Bishop of Oxford in the *Quarterly*. Agassiz, summarizing the evidence, maintained that "the animals known to the ancients are still in existence, exhibiting to this day the characters they exhibited of old."[36] Darwin's observation that three or four thousand years were almost as nothing on the geological scale was treated with sarcasm: "To the progressionist, a few thousands or millions of years more or less are of no moment."[37] Anti-Darwinians naturally stuck as long as possible to the traditional chronology,[38] but the argument could in fact be made independent of the length of the geological time-scale, since as far as *specific* change was concerned, it was always possible to maintain that it had not occurred

---

[34] *Annals and Magazine of Natural History*, 5, 1860, 348, reproducing letter by J. O. Westwood to the *Gardener's Chronicle*, 1860, Feb. 11. See also *Edinburgh New Philosophical Journal*, 11, 1860, 287.

[35] *Quarterly Review*, 108, 1860, 237. See also: *Illustrated London News*, 58, 1871, Mar. 11, 143, *Daily News*, 1861, Nov. 7, p. 2 (rep. D. Brewster).

[36] *Annals and Magazine of Natural History*, 6, 1860, 221.

[37] *Eclectic Review*, 3, 1860, 224.

[38] See e. g. *Globe*, 1864, Sep. 16.

at all. The rate of specific change was not only very slow, it was absolutely *nil:* "If 1,000 years, or 4,000, or 10,000 years, or 100,000 . . . have effected no appreciable change, it is reasonable to believe that multiplying any of these sums by a million would yield nothing but the same cipher."[39])

Darwinians, who naturally did not accept the existence of a fundamental distinction between specific and varietal modifications, simply asserted that even on the very shortest time-scale admitted by modern geologists, the 3,000 years of recorded history was an insignificant portion of the whole, and pointed with good effect to an obvious geological parallel: "Would it not excite a smile of ridicule in an individual to announce the discovery that Mont Blanc 3,000 years ago, had the same altitude as to-day,"[40]) asked Fawcett at the British Association in 1861, and he was not alone in pointing out that the argument of the anti-Darwinians would "be equally valid against the upheaval of mountain-chains."[41])

There was one other answer to the argument of specific permanence during historical times. Darwin often insisted that his theory was not one of necessary development. He only assumed that species would change if surrounding conditions, of whatever sort, also changed. Now in many cases there was no reason to think that conditions had materially changed during historical times. Therefore one should not expect any change in the species either.[42]) This negative argument, however, could only establish that the facts were not inconsistent with the theory: the positive argument was of course much more effective, by allowing the small but undeniable observed changes (in domestic animals) to be multiplied almost indefinitely, thus preparing the mind for receiving the transmutation theory.

Paleontological Evidence

By far the most important of the different forms of the missing link argument consisted in emphasizing the absence of transitional forms in

[39]) *Times*, 1863, Aug. 31, p. 7, reporting Crawfurd. See also *Temple Bar*, 5, 1862, 217, *Edinburgh Review*, 115, 1862, 76, *Good Words*, 1862, 6, *Annals and Magazine of Natural History*, 11, 1863, 426.

[40]) Quoted from *Nonconformist*, 1861, Sep. 18, 752.

[41]) Letter in *Spectator*, 1860, 380. See also *Leader*, 1860, 927, *Temple Bar*, 15, 1865, 237.

[42]) See also *Fortnightly Review*, 3, 1868, 364—6.

the geological record. It was this difficulty that worried Darwin himself more than any other, and he wrote in the *Origin*, "Geological research, though it has added numerous species to existing and extinct genera, and has made the intervals between some few groups less wide than they otherwise would have been, yet has done scarcely anything in breaking down the distinction between species, by connecting them together by numerous fine, intermediate varieties; and this not having been effected, is probably the gravest and most obvious of all the many objections which may be urged against my views."[43]) His answer to the difficulty may be summarized in the words of the heading of the chapter devoted to it, namely, "The Imperfection of the Geological Record."

Traditionalists did not appreciate Darwin's explanation. They were naturally loth to admit that what was picturesquely called the "great stone-book" was so extremely fragmentary as it must be if Darwin was right. This defensive attitude seems to have been especially widespread among amateur geologists and laymen: we should consider that geology had become the science *à la mode* in early 19th century Britain. A smattering of geological knowledge was to be found in fairly wide circles. Geology was popularised, and in the process it was almost inevitable that many of its findings should come to be represented as much more securely based, much more conclusive, than was really the case. Thus a dogmatic view of the geological record as giving a complete history of the past of our globe had gained ground among the general public. It was against this prejudice that Darwin had to contend.

Moreover, and perhaps more important, the picture of the earth's history that geologists had managed to piece together had been closely fitted to the prevalent theological-idealistic view of organic species as designed and directly created by God. The discontinuous series of organisms revealed by the positive evidence of paleontology was exactly what was to be expected on the separate creation theory: and from the negative evidence of the lack of transitional forms it was notoriously easy to draw the positive conclusion that no such transitional forms had ever existed. It was thus on the joint influence of three factors — the positive evidence of discontinuity, the traditional Design and separate creation thory, and the natural tendency to draw positive

---

[43]) *Origin* 1 ed. 1., 255.

conclusions from negative evidence — that anti-Darwinians could chiefly rely in order to make the missing link argument effective.

The pre-Darwinian traditionalist view was well expressed by the German naturalist H. G. Bronn: "All the successive modifications of the animal and vegetable population of the surface of the globe have been effected by the annihilation of the older species and the continual appearance of new species, without there having ever been any gradual passage from one species to another"[44]) A religious organ made it clear how this result of paleontological research was assimilated with the Design argument: "In all the long chain of beings, from monad to man, we see no evidence that one link has ever been other than it now is, or that there has ever existed a tendency, in a creature fitted for one sphere, to usurp that of another."[45])

After the publication of the *Origin*, traditionalists still found no reason to modify their interpretation of the paleontological facts. Agassiz, for instance, said, "The geological record, even with all its imperfections exaggerated to distortion, tells . . . that the supposed intermediate forms between the species of different geological periods are imaginary beings, called up merely in support of a fanciful theory." "Species appear suddenly, and disappear suddenly, in successive strata. That is the fact proclaimed by palaeontology." "Throughout all geological times each period is characterized by definite specific types, belonging to definite genera . . . built upon definite plans."[46]) Agassiz' argument displays all the features we have pointed out: first, refusal to consider the possibility that the negative evidence might support any other conclusion than that the undiscovered forms had never existed; second, heavy reliance on the positive evidence, and third, a reference to Design in the implied parallel between definite and distinct species on one hand, and definite and distinct type-plans on the other.

Many of the more violent anti-Darwinians, especially in the religious press, adduced the positive evidence as sufficient in itself to refute Darwin. "His theory is inconsistent with the very facts upon which he has rested it . . . we have absolute proof, indeed, of this immutability of species,"[47]) said the Scottish naturalist Sir David Brewster. "Geology

[44]) *Annals and Magazine of Natural History*, 4, 1859, 82.
[45]) *Eclectic Review*, ɪ, 1859, 557.
[46]) *Annals and Magazine of Natural History*, 6, 1860, 221, 223, 232.
[47]) *Daily News*, 1861, Nov. 7, p. 2, rep. Brewster.

not only does not help him, but absolutely opposes him,"[48]) wrote
another critic, turning Darwin's own words against him. Paleontology
yields "positive and incontestable evidence against his theory,"[49])
wrote a third. Others put the matter a little less categorically.
Amplifying Darwin's admission that the paleontological record
presented a grave difficulty on his theory, one might declare that
"the evidence of geology is strong against it,"[50]) or that "the facts
of geology are opposed to his theory,"[51]) or "seem at once to contradict
any theory of transmutation of species,"[52]) and so forth.[53]) All this,
of course, was going beyond the conclusions that the facts strictly
warranted. To conclude that the facts really contradicted the theory
it was necessary to consider the negative evidence as well: but to do
this would have weakened the effect of the assertion, and it was seldom
done in the anti-Darwinian press.

Even when no more was asserted than was strictly correct — namely,
that the evidence did not positively support the Darwinian theory —
the wording was often such that the reader might himself draw a more
far-reaching conclusion. "Geology ... reveals no single fact in confirmation of Darwin's hypothesis,"[55]) "What particle of evidence is
there,"[56]) "there is an entire absence of all evidence,"[57]) are typical
phrases.[58])

In the more popular press the missing link argument sometimes

---

[48]) *John Bull*, 1863, 127.

[49]) *Patriot*, 1860, Jan. 19. 45 See also: *Recreative Science*, 2, 1860—1, 276.

[50]) *Freeman*, 1860, Jan. 18. 46.

[51]) *Annals and Magazine of Natural History*, 11, 1863, 424.

[52]) *Athenaeum*, 1859: 2, 660.

[53]) See e. g. *London Quarterly Review*, 14, 1860, 300—1, *Daily News*, 1859, Dec.
26. p. 2.

[54]) — — —

[55]) *Watchman*, 1861, Sep. 18, 309.

[56]) *Contemporary Review*, 4, 1867, 48.

[57]) *Zoologist*, 3, 1868, 1394.

[58]) See also *Eclectic Review*, 3, 1860, 238, *Edinburgh Review*, 111, 1860, 530,
North British Review, 32, 1860, 481, *Blackwood's Magazine*, 92, 1862, 90, *Athenaeum*, 1862: 2, 54, *British Quarterly Review*, 38, 1863, 493, *Westminster Review*,
25, 1864, 586, *Manchester Guardian*, 1866, Aug. 28, p. 4, *Lancet*, 1868: 1, 622,
Family Herald, 1867—8, 380, *Freeman*, 1868, Aug. 28, 681, *North British Review*,
51, 1869—70, 601, *Record*, 1870, Sep. 23, p, 2. *John Bull*, 1872, Nov. 30, 824, *London
Quarterly Review*, 1872—3, 107—111.

appeared in rather curious shapes. One writer, for instance, objected that there was no geological evidence of "tentative and transitional forms,"[59]) another that "we do not learn from the geologists that they have detected any one species in the act of transforming itself into any other."[60]) None of these critics can have had a very clear idea of what sort of evidence he was demanding. On the Darwinian theory any individual organism might in some respects qualify as transitional, and any individual birth might be an instance of the species transforming itself: there was no reason to expect the missing links to be any more "tentative" *in themselves* than the actually existing forms. Such assertions, therefore, as that there was a *"thorough and complete absence* of that countless host of transitional links,"[61]) or that "in all the long chain between Ascidian and the man, [Darwin] has not certainly established one link,"[62]) hardly made any sense.

It seems in fact to be clear that even if transitional links had been forthcoming, Darwin's opponents would still hold to their assertions. There would remain two ways of saving the traditional thesis. Either the series of forms could be accepted as connected by descent: but then the forms themselves might be deprived of their specific status, and declared to be mere varieties which had before been regarded as species by mistake. Or else the meaning of the term "transitional link" could be restricted: a newly discovered intermediate form between two related species would still be declared distinct from either, since separated by the insurmountable specific barrier. In this way the Darwinians would have been faced with the necessity of finding a complete graduated series of possible parents and offspring embracing the whole organic world — a demand which was obviously impossible to fulfil.

Yet it was in these terms that the anti-Darwinians regarded the question. Thus when intermediate forms in the wide sense — forms which might be placed *somewhere* between others already known — were in fact produced, they were dismissed as not to the point. For instance, when Huxley drew attention to the remarkable series of fossil horses found in America, one writer objected: "He does not

[59]) *Good Words*, 1865, 303.
[60]) *Illustrated London News*, 58, 1871, Mar. 11, 243.
[61]) *Annals and Magazine of Natural History*, 5, 1860, 140.
[62]) *Leisure Hour*, 1872, 364.

produce a single fact in support of his bold assertion. All the facts
mentioned by him are isolated, and fail to supply the required connecting links. We still have nothing but discontinuity."[63])

The most widely discussed of these finds of transitional forms was
the famous fossil bird from a German coal seam. The fossil exhibited
precisely those characteristics which on the Descent theory were to be
expected from a transitional form between reptile and bird: it had a
lizard-like tail, wings furnished with claws, and a beak with teeth.
Anti-Darwinians, however, went out of their way to assert that it
was not "an intermediate creature, engaged in the transition from the
saurian to the bird."[64]) Of course there were still numerous missing
links. Among traditionalists the discussion concerned the question
whether the fossil should be classed as a bird or as a reptile. Some,
with the discoverer, held it to be undoubtedly a "reptile of the order
of Sauria,"[65]) but the majority in Britain followed Owen in maintaining
its "ornithic nature,"[66]) while a few declared that the question was still
in the balance.[67]) Everybody agreed that it must be either one or the
other. It is interesting to note that when the creature was represented
as a true bird it might even be used *against* the Darwinian theory, by
proving the existence of birds at a very early date, thus reducing the
time allowed for its supposed evolution out of lower forms:[68]) this
was the argument of the perfection of the early forms.[69])

The simple and natural Darwinian explanation was that the fossil
was neither a reptile, nor a bird, but an intermediate form between
the two.[70]) This straightforward solution of the problem could not be

---

[63]) *London Quarterly Review*, 36, 1871, 270.

[64]) Professor Wagner, the discoverer, quoted in *Chambers's Journal*, 1862: 1,
351.

[65]) *Annals and Magazine of Natural History*, 9, 1862, 266.

[66]) *Popular Science Review*, 2, 1862—3, 360. See also: *Critic*, 1862, June 7, 558,
Geologist, 1863, 4, *Quarterly Journal of Science*, 1, 1864, 127—9, *British and Foreign
Evangelical Review*, 15, 1866, 371.

[67]) *Popular Science Review*, 1, 1861—2, 524, *Edinburgh New Philosophical
Journal*, 17, 1863, 326.

[68]) *Nonconformist*, 1863, 730, quot. Lyell. *Edinburgh New Philosophical Journal*,
17, 1863, 326, *British and Foreign Evangelical Review*, 15, 1866, 371, *Chambers's
Journal*, 1862: 2, 414 (quoting Owen).

[69]) See below, p. 232—4.

[70]) *Chambers's Journal*, 1868, 143, quoting Huxley.

resisted indefinitely, which was at length, and somewhat ungraciously, recognized in the anti-Darwinian organs. "The general tendency of naturalists to seek for, and perhaps too readily to accept, any evidences, supposed or real, of a transmutation of species, have in the present case fostered the original idea of a 'feathered reptile' and led to a widespread readiness to accept the, as yet, untenable doctrine of the transmutation of flying reptiles into birds.[71 A])

The attitude of the traditionalists with regard to the discoveries of transitional forms only confirms that their opposition to the transmutation theory was based on *a priori* rather than factual grounds. As so often, the difference could be carried back to the fundamental opposition between the empiricist and the idealistic philosophical position. Indeed, we even find statements implying that no possible evidence would be accepted as refuting the anti-transmutationist position: "It is conceivable, though improbable in the highest degree, that scientific research may discover what has presumptuously been called the 'missing link' . . . But what will have been gained by such a revelation? . . . the step from resemblance to filiation is one that can never be made legitimately."[71 B]) These remarks chiefly concerned the question of the descent of man, a question on which prejudices were naturally especially strong, and absurdities less easily discovered, but they were general in their application. By such an argument the anti-Darwinian position could be made impregnable, but at the same time the whole question was brought out of the realm of science and common sense.

In order to sustain the argument that the positive evidence not only failed to confirm, but actually contradicted the Darwinian theory, it was necessary to assume that the negative evidence of the missing links could only support the inference that no such links had ever existed. Traditionalists did not hesitate to make this assumption: it was, indeed, implied in their philosophy of science, according to which no hypotheses were allowed which were not "close to the facts". And in connection with the argument of the missing links the popular prejudice against hypotheses could be appealed to with more effect than usual. To assume that transitional forms had existed, but had not been discovered because geologists had not yet investigated all possible

---

[71 A]) *Popular Science Review*, 2, 1862—3, 360. See also *Zoologist*, 1, 1866, 235.
[71 B]) *News of the World*, 1862, Oct. 12 (Leader on British Association).

strata was declared to be "baselessly hypothetical,"[72] "a gratuitous and dangerous hypothesis, by which any conceivable theory might equally be supported."[73] Indeed, said one writer, "in thus maintaining 'the imperfection of the geological record' ... Mr Darwin rejects all the leading truths of the science."[74] Or as another critic put it, "Doubtless there is much to be learnt, but we must infer the unknown from the known, and the known proclaims to all who love logic and hate sophistry that species are not transmuted one into the other."[75] These outbursts, expressed with more or less elegance, were exceedingly common in the early years of the controversy,[76] and the *Athenaeum*, steadfastly opposing Darwin even efter most informed people had come to accept at least the Descent theory, wrote as late as 1868 that "it is only by assuming that the unknown when revealed will help him, that he postpones the complete confutation which many think inevitably awaits his theory."[77]

Whenever a full discussion was devoted to the missing link problem, it was naturally impossible to maintain that the paleontological record was complete. A less vulnerable form of the argument was to concentrate on a limited number of points in the geological series where, one said, it could be reasonably asserted that the series of forms was unbroken. Sedgwick, for instance, insisted that in many sedimentary deposits there was an insensible gradation from the point of view of the rocks, but not from the point of view of the fauna,[78] and another well known geologist, Phillips, was of the same opinion.[79] This argument was repeated again and again by anti-Darwinian writers.[80]

Another of the more specific forms of the missing link argument was

---

[72] *Dublin Review*, 48, 1860, 80.

[73] *Eclectic Review*, 3, 1860, 242.

[74] *Good Words*, 3, 1862, 7.

[75] *Recreative Science*, 2, 1860—61, 277.

[76] See e. g. *Saturday Review*, 8, 1859, 775—6, *Christian Observer*, 1860, 571, *Quarterly Review*, 108, 1860, 250, *Annals and Magazine of Natural History*, 6, 1860, 221, *Good Words*, 6, 1865, 303.

[77] *Athenaeum*, 1868, May 28, 455.

[78] *Spectator*, 1860, 285.

[79] *Spectator*, 1860, 1172—3, *Nonconformist*, 20, 1860, 1034, *Annals and Magazine of Natural History*. 7, 1861, 402.

[80] *British and Foreign Evangelical Review*, 9, 1860, 430, *Fraser's Magazine*, 62, 1860, 80, *Quarterly Review*, 108, 1860, 242, *Annals and Magazine of Natural History*, 6, 1860, 222, *Intellectual Observer*, 1, 1862, 331, 392, *Quarterly Review*, 114, 1863,

to allege that the forms found in the very earliest strata were so highly developed, even perfect, that a very long series of predecessors must be assumed on any development theory. Yet none of these preceding forms had been found: indeed, the pre-silurian strata where they ought to occur yielded no evidence of life at all. A similar argument was applied to the later strata as well: one often asserted that whenever a new class of animals appeared, its first representatives were so perfectly constituted that they presupposed a very long preparatory stage. Illustrations were such things as the perfection of the eye of the Silurian Trilobites, declared to be quite as highly developed, or even more highly developed, than any to be found at the present time,[81] further, the highly organized reptiles of the mesozoic period,[82] the existence of Cephalopods among the oldest mollusks,[83] and the fact that the highest fishes were among the earliest of their class,[84] and so on.[85] Sometimes it was thought to be enough to adduce the very abundance of new species at the beginning of each geological period as sufficient to refute Darwin.[86]

The more violent anti-Darwinians sometimes put this argument more strongly. The early forms of each period were not only highly developed: they were actually more highly developed than the later forms. "There is no evidence anywhere of the development of higher from lower forms. On the contrary, it appears that the higher tribes of any given race first appeared, and that the type afterwards dwindled or was 'degraded', before the advent of a higher order."[87] "Any actual

412, *Examiner*, 1869, Apr. 17, 245—6, quoting Argyll. *Dublin University Magazine*, 74, 1869, 585, quoting Argyll, *Nature*, 3, 1870—1, 272, quoting Mivart, *Popular Science Review*, 10, 1871, 188, quoting Mivart, *Zoologist*, 6, 1871, 2617.

[81] *Annals and Magazine of Natural History*, 6, 1860, 222, *Contemporary Review*, 7, 1868, 138, *London Quarterly Review*, 39, 1872—3, 107—111.

[82] *Spectator*, 1860, 285.

[83] *Annals and Magazine of Natural History*, 14, 1864, 221.

[84] *Edinburgh New Philosophical Journal*, 19, 1864, 80.

[85] See e. g. *John Bull*, 1859, 827, *Rambler*, 2, 1859—60, 367, *Recreative Science*, 2, 1860—1, 275, *Nonconformist*, 1863, Sep. 9, 730, quot. Lyell, *British and Foreign Evangelical Review*, 15, 1866, 367, *Lancet*, 1872: 2, 297 (ref. C. R. Bree), *Guardian*, 1870, Sep. 28, 1137.

[86] See e. g. *Annals and Magazine of Natural History*, 6, 1860, 223, *Athenaeum*, 1861: 1, 866.

[87] *Eclectic Review*, 3, 1860, 242. See also: *Annals and Magazine of Natural History*, 11, 1863, 421, *Zoologist*, 3, 1868, 1385. *Recreative Science*, 2, 1860—1, 276.

knowledge," declared another traditionalist, "is in favour of a deterioration, not of a development theory.[88]) These opinions, needless to say, linked up both with the ancient belief in the Golden Age, and with the Christian view of the present world as fallen from the perfect state of the primeval Paradise. But it is worth noticing that one usually did not postulate a *general* deterioration. The deterioration occurred only *within* each geological period. According to the popular anti-Darwinian view, each such period was marked by a more or less complete renewal of organic species. The new organic world created by God was an improvement in relation to the preceding period: but as soon as the forms were left to fend for themselves, it was natural to suppose that they would fail to keep up to the perfect standard set by the Creator. This attempt to combine the prevalent 19th century belief in progress with the traditional belief in deterioration, however, became increasingly difficult to uphold as more and more paleontological material was collected, and catastrophist views were more and more abandoned by geologists. Moreover, the two leading scientific anti-Darwinians, Owen and Mivart, showed by their example that it was quite possible to combat the Darwinian theory, and even Descent, without encumbring oneself with any deterioration theory.

On the Darwinian side the argument of the missing links was constantly countered by a reference to the imperfection of the geological record. "The seeming breaks in the chain of being are not absolute, but only relative to our imperfect knowledge,"[89]) wrote Huxley in his *Times* review of the *Origin*, and in the *National Review* Carpenter amplified the argument: "As the accumulation of each deposit has often been interrupted, and as long blank intervals have doubtless intervened between successive deposits, we have no right to expect to find in any one or two of them all the intermediate varieties between the species which appear at the commencement and at the close of those periods; but we ought to find after intervals ... closely allied forms, or, as some authors term them, representative species. Now, as this is just what we do find, the theory of descent with modification, is so far conformable to positive facts, that it must be admitted to have at least as valid a foundation in a broad basis of phenomena as the theory

---

[88]) *Patriot*, 1863, Sep. 10, 594
[89]) *Times*, 1859, Dec. 26, p. 8,

of successive creations."[90]) As leading geologists, and notably Lyell, were inclined to agree with the imperfection view, the argument could not but be effective.[91]) Further, confirmation of the view that geological knowledge was still far from complete was obtained every time new fossil species were discovered. It is true that, as we have seen, these finds were sometimes used by anti-Darwinians to attack the development theory, and by religious people in order to emphasize the insecurity of any conclusions drawn from scientific data,[92]) but as many of the new finds were clearly precisely such intermediaries as the Descent theory demanded, they were in the main a powerful support for that theory.[93]) For the main argument in favour of transmutation was of course to point out the generally gradual progression over the whole geological history. Once this had been established, and in addition the scientific justification of a naturalistic, not to say natural, hypothesis had been granted, the Darwinian theory was no longer to be resisted.

The Insufficiency of Geological Time

One of the distinctive features of the Darwinian theory was its insistence on small, almost imperceptible variations as the vehicle of progressive change. As, moreover, these small changes were not assumed to occur in all individuals at once, through the direct influence of external conditions or otherwise, but only to become general by differential survival in the struggle for life, it was obvious that the theory demanded a very extended time-scale. The justification of such a time- scale therefore naturally became a matter of controversy in the debates around the theory. Darwin's opponents tended to maintain that the duration of the inhabitable earth had been much shorter than his theory demanded; in fact altogether too short for the production of the innumerable missing links.

To estimate the rate of change in the organic world on general grounds — from the observed range of individual variation from parent

---

[90]) *National Review*, 10, 1860, 212.

[91]) See e. g. *London Review*, 1, 1860, 58, Lyell at the British Association reported everywhere, *Edinburgh Review*, 128, 1868, 446, Wallace in *Quarterly Review*, 126, 1869, 386—7.

[92]) See above, Chapter 8, p. 169.

[93]) See e. g. *Examiner*, 1867, Nov. 30, 757, *London Review*, 16, 1868, 179, *Edinburgh Review*, 128, 446, *Westminster Review*, 35, 1869, 208.

to offspring, and the estimated efficiency of the differential survival, could only yield very vague results. The only experimental evidence that could be adduced was provided by the changes that had been effected by artifical selection. It might, indeed, be possible to draw some conclusions from the observed amount of change within geological periods of known length: but geologists were on the whole hardly in a position to translate their time-scale into absolute figures.

The difficulty of determining the probable rate of change was matched by the difficulty of determining the age of the world, and so the time at the disposal of evolution. Lyell and the uniformitarians were inclined to regard the world as for practical purposes infinitely old. Darwin, as a true disciple of Lyell, agreed with him in adopting a long time-scale. In the first edition of the *Origin* he attempted to give some substance to his estimate, concluding that probably "a far longer period than 300 million years has elapsed since the latter part of the Secondary period."[94 A]) When it was later pointed out that his calculation had contained some serious arithmetical slips, he cut out the whole passage; but it shows what time-scale he had in mind. Thus when discussing the origin of the eye, he spoke of the process of selection as going on for "millions on millions of years."[94 B])

Darwin's opponents, among whom catastrophism was for long the prevailing geological doctrine, tended to adopt a much shorter geological time-scale, agreeably both with their traditionalism and with the ability of their theory to account for great changes by marvellous agencies rather than drafts upon time. A definite figure was thrown into the debate in 1861, when Sir William Thomson, in a paper read at Edinburgh,[95]) gave his estimate of between 100 and 200 million years as the probable age of the habitable earth. The estimate was based on astronomical and physical considerations concerning the cooling of the earth, and though its originator advanced it with many reservations, his figure — that is, the lower one — was again and again referred to in the press as the current scientific estimate. Thomson, whose opposition to Darwinism came out clearly in his presidential address to the British Association in 1871, himself pointed out the bearing of his conclusions on the Darwinian theory, and this was eagerly advertised

---

[94 A]) *Origin* 1 ed. r., 245.

[94 B]) *Origin*, 1 ed. r., 162.

[95]) *Transactions* of the Royal Society of Edinburgh, 23, 1861, 157.

in the anti-Darwinian press: "In regard to Mr. Darwin, Dr. Thomson concludes that his geological estimates of time necessary to the postulates on which he reasons, are entirely inconsistent with the chronology of the sun to which cosmical laws induct us,"[96]) wrote one traditionalist organ, and the *Guardian* declared in its review of the *Descent* that "the famous argument of Sir William Thomson has never been refuted on its own ground ... perhaps because very few have the requisite knowledge for appreciating it."[97]) Clearly this matter was one of the instances where it could be said, and was said, that true science was in perfect harmony with true religion.[98])

Towards the end of our period the argument was worked out more fully by Mivart in his *Genesis of Species*. Mivart presented his own calculation of the time a Darwinian evolution would demand, namely, 2,500 million years, which was, as he said, 25 times as much as Thomson would grant as having elapsed.[99])

The Darwinians had to be very much on the defensive in these matters. They had no estimate of their own to put forward in opposition to Thomson's. His argument therefore had to be taken seriously. It was counted in two ways. In the first place, the uncertainty of the calculation was stressed. This was done most effectively by Huxley in his address to the Geological Society in 1869: "Is the earth nothing but a cooling mass, [like "a hot-water jar such as is used in carriages," or "a globe of sandstone"] and has its cooling been uniform? An affirmative answer to both these questions seems to be necessary to the validity of the calculations on which Sir William Thomson lays so much stress."[100]) In the second place, Darwinians argued that Darwin's theory did not assume a fixed and constant rate of change, since the changes depended on the changes in the external conditions, rather than on internal causes, and that, further, the rate of change had to be chiefly estimated from

---

[96]) *Intellectual Observer*, 1, 1862, 4.

[97]) *Guardian*, 1871, 936.

[98]) See e. g. *Nonconformist*, 1864, Sep. 21, 767,*Standard*, 1864, Sep. 16, p. 4, *North British Review*, 46 1867, 304—5, *Guardian*, 1870, Sept. 28, 1871, 137, June 7, 682 (letter), *London Quarterly Review*, 39, 1872—3, 107—111.

[99]) Mivart, *Genesis of Species*, London, 1870, 160. See also: *Month*, 14, 1871, 528. *Dublin Review*, 16, 1871, 483—5, *Fortnightly Review*, 9, 1871, 414—15, *Popular Science Review*, 10, 1871, 189.

[100]) Quoted from T. H. Huxley, *Lay Sermons*, 270. See also *Quarterly Journal of Science*, 4, 1867, 487 (Wallace), *Saturday Review*, 31, 1871, 180.

geology itself. Here again Huxley came forward: "Biology takes her time from geology. The only reason we have for believing in the slow rate of the change in living forms is the fact that they persist through a series of deposits which, geology informs us, have taken a long while to make. If the geological clock is wrong all the naturalist will have to do is to modify his notions of the rapidity of change accordingly."[101]

Darwin himself also dealt with Thomson's argument in the later editions of the *Origin*. After stressing the uncertainty of the estimate, he considered in more detail some of the paleontological implications of the shorter time-scale. It has been estimated, he said, "that about 60 million years have elapsed since the Cambrian period, but this, judging from the small amount of organic change since the commencement of the Glacial epoch appears a very short time for the many and great mutations of life, which have certainly occurred since the Cambrian formation; and the previous 140 million years can hardly be considered as sufficient for the development of the varied forms of life which already existed during the Cambrian period. It is, however, probable, as Sir William Thompson insists, that the world at a very early period was subjected to more rapid and violent changes in its physical conditions than those now occurring; and such changes would have tended to induce changes at a corresponding rate in the organisms which then existed."[102]

Darwins' judgment of the biological questions involved was sound enough. 60 million years *is* too short a time for the organic evolution that has taken place since the Cambrian era, and 140 million too little for the preceding evolution. Modern estimates give c. 500 million years for the former, and at least as much for the latter, with possibly as much as 1,000 million years of organic evolution before the Cambrian.[103]

The Bear becoming a Whale

The object of the missing link argument in its various forms was to show that the transitional forms which Darwin's theory presupposed could not be found anywhere, and that therefore the theory was to be rejected. There were no links in the present organic world, none had

---

[101] T. H. Huxley, *Lay Sermons*, 270. See also *Nature*, 1 1869—70, 455.

[102] *Origin*, 6 ed., 447—8.

[103] Frederick E. Zeuner, *Dating the Past*, especially p. 337.

been produced artifically, none had been recorded since the dawn of history, none in the geological strata, and the time-span granted by physicists for the existence of organic life on the earth seemed to be too short for their production in sufficient numbers. All these were quite fair objections, though, as we have seen, the conclusions drawn from them often went beyond what was justifiable. Because of the deficiency of the evidence, the theory was to that extent unconfirmed, but it was neither refuted nor illegitimate, unless an idealistic philosophy of science was taken for granted.

Now in the extreme anti-Darwinian organs it was not uncommon to turn the argument round, and to maintain that *since* there were no transitional forms, therefore the Descent theory must imply that the existing gaps between species in the fossil and recent series were covered, so to speak, in one jump. In other words, one imputed to Darwin the view that, for instance, a donkey could beget a horse, or a monkey give birth to a human child, whereupon the theory was naturally declared to be absurd.

The starting-point of these attempts to ridicule the theory was very often a passage in the first edition of the *Origin:* "In North America the black bear was seen by Hearne swimming for hours with widely open mouth, thus catching, like a whale, insects in the water. Even in so extreme a case as this, if the supply of insects were constant, and if better adapted competitors did not already exist in the country, I can see no difficulty in a race of bears being rendered, by natural selection, more and more aquatic in their structure and habits, with larger and larger mouths, till a creature was produced as monstrous as a whale."[104]) Owing to the comments that this passage gave rise to, Darwin struck out its latter part, from "Even in so extreme a case . . ." in subsequent editions. Still, the cat was out of the bag, and the bear becoming a whale was one of the favourite illustrations of the absurdity of the theory.

It is interesting to note that anti-Darwinian writers seem to have deliberately played on the ambiguity of "a bear" and "a whale". The words can be used both to denote the individual, and to denote the class, or species. Darwin of course used them in the latter sense. A *race* of bears, as he said, might by natural selection become transmuted into whales through successive generations. His opponents, bent on

---
[104]) *Origin,* 1 ed. r., 158.

ridiculing the theory, instead took the words as referring to individuals, which made the proposition patently absurd. "Darwin seems to believe that a white bear, by being confined to the slops floating in the Polar basin, might be turned into a whale; that a lemur might easily be turned into a bat, that a three-toed Tapir might be the great grandfather of a horse . . ."[105]) wrote Sedwick in the *Spectator*. In popular organs, such assertions occurred without any qualifications. "Mr. Darwin has, in his the most recent and scientific book on the subject, adopted such nonsensical 'theories' — as that of a bear swimming about a certain time till it grew into a whale, or to that effect," wrote the *Family Herald* in an answer to a correspondent.[106])

The effect of this argument could be further heightened by inventing various ludicrous genealogies or preposterous imaginary links. "With such a range and plasticity . . . we know not where to stop — centaurs, dryads, and hamadryads . . . [and perhaps] mermaids once filled our seas,"[107]) wrote one naturalist sarcastically. The mermaid was a popular illustration. The *Family Herald* used it repeatedly: "We defy any one, from Mr. Darwin downwards, to show us the link between the fish and the man. Let them catch a mermaid, and they will find the missing link."[108]) Elaborating the Darwinian theme of our aquatic origin, the *Pall Mall* once joked about "the stages by which a polypus is finally developed into a professor, and a bed of oysters by the force of 'natural selection' into a gathering of the British Association."[109]) This was one better than the *Annals and Magazine*, which — evidently in the interest of the alliteration — stopped short at "a law of sufficient force to convert an eel into an elephant, or an oyster into an orangoutan,"[110]) but perhaps equalled the *Weekly Review*, where Darwin was taunted for

---

[105]) *Spectator*, 1860, 286.

[106]) *Family Herald*, 1860—61, 364. See also: *North British Review*, 32, 1860, 465, 473, *Eclectic Review*, 14, 1868, 346, *Edinburgh New Philosophical Journal*, 11, 1860, 281, *Blackwood's Magazine*, 89, 1861, 616. 92, 1862, 91.

[107]) *Edinburgh New Philosophical Journal*, 11, 1860, 281. See also: *Tait's Edinburgh Magazine*, 27, 1860—1, 526, *Record*, 1862, Oct. 8, p. 2, *Recreative Science*, 2, 1860—1, 271, *Daily News*, 1861, Nov. 7 p. 2, (rep. Sir David Brewster), *Standard*, 1867, Sep. 12, p. 3, (reporting Crawfurd), *London Society*, 18, 1870, 476, *Month*, 15, 1871, 80, *Tinsley's Magazine*, 8, 1871, 399.

[108]) *Family Herald*, 1861—2, 268; 1867—8, 188.

[109]) *Pall Mall Gazette*, 1865, Sep. 12, 381.

[110]) *Annals and Magazine of Natural History*, 11, 1863, 424.

fancying that "tadpoles could develop into philosophers."[111]) Such facetiousness of course meant nothing to informed readers, but among the uneducated it was probably an effective argument. It was certainly very popular.[112])

Pro-Darwinian organs noted this with some bitterness: "We are aware that this view of the Darwinian theory is quite common, and that people are accustomed to demolish the theory by triumphantly asking how a whale ever became a monkey."[113]) The *Pall Mall* analysed the argument more closely: "The majority of those who maintain the fixity of Species do not, apparently, bear in mind this distinction between the modifications possible to an individual, and the modifications multiplied to each other of successive generations; yet this is the cornerstone of the development hypothesis."[114]) Since the argument was founded on a misunderstanding of the nature of the theory, the only effective answer to it was more and better information. That, of course, was not provided, since, as we pointed out in Chapter 2, the more anti-Darwinian an organ was, the less information on the theory could be expected from it. This was a guarantee that attitudes would not change too quickly. Anti-Darwinianism led to the Darwinian theory being caricatured, and the caricature perpetuated the anti-Darwinianism

---

111) *Weekly Review*, 1863, Aug. 29, 514.
112) See note 107.
113) *London Review*, 16, 1868, 644.
114) *Pall Mall Gazette*, 1868, Feb. 15, 636.

CHAPTER 12

The Battle against Natural Selection

The missing link arguments, and the arguments designed to establish the existence of a permanent and fixed specific type, were used primarily against the transmutation theory as such. They were not new: they had been used against pre-Darwinian evolutionists as well. Many anti-Darwinians, however, were not absolutely opposed to the transmutation theory, and instead directed their arguments against the Natural Selection part of the Darwinian doctrine. They were inclined to grant that new species might arise through the accumulation of gradual modifications, but they denied that the variations were random, and that they could be preserved and accumulated by purely natural means. Several lines of reasoning were employed to support this contention.

One general objection was that Darwin's Natural Selection could not be effective, since under nature no effective bar existed against the intercrossing of those incipient species which arose, and that therefore the favourable variations that occurred would soon be diluted and swamped, before the next step in the onward direction had time to occur.

Other arguments were directed against specific applications of the Natural Selection theory. One granted that Natural Selection might explain many structures and organs, but denied that it could explain all. For instance, in many cases an organ might be of undoubted service to its possessor in its fully developed state, but in its incipient stages, through which it had to pass, on Darwin's theory of evolution by minute variations, the organ might be useless, and perhaps even directly harmful. Such an organ could obviously not be naturally selected.

Moreover, there existed in nature several organic structures which, even when fully developed, did not seem to be of any use. These as well were inexplicable on the theory of Natural Selection.

Another supposed difficulty was the persistence of many forms throughout long geological periods, and in particular, the persistence of very low and undifferentiated forms. If Natural Selection continually favoured the improved races, how was it that some species were thus left behind?

Closely similar structures appearing in unrelated species presented yet another difficulty. It was highly unlikely that similar results could have been produced in two or more altogether separate lines of descent by the accumulation of purely random variations.

In all these cases the underlying idea was that if there had been development, that development was due to other forces than purely natural selection. More particularly, Darwin's opponents wished to emphasize that no explanation of the origin of the organic world could be complete without introducing final causes into the picture. Indeed, anti-Darwinians refused to accept Natural Selection as a causal explanation at all, since Darwin, as he himself admitted, could not explain the causes of the individual variations which Natural Selection worked upon. These causes, the opponents said, were what science had to concern itself with. As long as Darwin could not account for the individual variations, his opponents considered themselves at liberty to interpolate final causes for them.

Dilution through Intercrossing

At the time when Darwin published the *Origin*, the laws of inheritance were little known, or, as Darwin himself said, "quite unknown."[1]) By and large one believed in "blending inheritance", i. e., that the characteristics of the offspring must in general be expected to be intermediate between those of both parents, and that it was the blended type which was in turn transmitted to the second generation. On the other hand, it was also known that peculiarities were sometimes inherited complete, and that they might sometimes skip several generations. But these phenomena seem in the main to have been reckoned as anomalies for which special explanations must be found, for instance, similar conditions recurring to produce similar effects, or a special prepotency associated with some peculiarities, or, in the case of reversion, the preponderant influence of the specific type. The general rule of blending was the natural assumption, which was also supported by the traditional belief in the "mixing of the blood" which people in general, and many naturalists, believed to form the material basis of the phenomena of inheritance.

---

[1]) *Origin*, 1 ed. r., 11.

Now on the assumption of blending, and on the further assumption that favourable variations were very rare, it was a valid objection against the Natural Selection theory that intercrossing between modified and unmodified individuals, which was bound to occur in nature, but was artificially prevented by domestic breeders, would fairly soon lead to the dilution and eventual extinction of those favourable variations which occurred. This criticism was quite a common one.[2] "In the wild state," wrote an early critic, "after a very few generations at the utmost, any accidental variation, whether apparently favourable or unfavourable, will be merged in a return to [the] original condition."[3] The analogy with artifical selection, on which Darwin depended so much, therefore broke down at a crucial point. "Animals will intermix, if left to themselves. There is small chance, therefore, that any minute element of divergence will be allowed to expand until it becomes the dominant quality of a race."[4]

The argument was developed with much force and effectiveness in an article by Fleeming Jenkins in the *North British Review* in 1867. The writer used a mathematical line of reasoning to show that the chances were practically *nil* of any favourable variation being preserved over a series of generations.[5] In the 6th edition of the *Origin* Darwin acknowledged the force of the criticism. Summarizing the argument, he said: "[The critic] shows that if a single individual were born, which varied in some manner, giving it twice as good a chance of survival as that of the other individuals, yet the chances would be strongly against its survival. Supposing it to survive and to breed, and that half its young inherited the favourable variation; still ... the young would have only a slightly better chance of surviving, and this chance would go on decreasing in the succeeding generations." Darwin concluded, "The justice of these remarks cannot, I think, be disputed."[6] His answer to the criticism was to change his ground a little. Formerly he had assumed that even very rare variations might be selected; he now

---

[2] See, in addition to those quoted below: *Fraser's Magazine*, 62, 1860, 76, *London Quarterly Review*, 14, 1860, 289, *Gardener's Chronicle*, 1860, Feb. 11, 122, *Guardian*, 1871, 936, *Tablet*, 1871, Apr. 15, 456.

[3] *Zoologist*, 1859, March, 6475.

[4] *British Quarterly Review*, 31, 1860, 417.

[5] *North British Review*, 46, 1867, 277—318.

[6] *Origin*, 6 ed., 112.

insisted more strongly on frequently recurring, but by the same token much less strongly marked variations.

On this ground, however, he met with another challenge. If he took his stand exclusively on extremely small variations, it was argued, then the advantage conferred by each in the struggle for life also became so small as not to count. This was the argument of the uselessness of the incipient stages of developing structures, which we shall deal with below.

To Darwin's opponents the objection raised by the intercrossing argument seemed unanswerable. One maintained that, in order to establish the development theory, Darwin needed to include further assumptions, and these would destroy its naturalistic character. For instance, he might postulate a tendency for the improved varieties to pair with similarly improved mates only: "If it could be proved that there is a predominant tendency in the more perfect and robust of each species to pair with individuals like themselves, for the transmission of their kind, then the necessary existence of natural selection as an operative cause must be admitted,"[7]) wrote one reviewer. But this appeared hardly realistic even in the case of the higher animals, and clearly absurd in the case of the lower animals and plants.[8]) Moreover, if the variation was a rare one, then the chance of two similar variations occurring at the same time and place was almost infinitesimally small, and the chance of these similarly improved individuals meeting and pairing smaller still. This was so popular an objection that we even find it in *Punch*.[9])

Another assumption which one suggested might save Darwin's theory was to endow the favourable variety, and that only, with a special power of being transmitted undiluted to the offspring. In other words, even though only one of the parents was improved, all the offspring would exhibit the favourable peculiarity in its full force. The phenomenon in itself was not unknown. Huxley pointed it out in one of his reviews of the *Origin:* "There seems to be, in many instances, a pre-potent influence about a newly-arisen variety which gives it what one may call an unfair advantage over the normal descendants

[7]) *Fraser's Magazine*, 62, 1860, 76.

[8]) *North British Review*, 32, 1860, 471.

[9]) *Punch*, 64, 1863, Mar. 21, 122. See also, *Dublin University Magazine*, 74, 1869, 587, *Tablet*, 1871, Feb. 25, 232.

from the same stock."[10]) But Huxley did not, of course, wish to imply
that this prepotency was any more associated with favourable than
with unfavourable variations. Yet this was clearly what Darwin's
opponents thought was needed. The *North British* reviewer elaborated
this argument: "If ... the advantage given by the sport is retained
by all descendants, independently of what in common speech might
be called the proportion of blood in their veins directly derived from
the first sport, then these descendants will shortly supplant the old
species entirely, after the manner required by Darwin. But this theory
of the origin of species is surely not the Darwinian theory ... What
is this but stating that, from time to time, a new species is created?
it offers no explanation of the cause of the divergence from the progenitors, and still less of the mysterious faculty by which the divergence is
transmitted unimpaired to countless descendants. It is clear that
every divergence is not thus transmitted."[11]) The argument impressed
theologically minded critics: "Darwin's theory requires a law of hereditary generation transmitting the (accidentally favourable) variations,"[12]) wrote a contributor to the *Spectator*, and in the Catholic *Tablet*
we read: "The only way we can imagine of escaping from the dilemma
would be, in parliamentary language, to saddle the law of heredity
with an 'instruction', that, whatever other variations it should permit,
it should permit all improvements. But this would suit neither Mr.
Darwin nor Mr. Wallace. For the law of heredity possesses no intelligence."[13]) The implication is clear enough. The argument was believed
to establish that the law of heredity was so constituted that it gave in
itself evidence of Design. It bore the marks of its Divine origin.[14])
Once that had been established, the theologically obnoxious features
of the Natural Selection theory were removed, and with them the
chief obstacle against accepting the transmutation theory.

The Preservation of Incipient and Useless Structures

It was an essential characteristic of Darwin's theory to assume not
only that the steps whereby the development of the organic world had

[10]) *Westminster Review*, 17, 1860, 548.
[11]) *North British Review*, 46, 1867, 289—92.
[12]) *Spectator*, 1867, 18.
[13]) *Tablet*, 1871, Sep. 9, 333.
[14]) See above, Chapter 6, p. 128 ff.

taken place were almost insensibly small, but also that each single step implied an advantage in the struggle for life for the population in which it occurred. Therefore it was not enough to show that an organ or structure, in its fully developed form, conferred an advantage on its possessors over those who did not have it: it was also necessary to show that the steps by which that organ or structure had come into being had all been similarly advantageous in comparison with the previous step.

One of Darwin's great difficulties was that it was impossible for him to provide empirical evidence of all the various stages through which the development had passed in each particular case. This difficulty became especially acute when the organ was such that no parallels for it could be found in related species, by which a clue could be given as to its mode of development. Failing such clues, Darwin necessarily had to be rather vague and hypothetical, and the common criticism that his theory was an imaginary speculation could be brought against him. His opponents therefore diligently sought for instances of such unique structures. "Certain organs", said one critic, "are special, and appear to have no prototype: such as the fangs and poison-secreting glands of venomous snakes; the electric battery of the torpedo; and the spinning apparatus of the spider."[15]) Similar difficulties arose in the case of highly complex structures. By far the most common instance cited on this score was the human eye. Darwin had anticipated this objection: he certainly had in mind the emphasis that Paley had placed on the perfect structure of the eye as evidence of Design. The eye was a very complex organ, whose usefulness depended on the harmonious development of a very great number of parts. If the development of any one of these parts was changed ever so little, the function of the whole organ might be seriously impaired. How, then, it was argued, could the eye arise by the accumulation of small individual variations? For if one part varied without a corresponding variation in the other parts, the result would almost inevitably be a deterioration. Darwin agreed that the difficulty was serious: yet, he said, "reason tells me that if numerous gradations from a perfect and complex eye to one very imperfect and simple, each grade being useful to its possessor, can be shown to exist ... then the difficulty of believing that a perfect and complex eye could be formed by natural selection, though insuperable by our imagina-

---

[15]) *Chambers's Journal*, 1862, May 31, 351, quoting Sir B. Brodie.

tion, can hardly be considered real."[16]) He then went on to indicate that gradations did really exist in nature. His critics, however, most of them steeped in Paley's thought, would not follow him. "He must furnish us with a very different logic from what he has used, a far more formidable series of fact than he has done, before any sober man of common sense will believe that the operation of any such law as he describes can produce such an organ as the human eye."[17]) This criticism concerning the eye remained on the list of objections against Darwin's theory all through our period, especially among theological writers.[18])

A somewhat similar illustration of the absurdity of Darwin's theory was the wonderful constructive instincts of the hive bee.[19]) Here again, Paley's influence was noticeable: the bees were supposed to be guided by unconscious mathematical principles in the construction of their cells.[20]) Darwinians countered this argument in two ways. In the first place, it was possible to point out that a whole gradation of cell-colonies, from the simplest to the most elaborate, existed within the bee family.[21]) In the second place, the mathematical accuracy of the building even of the best colonies had been exaggerated: and the results which were produced could be accounted for rather simply on the principle of Natural Selection, since the better constructed colonies achieved a great saving of wax.[22])

The argument of the uselessness of incipient stages was elaborated very fully by Mivart in his *Genesis of Species*. Among the instances he cited were, in addition to the eye, such things as the neck of the giraffe, the position of the eyes of certain flat-fishes, the formation of whalebone, the larynx of the kangaroo, the rings of the rattlesnake, etc. Many of these instances were repeated in the numerous and largely favourable press reviews of Mivart's book.[23 A]) In the 6th edition of the *Origin*

---

[16]) *Origin*, 1 ed. r., 160.

[17]) *Daily News*, 1859, Dec. 26; p. 2.

[18]) *Good Words*, 3, 1862, 7, *British Quarterly Review*, 38, 1863, 494, *Record*, 1870, Sep. 23, p. 2, *London Quarterly Review*, 36, 1871, 275—9, *Lancet*, 1872:2, 297.

[19]) *Record*, 1870, Sep. 23, p. 2, *Daily News*, 1859, Dec. 26 p. 2.

[20]) *Annals and Magazine of Natural History*, 11, 1863, 415—429, article by Samuel Haughton.

[21]) *Origin*, 1 ed. r., 193.

[22]) *Annals and Magazine of Natural History*, 12, 1863, 304 (Wallace).

[23 A]) *Tablet*, 1871, Feb. 25, 232, *Popular Science Review*, 10, 1871, 188, *Nature*, 3, 1870—1, 272, *Fortnightly Review*, 9, 1871, 414—15, *Record*, 1872, Feb. 28, p. 4,

Darwin sought to answer Mivart's specific objections. To do this he had to go into rather abstruse anatomical details, and such a discussion necessarily passed over the heads of the general public, who only got the impression that the authorities disagreed. Those who were not experts therefore took their stand according to preconceived notions. The *Tablet* admitted that Darwin put his case plausibly enough. "But the controversy has gone far enough to make every reader feel that just as he has put a fresh complexion on the case, by fresh facts and fresh hypotheses, so Mr. Mivart, when his turn comes again, will have other facts, quite as good, and other hypotheses, quite as sound, and the thing will stand where it was."[23 B])

The argument we are now dealing with was sometimes put in a slightly different form. Instead of insisting on the impossibility of imagining a gradual series of advantageous variations, one declared that the gradual steps which Darwin postulated were so small that the advantage each conferred must be considered infinitesimal or none. This objection was effective with a public unused to statistical reasoning. "What advantage", wrote an early reviewer of the *Origin*, "could it afford an insect that was about to be swallowed by a bird, that it possessed a thousandth fragment of some property?"[24]) Another said: "So trifling are all modifications when they first appear — so many ages required to give them any prominence by means of natural selection alone — that to expect any immediate results would be like saying that a single penny added to the capital of a merchant, should enable him to outstrip all his brother merchants in the race for opulence."[25]) This sort of criticism was worked out mathematically in a paper by A. W. Bennett, read at the British Association in 1870, and reprinted in *Nature*.[26]) Using a case of mimicry in butterflies as an illustration, the author argued that it might be assumed that at least twenty variations in a definite direction were needed before any advantage at all would arise. Assuming that each single variation might occur in 20

Guardian, 1871, 935; (somewhat differently:) *Annals and Magazine of Natural History*, 7, 1871, 418. A similar argument was also advanced earlier in *Month*, 11, 1869, 41—2.

[23 B]) *Tablet*, 1872, May 25, 648.

[24]) *Eclectic Review*, 3, 1860, 227.

[25]) *British Quarterly Review*, 31, 1860, 415. See also: *Christian Remembrancer*, 40, 1860, 257, *Zoologist*, 18, 1861, 7600.

[26]) *Nature*, 3, 1870, Nov. 10, 30—33.

different directions, one of which only was favourable, the probability of that favourable variation occurring was evidently 1/20. On this basis he concluded that the chance for the modification to become big enough for Natural Selection to begin to operate was only $1/20^{20}$, which seemed hardly removed from sheer impossibility.

The Darwinian answer to this argument was to point out that there was no need to assume so very small basic variations, and that, further, it was a mistake to equate a small advantage with no advantage at all. Provided the variations were numerous — and the smaller they were, the more numerous were they likely to be — even a small advantage might establish a modification in the long run.[27]

The argument from the uselessness of incipient structures could, like that of dilution, be used both against the transmutation theory as such, and against the Natural Selection theory alone. Darwin was faced with the dilemma of either abandoning the transmutation theory, or else admitting that Natural Selection had to be eked out by some principle of directed evolution. In either case the anti-Darwinians were able to save their main position: the argument of Design.

Useless Organs

The Natural Selection theory was thoroughly utilitarian. It was solely the usefulness of a variation in the struggle for life which determined whether it was to be preserved or not. Organs or structures which did not increase their possessors' chances of survival could not be developed by Natural Selection. Darwin was very emphatic on this point; he even maintained that if it could be shown that any structures had developed for some other purpose, it "would be absolutely fatal"[28] to his theory.

Darwin's critics therefore naturally tried to find instances of this. However, as practically all the anti-Darwinians entertained teleological views themselves, it was not easy for them to acknowledge that any organ was really useless. Had not the Creator adapted every organic being to its station? To argue that something in the organic world was completely useless therefore appeared almost sacrilegious. But the dilemma could be solved by means of an anthropomorphic development

---

[27] See letter by Wallace in *Nature*, 3, 1870, Nov. 17, 49.

[28] *Origin*, 1 ed. r., 171.

of the Design argument. Many structures in the organic world, one held, could be explained not with reference to their usefulness to the possessors themselves, but to man as the lord of creation. We have already seen how Professor Owen adhered to this doctrine in the case of the horse genus.[29]) Other naturalists had given the principle quite a wide application. For instance, there were those who seriously maintained that the rings of the rattlesnake were created to warn its unfortunate victims off[30]) — the snake itself was apparently not a worthy object of the Creator's benevolence.

A variant of this argument was that many structures in the organic world had been created solely for the sake of beauty, wholly irrespective of use. Darwin mentioned this idea in the *Origin:* it is interesting that whereas in the first edition he describes it by the words, "created for beauty in the eyes of man," in the sixth edition he adds, "or the Creator (but this latter point is beyond the scope of scientific discussion)".[31]) He probably had in mind the Duke of Argyll's *Reign of Law*, where this line of reasoning was vigorously pursued. "We shall never understand the phenomena of Nature unless we admit that mere ornament or beauty is in itself a purpose, an object, and an end,"[32]) wrote the Duke, and many of his reviewers assented.[33])

In addition to these instances, some of Darwin's critics adduced others where neither the principle of utility, nor that of beauty could come into play. "A vast number of 'modifications' have apparently no reference whatsoever to the 'good', or advancement of the species . . . but are often, as it were, fantastic, or grotesque, having no connexion with either its well-being or mode of life, and the final cause of which it is utterly hopeless to discuss."[34]) Chief among these was the sting of the bee, which destroyed the bee itself. Natural Selection, it was contended, could not endow a race with a quality injurious to every individual which possessed it.[35]) A subsidiary argument in this case was

---

[29]) *Gardener's Chronicle*, 1868, Dec. 26, 1343. See above, p. 223.

[30]) Darwin, *Origin*, 6 ed. p. 254. Mivart, *Genesis of Species*, 56, mentions the rings but does not try to explain them.

[31]) *Origin*, 1 ed. r., 170, 6 ed. p. 249—250.

[32]) The Duke of Argyll, *The Reign of Law*, 188.

[33]) E. g. *Guardian*, 1867, Apr. 24, 455, *Eclectic Review*, 12, 1867, 56—7, *Tablet*, 1868, Nov. 14, 49, *Contemporary Review*, 5, 1867, 78.

[34]) *Annals and Magazine of Natural History*, 5, 1860, 137.

[35]) See *Eclectic Review*, 3, 1860, 240, *North British Review*, 46, 1867, 283.

that a variation in a sterile individual could not become naturally selected, since it could not transmit any improvement to offspring. Moreover, it was natural to hold that the sterility as such was hardly to be reckoned as a useful property: "[Darwin] has told us that only advantageous variety will be perpetuated; but here we see the extreme imperfection of sterility made permanent,"[36]) said one reviewer of the *Origin*.

In regard to the difficulty presented by the bee, Darwinians had easy work: as the bees were social animals, a variation that brought advantage to the group could be naturally selected although it might under certain circumstances be disadvantageous to each individual which possessed it. A well-defended society might produce more offspring, and thus sustain the loss of more warriors: "If on the whole the power of stinging be useful to the community, it will fulfil all the requirements of natural selection, though it may cause the death of some few members,"[37]) explained Darwin.

As to the scope of the utility principle, Darwin gradually came to admit that he might have overworked it. He explained it as a survival of his former acceptance of Design. "I was not . . . able to annul the influence of my former belief, then almost universal, that each species had been purposely created; and this led to my tacit assumption that every detail of structure, excepting rudiments, was of some special, though unrecognised, service."[38]) He was becoming increasingly willing to ascribe some variations to mere chance fluctuations, or to what he called "correlation of growth": the fact that, in many cases, variations in one organ may be automatically accompanied by a variation in a different part of the organism. One of these variations might be useful, and thus naturally selected, the other, concomitant one, might be wholly useless, or even slightly injurious. The selection depended on the total effect.[39])

Persistence of Unimproved Species

At the very end of the *Origin* Darwin summarized the operation of Natural Selection in the following words: "As natural selection works

---

[36]) *London Quarterly Review*, 14, 1860, 305. See also: *Daily News*, 1859, Dec. 26, p. 2, *Edinburgh Review*, 111, 1860, 524—5.

[37]) *Origin*, 1 ed. r., 173.

[38]) *Descent*, 92.

[39]) *Origin*, 1 ed. r., 124, 168.

solely by and for the good of each being, all corporeal and mental endowments will tend to progress towards perfection."[40]) The interpretation of the Darwinian theory implicit in that short sentence served as a basis for a fairly common objection. If natural selection was constantly at work to perfect each organic being, how was it that even at the present day millions and millions of lowly organized creatures were still in existence? The most popular instances quoted against Darwin were certain mollusks, declared to be exactly the same through all the geological periods to the present day. That, said one naturalist organ, "does not tally with that steady movement towards perfection, that certain progress, of some kind or other (even though slow), of organic forms, which a reception of this 'natural selection' idea so loudly and positively demands."[41])

The fact of the permanence of ancient forms could of course be taken as support for the theory of the fixity of species, and thus as evidence against any transmutation theory. But it was also an argument against Natural Selection. As such it was particularly effective when it could be shown that the unimproved and improved forms existed side by side. "We find," wrote Professor Owen, "that every grade of structure, from the lowest to the highest, from the most simple to the most complex, is now in being — a result which it is impossible to reconcile with the Darwinian hypothesis of the one and once only created primordial form."[42]) One writer used the bee family as an illustration: "The humble bee and the hive-bee coexist together, and the latter is supposed to be developed from the former by the law of natural selection."[43])

Worse still: not only were the unprogressive forms seen to coexist with the progressive ones: there were even cases, one alleged, where the latter had disappeared, while the former remained. One did not assert that this was the general rule — which would have been to embrace the deterioration theory — but that the occurrence of even

[40]) *Origin*, 1 ed. r., 414.

[41]) *Annals and Magazine of Natural History*, 5, 1860, 141. See also: 6, 1860, 222, 7, 1861, 403, *Edinburgh New Philosophical Journal*, 11, 1860, 287, *Edinburgh Review*, 111, 1860, 512, *Good Words*, 3, 1862, 6, *Blackwood's* 92, 1862, 90, *Parthenon*, 1863, 174, *British and Foreign Evangelical Review*, 15, 1866, 365, *Guardian*, 1871, 977.

[42]) *Edinburgh Review*, 111, 1860, 515.

[43]) *Annals and Magazine of Natural History*, 11, 1863, 427. See also: *Zoologist*, 18, 1861, 7597.

one such case was incompatible with Natural Selection. "The grandest and strongest types of animal life have become extinct, while dwindled specimens of the same groups survive amongst us,"[44] wrote one early reviewer.

To a large extent this argument was due to a misunderstanding of the principle of Natural Selection. One assumed that it meant the selection of the strongest, or biggest, or most highly organized animals or plants. In short, one replaced "survival of the fittest", which meant only the survival of those races which, for whatever reason, were able to produce most offspring able to survive to a mature age in a certain environment, by "survival of the best", which meant the survival of those forms which, from some arbitrarily chosen, often anthropomorphic point of view, were considered worthy of such an epithet. Such a misunderstanding, implying a confusion of values with facts, was quite common,[45] and its popularity went on increasing: the whole of the movement of Social Darwinism may be said to be based on it.[46] It certainly prompted, for instance, the writer who asserted that "Selection implies choice, the choice of the better," and asked how a "blind agency" could know what is better.[47] This point of view was put quite clearly by James Martineau in a *Contemporary* article: "Even the prevalence of the better (or 'fitter to live') it would not account [for] except on the assumption that whatever is *better* is *stronger* too; and a universe in which this rule holds already indicates its divine constitution."[48] It is obvious that Martineau read a moral sense into "fitter to live" — apparently "more worthy to live" — since he so unhesitatingly identified it with "better". Further, he apparently took "stronger" to be synonymous with the Darwinian "fitter to live." In fact, of course, Darwin's theory said nothing about "better" in Martineau's sense; Darwin's terms moved wholly in the naturalistic sphere. Herbert Spencer replied to Martineau: "Under its rigorously-scientific form, the doctrine is expressible in purely physical

---

[44] *John Bull*, 1859, Dec. 24, 827. See also: *Edinburgh New Philosophical Journal*, 11, 1860, 288.

[45] See in addition to those quoted separately: *Westminster Review*, 37, 1870, 540, *Nature*, 4, 1871, 161 (letter).

[46] See R. Hofstadter, *Social Darwinism*.

[47] *British and Foreign Evangelical Review*, 19, 1870, 111.

[48] *Contemporary Review*, 19, 1871—2, 619.

terms, which neither imply competition nor imply better and worse."[49])
Darwin himself would have agreed with part of this assertion; Spencer's
denial of the necessity of competition, however, was peculiarly his own.
Spencer had formulated a law of evolution according to which everything must develop from incoherent homogeneity to coherent heterogeneity — a "law" which to some extent came into the "predetermined
evolution" class, though Spencer substituted the Unknowable for God
— and regarded Darwin's Malthusian Natural Selection theory as a
partial explanation only.[50])

One ramification of this argument was that on the Darwinian theory
one should have expected further evolution than had actually occurred
in many forms. Why, indeed, have not other animals developed as
far as man? Darwin, one said, gives no reason to "account for the
fact that a similar brain-growth does not appear to have taken place
in any other of the numerous animal species of the earth, though so
many have, for countless ages, exercised intelligence and cleverness in
their respective struggle for existence."[51])

The Darwinians defended their view by emphasizing that theirs
was not a theory of necessary or pre-determined evolution. The improvement effected by Natural Selection had to be seen in relation to the
environment and conditions of each species. And the environment
was by no means to be equated with such external conditions as climate
and soil conditions. The struggle for life was as much, or more, against
other species living in the same physical environment. Thus in a sense
a lion may be called an "improved" cat — but since the two species live
on different sorts of prey, a country may have both lions and cats.
Above all, it would be no advantage for the cats to vary in the direction
of the lions, since they would then come into competition with them,
and naturally at first be at a disadvantage. To those critics who asked
why only the giraffe had developed such a long neck, when several
other grazing animals would have derived benefits from a similar
modification, Darwin replied by pointing to the look of the trees in
English parks, whose lower branches were never allowed to reach below
a certain level, because cattle nibbled off all leaves and shoots within

---

[49]) *Contemporary Review*, 20, 1872, 147.

[50]) Spencer, *First Principles*, II, Ch. 14, 17, 18, 24.

[51]) *Contemporary Review*, 17, 1871, 280. See also *Edinburgh Review*, 134, 1871,
212.

their reach. The level was determined by the reach of the larger cattle: sheep, for instance, would derive no advantage from getting slightly longer necks in such an environment. The possibility of development depended on the niche in the economy of nature not being already occupied by another species.[52])

Independent Homologies

One of the more recondite arguments against the Natural Selection theory was that it was unable to account for the development of very similar structures independently of each other in quite distinct branches of the organic world. The chance of a long series of small, purely accidental variations taking the same course more than once would seem to be infinitesimally small. Yet it had apparently happened: the most common instance advanced was the closely similar eyes in mollusks like cuttlefish on one hand, and vertebrates on the other. The argument was urged very strongly by Mivart in the *Genesis of Species*.[53 A]) But it first occurred, it seems, in an article in the *Saturday Review* in 1860: "Whence are derived the analogies that connect this descendant of the original mollusk with the offspring of the first radiate, the first articulate, and the first vertebrate animal? Do they not give distinct intimation of the presence of another law of structure, another *vera causa* regulating the forms of the animated world, besides ... natural selection?"[53 B]) Mivart was prepared to accept the Descent theory: what he wished to establish by his argument was that Natural Selection was not enough. Other writers used the same argument to support the separate creation view. "Where correlation of growth through genealogical affinity is out of the question, it is not unreasonable to suppose that the uniformity of modes, upon which the more important parts are formed, is the result of special creation working out a consistent proportion of the integral parts,"[54]) wrote one orthodox naturalist. Mivart's view, however, was the prevailing one among those who advanced this argument, which was fairly common in the better class

---

[52]) *Origin*, 6 ed. 153, 279.
[53 A]) Mivart, *Genesis* of *Species*, Chapter III.
[53 B]) *Saturday Review*, 1860, 574. Mivart?
[54]) *Zoologist*, 1861, 18, 7587.

periodicals, especially towards the end of the 'sixties and early 'seventies: Mivart devoted a whole chapter to it.[55])

Darwin's answer to this difficulty was, first, that the similarity of the structures had been exaggerated, and secondly, that similarity of external and internal conditions was likely to produce similar results.[56])

Natural Selection not a Causal Explanation

The arguments against the Natural Selection theory which we have dealt with so far were all intended to show that it was contrary to the facts: either the required variations were said not to occur, or natural selection was denied to be effective in accumulating them. But there existed also an objection of another nature, which concerned the logical status of the Natural Selection theory. Many of Darwin's critics maintained that, assuming that evolution had taken place, the Darwinian theory did not really explain it, since it did not indicate how the variations themselves arose. Darwin did not say what were the causes of the variations which were gradually giving rise to new forms of life.

Darwin himself repeatedly asserted that he knew nothing of the causes of the individual variations. Summarizing his chapter on the "Laws of Variation," he said, "Our ignorance of the laws of variation is profound. Not in one case out of a hundred can we pretend to assign any reason why this or that part differs, more or less, from the same part in the parents."[57]) But he naturally did not admit that ignorance in this respect implied that his explanation of evolution was incomplete. Knowledge of the immediate causes of the individual variations would not explain more fully why evolution had taken the course it did.

Previous theories of development had taken into account almost only the actual variations by which progress had been effected, seeking to find out by what conceivable mechanism such undoubtedly adaptive changes could take place. What Darwin did was to consider the adaptive variations in their wider context: the total variability of the organism.

---

[55]) See e. g. *Good Words*, 3, 1862, 7, 8, *Fortnightly Review*, 3, 1868, 625—8, *Month*, 11, 1869, 148, 153, *Nature*, 1, 1869—70, 106—7, *Gardener's Chronicle*, 1871, 649, *Popular Science Review*, 10, 1871, 188, *Dublin Review*, 16, 1871, 483—5, *Nature*, 3, 1870—2, 272.

[56]) *Origin*, 6 ed., 237.

[57]) *Origin*, 1 ed. r., 144.

Since that variability clearly included both favourable and unfavourable
— adaptive and non-adaptive — variations, the question of the direction
of the variations was separated from the question of the causes of the
variability. Whatever were these causes, the direction of evolution
was not determined chiefly by them, but by the conditions which made
one variation survive in greater numbers than others. As long as
naturalists had only kept in view the actual forms that were to be
found in the present and fossil organic series, and at most also the
forms that could reasonably be interpolated in the direct line of development between neighbouring forms, the question had to be, what are
the causes of the undoubtedly favourable variations making up this
progressive series? And the answer to that question had to explain
why the individual variations were adaptive. Now Darwin's point
was that if variability was random and indefinite, favourable variations must necessarily occur with a certain ascertainable frequency.
Therefore the question had to be, not why do these variations occur?
but why are the favourable variations, which are bound to occur,
preserved and accumulated in preference to others? To that question
Natural Selection provided an answer.

This distinction between explaining the variations themselves, and
explaining the course of evolution, was one of the revolutionary features
of Darwin's theory, and one which it took time to understand and
accept. It must even be admitted that Darwin himself, and other Darwinians of the time, did not bring out very clearly this fundamental
difference between the new and the old explanations of the evolution
of organic life. The term *random* variation was not employed by Darwin,
who preferred *indefinite*, which might mean both indefinite in direction
and indefinite in extent. Darwinians, like everybody else at the time,
were also extremely loth to use the terms *chance* or *accidental*. For
instance, Darwin's chapter on Laws of Variation begins: "I have
hitherto sometimes spoken as if the variations — so common and
multiform in organic beings under domestication, and in a lesser degree
in those in a state of nature — had been due to chance. This, of course,
is a wholly incorrect expression, but it serves to acknowledge plainly
our ignorance of the cause of each particular variation."[58]

Darwin's reluctance to employ words implying that any process
in the course of nature might be described as purely random was evident-

[58] *Origin*, 1 ed. r., 114.

ly bound up with the fact that he probably — and Huxley certainly[59]) — held a deterministic view of nature, according to which it would be possible, given the knowledge of the state of the universe at one stage, and of the laws of nature in that universe, to calculate any future state. Though one of course recognized that the feat was impossible in practice, one assumed as a working hypothesis that each event had a definite and at least in principle ascertainable cause.[60])

In spite of Darwin's frequent references to the "laws of variation," he expressly repudiated the view that there were any laws of necessary development.[61]) Indeed, the whole bearing of his argument was that the variations that occurred were both progressive and unprogressive, both adaptive and non-adaptive. But he never clearly brought out the fact that on his theory the non-adaptive variations must be assumed to be at least as frequent as the adaptive ones. Nor was the pro-Darwinian press any more explicit on this point.[62])

In the first edition of the *Origin* Darwin usually employed the neutral term variability; only in the later ones did he elaborate this expression, e. g. "*fluctuating*" variability.[63]) Likewise, it is only in the later editions that the following passage occurs, where the distinction is made between explaining the individual variations, and explaining the course of evolution: "When man is the selecting agent, we clearly see that the two elements of change are distinct; variability is in some manner excited, but it is the will of man which accumulates the variations in certain directions; and it is this latter agency which answers to the survival of the fittest under nature."[64])

Darwin never made any consistent attempt to bring out the randomness of the variations by actually recording and counting the variations in different directions and of different extent which occurred in a specific population. He did not prove experimentally, i. e. statistically, that the variations were really indefinite in direction in the sense of neutral from the point of view of adaptation. Moreover, he even admitted that they were probably definite in some instances: he was

---

[59]) See above, Chapter 6, p. 140.
[60]) *Origin*, 1 ed. r., 146.
[61]) *Origin*, 6 ed., 153.
[62]) See, however, *Gardener's Chronicle*, 1868, 184.
[63]) *Origin*, 6 ed., 204.
[64]) Origin, 6 ed., 167.

prepared to accept, and increasingly as time went on, that external conditions, and also use and disuse, had direct effects.

It is therefore small wonder that many readers of the *Origin* did not at first grasp the essential feature of his argument: the randomness of the variations which were the building stones of Evolution. And when Darwin wrote about our profound ignorance of the laws of variation,[65] his opponents naturally insisted that in such a case he had no right to exclude the possibility that the variations might be in fact directed by some Intelligence. They therefore thought themselves fully justified in asking the old question, what causes the adaptive variations to occur? instead of the new one to which Darwin provided an answer, what causes any variation that happens to be adaptive to be preserved?

As long as Darwin's critics looked at the problem in this way they were unable to see the point of the Natural Selection theory at all. If Darwin could not indicate the causes of the individual adaptive variations, what possible help could he get from bringing the non-adaptive variations into the picture? He was precisely as ignorant of *their* causes. In this manner, the critics could dismiss the Natural Selection theory altogether, and treat the Darwinian doctrine as no more than a description of the course which evolution perhaps had followed, and by no means as an explanation of why it had followed that course. As long as they refused to see that the Darwinian assumption of random and indefinite variation implied the necessary occurrence of at least some favourable variations, they remained convinced that he committed a logical fallacy by offering a description of gradual development as an explanation of that development. Darwinians, declared a theological critic, "have confounded progress in a series with growth from a germ . . . The argument here is, that though a perfect eye seems to indicate indubitably that it is the product of a designing mind, this conclusion is invalidated by the fact, that a *gradation* of eyes from the perfect to the imperfect is discoverable."[66] Such statements read as if their authors had not heard of the Natural Selection theory. Yet the following, for instance, occurred in a review of *Descent:* "The hypothesis of 'evolution' is to our minds a mere hypothesis of gradual accession and rise, but the addition of new power is not the

---

[65]) *Origin*, 1 ed. r., 144.
[66]) *Eclectic, Review* 8, 1865, 392.

less real because it is gradual: and it seems to us to be no causal explanation of the high intelligence of man to show that a much lower form of
intelligence is found in the animals from which his stock originally
diverged."[67])

This argument that Natural Selection did not explain evolution
because it did not explain the *origin* of the modifications was extremely
common. "Whence, then, these newer qualities and higher functions?
Clearly not from the predecessor, who did not possess them, nor from
the law, which is simply a mode of operation, but from the Lawgiver,
who ordained, and continues to sustain the method of development."[68])
A more popular[70]) way of expressing this was to assert that "We cannot
inherit more than our fathers *had*."[69]) But on the principle of Natural
Selection, of course, the whole point was that offspring sometimes did
inherit more than their parents had: and sometimes less.

Many critics attacked Darwin from another angle also. They seized
upon his statement that the variations were accidental. Forgetting
the implication that if so, favourable variations were bound to occur
at least sometimes, and disregarding Natural Selection, they were able
to represent the theory as absurd, and sacrilegious into the bargain. "If
Chance . . . really governs the universe, we must hold that she is capable
of performing a host of miracles every minute . . ."[71 A]) wrote a theological critic. This sort of argumentation, which of course bore directly
on the Design argument, was common enough, especially in the religious
press.[71 B]) The mere mention that the Darwinian theory depended on
Chance was generally thought effective enough to condemn it as

[67]) *Spectator*, 44, 1871, 289.

[68]) *Edinburgh New Philosophical Journal*, 19, 1864, 145, quot. Page.

[69]) *Edinburgh Review*, 134, 1871, 220. See also, similarly worded, *British and Foreign Evangelical Review*, 21, 1872, 24.

[70]) *Contemporary Review*, 14, 1870, 186. *British Quarterly Review*, 54, 1871, 479, *Spectator*, 37, 1864, 1413. *Christian Remembrancer*, 54, 1867, 147, *Tablet*, 1872, May 25, 648.

[71 A]) *Eclectic Review*, 3, 1860, 414.

[71 B]) *Freeman*, 1860, Jan. 18, 45, *Eclectic Review*, 3, 1860, 241—2, *Popular Science Review*, 2, 1862—3, 397, *Spectator*, 37, 1864, 1413, *Good Words*, 6, 1865, 302, *Times*, 1866, Aug. 24, p. 8, *Pall Mall Gazette*, 1866, Aug. 25, 605, *Contemporary Review*, 5, 1867, 78, *North British Review*, 46, 1867, 294, *Blackwood's Magazine*, 101, 1867, 687, *Guardian*, 1867, Apr. 24, 455 *Good Words*, 9, 1868, 250, *Eclectic Review*, 15, 1868, 478, *Edinburgh Review*, 128, 1868, 436, *Edinburgh Review*, 134, 1871, 207, *Month*, 14, 1871, 529.

scientifically illegitimate.[72]) Sometimes, however, critics found it convenient to emphasize the inadmissibility by an *a priori* argument from the supposedly known principles of nature: "It cannot surely be the method of nature to give out blindly, as it were, from time to time, all possible varieties without any law of successive or progressive development."[73])

Pro-Darwinian writers often tried to defend Darwin by denying that his theory did include an element of chance. To support this interpretation they could indeed appeal to Darwin himself. But like Darwin, they were apt to obscure the difference between the Natural Selection theory and its predecessors by using this sort of argument: and it appears likely that some of Darwin's defenders were not themselves quite clear about the difference. "What are *accidental* varieties — in what sense can we see the word accidental, but in that of belonging to some unknown law?"[74]) asked one writer. Another commentator, even more ambiguously, declared that the word *fortuitous* was taken by Darwin, "not in the gross sense of blind chance, but as denoting the intervention of agents beyond our powers to generalise or to calculate beforehand."[75]) The religious implication could be made clearer still: "How little Mr. Darwin supposes [the power behind evolution] to be acting blindly or by chance is shown when he tells us that he sees no good reason why his views 'should shock the religious feelings of any one'."[76])

It is of course true in a sense that Darwin, when he said that the variations were accidental, only meant that they were due to unknown

---

[72]) See e. g. *Christian Observer*, 1860, 571, *Geologist*, 1860, 471, *Eclectic Review*, 3, 1860, 224—230, *Annals and Magazine of Natural History*, 5, 1860, 137, *Edinburgh New Philosophical Journal*, 11, 1860, 288, *Recreative Science*, 2, 1860—1, 274, *Press*, 1860, 1207, *Athenaeum*, 1861: 1, 536, *Edinburgh New Philosophical Journal*, 15, 1862, 255, *British Quarterly Review*, 38, 1863, 495, *Eclectic Review*, 6, 1864, 202, *London Review*, 9, 1864, 462, *Christian Observer*, 1866, 9, *Guardian*, 1868, Mar. 11, 303, *Eclectic Review*, 14, 1868, 349, *Quarterly Review*, 127, 1869, 165, *Christian Observer*, 1871, 889, *John Bull*, 1871, Aug. 12, 550, *Nonconformist*, 1871, Aug. 9, 785, *Tablet*, 1871, Feb. 25, 232, *Freeman*, 1871, Apr. 28, 199, *London Quarterly Review*, 36, 1871, 275, *Record*, 1871, Aug. 9 p. 2.

[73]) *Blackwood's Magazine*, 101, 1867, 686—7. See also *British Quarterly Review*, 45, 1867, 573, *Times*, 1867, Jan. 31, p. 5, and above, Chapter 6.

[74]) *Macmillan's Magazine*, 4, 1861, 242.

[75]) *Saturday Review*, 31, 1871, 180.

[76]) *Inquirer*, 1868, Aug. 29, 549. See also: *Saturday Review*, 18, 1864, 605, *Manchester Guardian*, 1866, Aug. 28, p. 4.

causes. But he did not mean this in the sense conveyed by many of his expositors: in particular, he refused to hypostatize the unknown causes into a mysterious tendency or a Divine plan. His theory was, as one commentator aptly put it, "a remarkable attempt to reduce the doctrine of what used to be called Chance to an orderly philosophical system."[77])

In their demand for a full statement of the causes of the individual variations as a preliminary to the explanation of Evolution, Darwin's opponents often appealed to the idealistic tradition in the philosophy of science, according to which no explanation of an event was to be considered complete until incorporated into a chain of causes, ending only in the First Cause, or God.[78]) As Darwin could not find any natural cause of the individual variations between parents and offspring, was it not possible that the cause was a supernatural one? And if this was really the case, why assume that there was any Natural Selection at all?[79 A])

It may be objected that this argument was theological and not scientific. But the difference was not always clear. Anyhow, the demand that Darwin should show up the causes of the individual variations did not only come from his non-scientific critics, but also from some eminent scientists. And instance in point was W. B. Carpenter, president of the British Association in 1872: "I cannot but regret that an undue importance ... should have been attached to the doctrine of 'Natural Selection' as a *vera causa*. For Natural Selection, or the 'survival of the fittest,' can do nothing else than perpetuate, among varietal forms already existing, those which best suit the external conditions of their existence; and the scientific question for the Biologist is, — what is the Cause of departure from the uniformity of type ordinarily transmitted by Inheritance, whereby these varieties come into being; and under what conditions does that Cause operate?"[79 B]) The implication was that Darwin did not indicate the *true* cause: the true cause was still wrapt in mystery, and would continue to be so until the complete causal chain had been laid bare. Carpenter made it sufficiently clear that he believed the variations entered as elements in a Divine Plan.[79 C])

---

[77]) *Nonconformist*, 1866, Feb. 21 156, quot. Acland.

[78]) Above, Chapter 9, p. 177 ff.

[79 A]) See above, Chapter 6, p. 132 ff.

[79 B]) *Contemporary Review*, 20, 1872, 742. See also *Guardian*, 1872, 1097.

[79 C]) See also above, Chapter 4, p. 90—92.

There is reason to ask whether Darwin's critics would have been satisfied if he had in fact been able to show what caused the individual variations. In accordance with the view that scientific explanation consisted in laying bare as much as possible of the chain of causation leading up to the First Cause, they would in all probability have gone on to ask, what caused the causes of the individual variations? and so on, *ad infinitum*. Thus when Huxley once used the analogy of a particle of rust on the pinion of a watch causing the watch to alter its rhythm, or even to stop, an anti-Darwinian writer commented: "Mr Huxley can scarcely mean to imply that the infinite divergencies between the human and the brute species may have originated in something as fortuitous and insignificant as a little rust on the pinion of a watch. Yet if he does entertain that opinion, it helps him not, for the rust on the pinion of a watch must have its cause also."[80])

Darwin's critics apparently did not consider the possibility that the same sort of causes could be reasonably assumed to lead both to unfavourable variations and to favourable ones. When discussing evolution, it was the favourable variations only that they kept in view; it was the causes of these that they demanded to know. "Natural Selection only destroys monstrosities, removes antecedent species, and favours useful variations, 'though it is impotent to originate them,'"[81]) is a statement which reveals the attitude. Natural Selection is clearly impotent to originate useful variations: but it is equally impotent to originate the monstrosities: and if one sort of variations were due to miraculous interference, then there was no valid reason for not assuming a similar cause for the other sort. Yet this inference was not drawn, for then God would have been represented as creating indiscriminately good, bad, and indifferent varieties, leaving it to the struggle for life to decide which were to survive and leave offspring.[82])

In short, the anti-Darwinian argument consisted in disregarding the non-adaptive variations, bringing out the adaptive ones, and asking, what causes the variations to be adaptive? The argument was not, however, put in this question-begging form. Critics evidently believed that there were valid reasons for setting the favourable variations apart from the others. In order to do this they tried to establish, as

---

[80]) *Edinburgh Review*, 117, 1863, 567.

[81]) *Month*, 14, 1871, 529.

[82]) See above, Chapter 6, p. 136 ff.

we pointed out in Chapter 6, that two kinds of variation could be discerned. There was "ordinary" variation, which might perhaps be random, but which, on the other hand, never passed beyond specific frontiers. But there was also, one maintained, another sort of variation, of a fundamentally different nature. These variations were much rarer, and they were also much more spectacular, and might lead to specific changes. These extraordinary variations were directed, designed. Thus one tried to turn the evidence of the big mutations — which undoubtedly sometimes occurred — against Darwin's theory.

Darwin himself recognized the existence of sudden big mutations,[83] but he never looked upon them as of any importance in the process of Natural Selection, for which he definitely maintained the maxim of continuity.[84] This point, however, was not really important in the Darwinian doctrine, and Huxley in fact declared that "Mr. Darwin's position might, we think, have been stronger than it is, if he had not embarrassed himself with the aphorism, Natura non facit saltus."[85] What was important was not the extent of the variation, but the principle of randomness. The number of adaptive variations, whether big or small, should not be greater than was to be expected on the assumption that the variations were indefinite in direction. On this principle it was evident that the bigger a mutation was, the smaller would be its likelihood of being favourable.

Here was the essential difference between the Darwinians and their opponents on this point: the Darwinians denied that the big mutations, any more than the smaller ones, were adaptive. As Huxley said in the review from which we have just quoted: "Doubtless there were determining causes for these as for all other phenomena, but they do not appear, and we can be tolerably certain that what are ordinarily understood as changes in physical conditions, as in climate, in food, or the like, did not take place and had nothing to do with the matter."[86] It is possible that Huxley implied that adaptation to conditions did not occur with more frequency than the randomness hypothesis allowed. But it must be admitted that the problem was never stated precisely in those terms during the period we are considering. One was content

[83] *Origin*, 1 ed. r., 25.
[84] *Origin*, 1 ed. r., 166.
[85] *Westminster Review*, 17, 1860, 569.
[86] *Westminster Review*, 17, 1860, 548.

with vaguer claims. Darwin, for instance, when he considered these matters in the later editions of the *Origin*, argued that the big variations which had been actually observed were *in general* far from advantageous: "Several may be attributed to reversion ... A still greater number must be called monstrosities."[87] Moreover, those few favourable big mutations that occurred would hardly have any chance of being selected and preserved, since, as Darwin said, "they would be liable ... to be lost by accidental causes of destruction and by subsequent inter-crossing."[88]

Anti-Darwinians did not at all accept the logic of this argument. They insisted that if Darwin and Huxley could not say what were the causes of the big mutations, then what right had they to exclude the possibility that the cause was an adaptive principle? The favourable mutations might be rare — but so are most miracles. And the objection that they would be diluted by intercrossing only affected the Natural Selection theory: it only established that if there had been evolution, it had not been due to natural causes, which was precisely what one sought to prove. Therefore Darwin's opponents often alleged that the mere existence of big variations was enough to refute the Darwinian theory, or at least to prove that it was incomplete. Thus Owen implied that instances of complete specific mutations had come under observation: "Has not Cuvier, in a score or more of instances, placed the parent in one class, and the fruitful offspring in another class of animals? Are the entire series of parthenogenetic phenomena to be of no account?"[89] Other writers, without asserting that the observed mutations were in themselves specific changes, maintained that they might at least "throw light upon specific origination."[90] The argument was further illustrated by Mivart in his *Genesis of Species;*[91] it became quite common

---

[87]) *Origin*, 6 ed., 314.

[88]) *Origin*, 6 ed., 314.

[89]) *Edinburgh Review*, 111, 1860, 502.

[90]) *Popular Science Review*, 10, 1871, 188. See also: *Edinburgh Review*, 128, 1868, 431.

[91]) Mivart, *Genesis of Species*, 115, quoting argument from Owen, *Anatomy of the Vertebrates*, III, 795.

[92]) *Annals and Magazine of Natural History*, 6, 1860, 152, 223; 12, 1863, 89 (ref. Candolle), *Quarterly Review*, 131, 1871, 54 (Mivart), *Month*, 14, 1871, 529, and several reviews of Mivart's *Genesis*.

in the late 'sixties and early 'seventies.[92]). The attitude prevalent at the time is well brought out in the following declaration by Mivart: "Mr. Darwin concedes *in principle* the very point in dispute, and yields all for which his opponents need argue, when he allows that beautiful and harmonious variations may occur *spontaneously* and *at once*, as in the dark or spangled bars on the feathers of Hamburgh fowls."[93]) Mivart evidently did not consider the question of the probability of such an event on the randomness hypothesis: the very occurrence of such a remarkable mutation needed, it seemed to him, a proper causal explanation; and he did not leave his readers in doubt as to the nature of that explanation.

The Darwinian answer to this argumentation was to appeal to empiricist principles. The *Examiner*, reviewing Mivart's book, first of all criticised Mivart for alleging an inherent tendency as an explanation. "This," it said, "is exactly on all-fours with the celebrated explanation of the power that opium has of producing sleep, — namely, that it is a drug possessing a soporific virtue. It is a mere statement of the fact to be explained, twisted, by the use of a metaphysical jargon, into the form of an explanation." Mivart's explanation, the reviewer went on, was acceptable under one condition only, and one which observation proved did not prevail: "The only case in which Mr. Mivart's position would be worth the trouble of defending, would be if there were no abortive or feeble attempts at evolution."[94]) Darwin himself, as we have seen, held that to argue as his opponents did in this matter was to "enter into the realms of miracle, and leave those of Science."[95]) Indeed, all his critics made it sufficiently clear that they demanded an explanation "beyond the ken of the naturalist,"[96]) as one of them put it.

Alternatives to Natural Selection

Most of the criticism directed against Natural Selection came from writers who were prepared to accept the Descent theory. This is easily explained: the transmutation doctrine had been practically granted by most informed people towards the end of the 'sixties, and it was chiefly

---

[93]) *Quarterly Review*, 131, 1871, 54 (Mivart).
[94]) *Examiner*, 1871, Feb. 4, 126.
[95]) *Origin*, 6, ed., 318.
[96]) *Edinburgh Review*, 134, 1871, 227.

in the better-class periodicals that the Natural Selection theory was discussed at all.[97 A])

Now if Natural Selection was denied while transmutation was accepted, it was necessary to suggest some alternative to Darwin's theory. Our illustrations above will have shown that what the critics chiefly objected to was the randomness element in Darwin's theory. The alternatives suggested therefore practically always implied some sort of directed or pre-determined evolution. One might, in the first place, emphasize the internal forces of the organism, or the fixed laws of development, which were said to push the forms inevitably along the road of progressive change. Secondly, it was possible to lay stress on the direct action of external circumstances on the organism. Thirdly, evolution might be ascribed to the action of some Vital Force.

These lines of explanation did not exclude one another. The external conditions, for instance, necessarily had to interact with the internal capacity of the organism, and the Vital Force was naturally one of the internal forces. Common to all the explanations was their *ad hoc* nature: the laws of internal development were derived from the actual development of each organism; the way external conditions operated had to be inferred from their supposed effect in each particular case; and the nature of the Vital Force had to be inferred from the changes it was supposed to produce. Thus the mysteriousness of the thing to be explained — the progressive and adaptive evolution — was carried over undiluted to these explanatory concepts. The need for a principle of Design was therefore felt quite as strongly for these concepts as for the phenomena they were intended to explain. It was of course this circumstance which made them acceptable to the theologically minded anti-Darwinians.[97 B])

The explanations were by no means new: previous evolutionists, from Demaillet and Erasmus Darwin to Lamarck and the author of the *Vestiges*, had all incorporated one or more of them in their theories. What was new — at least in England — was that these arguments were now taken up by defenders of the religious interpretation of nature. Before Darwin nearly all scientists, as well as theologians, had treated them with contempt. Now many theologians, as well as some scientists, changed front, adopting some of the arguments of their former opponents

---

[97 A]) See above, Chapter 2, p. 24, and App. I, p. 346—7.

[97 B]) See above, Chapter 6, especially p. 132 ff.

in order to combat the theologically even more obnoxious theory of Natural Selection.

What had roused the theological opposition against Lamarck — the best known of the pre-Darwinian evolutionists — was not only his unorthodox theory of Creation, but above all his materialism. Lamarck sought to explain Evolution by means of tendencies inherent in matter itself, without the interposition of a watchful Providence. As to the rejection of supernatural interpositions, Darwin and Lamarck could join hands. But they differed fundamentally on the nature of the explanation that was offered for evolution. Lamarck's inherent tendencies were no more than names for the thing to be explained, while Darwin's Natural Selection placed the facts of development in a wider context of other observed facts and regularities. It was because he offered a truly scientific explanation that Darwin made the Descent theory acceptable in the scientific world, where it had been repudiated as long as no better explanation than Lamarck's was available.

But when theologians and theologically minded scientists accepted the Descent theory, it was obviously not because they accepted Darwin's explanation of it. They did not recognize the validity of the explanation. The remarkable thing is that they substituted for it precisely those Lamarckian and other explanations which they had previously found inadequate. It is hardly possible to avoid the conclusion that their acceptance of the Descent theory was a concession to the prevalent informed opinion of the age, and that their previous opposition to the Lamarckian explanation of Evolution had been due less to its scientific deficiencies than to its religious implications.

Lamarck's name naturally remained suspect in the eyes of the religious. Though it was possible to use his explanations in support of the principle of Design, Lamarck himself had failed to do so. It was obviously desirable to refute Lamarck's materialistic metaphysics, though the quasi-scientific explanatory concepts he had suggested were used to the best advantage. And it was not difficult to show that these concepts really did not explain anything without the assumption of a guiding Providence.

Many of Darwin's critics did not see, or did not wish to point out, the difference between his theory and Lamarck's. In the press Darwin's name was often, especially during the first years, linked with Lamarck's, in order that his views might stand out as old and long since refuted ones.

We therefore meet with such expressions as "the Lamarckian theory of development, and . . . that most recent modification of it which Mr. Darwin has put forth."[98]) Samuel Haughton declared that the "Law of Natural Selection forms the only *bona fide* addition made by Darwin to Lamarck's famous theory of Progression," but immediately withdrew even this admission by insisting that Natural Selection was "implicitly involved"[99]) in Lamarck's doctrine. Professor Owen read a paper at the British Association, entitled, "The Aye-Aye as a Test of the Lamarckian and Darwinian Hypothesis."[100]) Religious organs declared that in the *Origin* "the old and, as it was once thought, exploded theory of Lamarck is substantially revived."[101]) As late as 1869 a speaker at the British Association stated that "the theory of Lamarck has not been proved by palaeontology or the elaborate researches of Darwin."[102])

Several critics obviously did not realise that Darwin's explanation of evolution differed from those provided by the pre-Darwinian evolutionists. They therefore used the same arguments against both, attacking the supposed materialism, and insisting that the explanations that were offered needed the addition of a principle of Design. "It is essential to Mr. Darwin's theory that he should have a force like that which can animalize vegetable matter, and which is further endowed with an inherent and inexplicable power of transmutation into higher structures,"[103]) wrote one orthodox naturalist. More instances of this kind of reasoning were given above, Chapter 6.

Most of the writers we are now concerned with, however, recognized the difference between Darwin and his predecessors, and expressly contrasted their own explanations with Natural Selection. The alternative most frequently advanced seems to have been the one postulating a special law of development. This view, for instance, was advocated by Professor Owen. In his *Edinburgh* review of the *Origin* his own theory was only vaguely foreshadowed: he did not yet wish to commit himself on the subject of transmutation: "It may be inferred: — that the primordial as well as all other forms of organic beings,

---

[98]) *Leader*, 1860, 378 and *Athenaeum*, 1863:1 221.

[99]) *Annals and Magazine of Natural History*, 11, 1863, 423.

[100]) See above, Chapter 4, p. 71.

[101]) *Good Words*, 1865, 303.

[102]) *Daily Telegraph*, 1869, Aug. 25, p. 3, report of British Association meeting.

[103]) *Recreative Science*, 3, 1861—2, 223—4.

originate, and have ever originated, from the operation of secondary
and continuously operating creative laws; and that the various grades
of organisms now in being, from the microscopic monad upwards,
indicate the various periods in time at which the first step of the series
they respectively terminate began. The monad that by 'natural selection' has ultimately become man, dates from the farthest point in the
remote past ... the monad which by its vibratile cilia darted across
the field of the microscope we were looking through this morning, is
the result of the collocation of particles which ... took place under
the operation of the heterogeneous organising force yesterday."[104])
It appears that Owen was prepared to consider the possibility of Evolution having taken place within each specific line of descent, each separate
species thus having developed, according to a fixed and predetermined
order, from a monadic germ. The view was elaborated more fully
in his *Anatomy of the Vertebrates*, especially in its third volume, in
which he gave full adherence to the doctrine of Evolution, while stating
more fully the difference between his own view, which he called Derivation, and Darwin's Natural Selection: "'Derivation' holds that every
species changes, in time, by virtue of inherent tendencies thereto"
"'Derivation' sees among the effects of the innate tendency to change
... a manifestation of creative power in the variety and beauty of the
results, and in the ultimate forthcoming of a being susceptible of
appreciating such beauty, evidence of the preordaining of such relation
of power to the appreciation... 'Natural selection' leaves the subsequent
origin and succession of species to the fortuitous concurrence of outward
conditions." "'Derivation' recognizes a purpose in the defined and
preordained course."[105])

In the traditionalist — though not in the extreme religious — press
Owen's ideas were naturally welcomed[106]) and an equally favourable
reception was extended to Mivart's *Genesis of Species*, which was written
with the express object of presenting a theory of Evolution to supersede
the Darwinian one. Mivart, like Owen, laid stress on an "internal law,"
whereby the organism was steered along a pre-determined course under

104) *Edinburgh Review*, 111, 1860, 514.

105) Owen, *Anatomy of the Vertebrates*, III, 808—9. Quoted in *Examiner*, 1868,
Dec. 26 821.

106) E. g. *London Society*, 13, 1868, 332, *Athenaeum*, 1866, Apr. 28, 560 1869,
Feb. 13, 244, *London Quarterly*, 36, 1871, 280.

the influence of external conditions: "The theory propounded in this work allows . . . a greater and more important share to external influences, it being believed by the author that these external influences equally with the internal ones are the results of one harmonious action underlying the whole of nature, organic and inorganic, cosmical, physical, chemical, terrestrial, vital, and social. According to this view, an internal law controls the actions of every part of every individual, and of every organism as a unit, and of the entire organic world as a whole."[107]

Various suggestions were made in order to elaborate the idea of a developmental law. The majority no doubt simply spoke of the necessity of assuming Design,[108] and many spoke vaguely about "progressive change" as a "principle inherent in animated nature"[109] and the like. Others attempted at least quasi-scientific explanations. One naturalist wrote a book on "The Origin of Species by Organic Affinity"[110] in which the gradual formation of specific characters was compared with the formation of chemical compounds according to laws of chemical affinity.[111] The most effective form of the argument, however, was to refer to the factual evidence of regulated changes in the organic world outside the sphere of evolution proper. There were, for instance, the phenomena of correlation of growth, to which Darwin himself drew attention. "Darwin . . . regards correlation of growth as the origin of what is commonly called 'type' in some instances; but if in some, why not in all?" wrote one naturalist. "If this be so, it implies the impression upon the primordial germ of the power of evolution under certain conditions of existence in certain directions, and in no other,; in other words, that germ must have been endowed with the potentialities of all the variations through which its progeny could pass; which, however, involves the notion of a plan in creation."[112] Mivart expounded this idea with much force: "It is, we think, evident from the facts of homology . . . that some innate and substantial cause exists in each organism, which may at the same time account for both specific resem-

---

[107] Mivart, *Genesis of Species*, 274, *Contemporary Review*, 19, 1871—2, 170, *Dublin Review*, 16, 1871, 486, 17, 1871, 15, *Record*, 1872, Feb. 28, p. 4

[108] See above, Chapter 6.

[109] *Edinburgh New Philosophical Journal*, 15, 1862, 255.

[110] H. Freke, *On the Origin of Species through Organic Affinity*.

[111] *Zoologist*, 18, 1861, 7579.

[112] *Annals and Magazine of Natural History*, 7, 1861, 400.

blance and specific divergence. In obedience to the law of parsimony it is more desirable to make use of one such conception than to imagine a number of, to all appearance, separate and independent 'laws of correlation' between different parts of each animal... If... it is still necessary to conceive a substantial form moulding each organic being... then the claim of 'Natural Selection' to explain by itself the evolution of each animal form, or the 'origin of species,' must fall to the ground."[113])

It was almost inevitable that the idea of a law of development should be illustrated by means of the development of an ovum into a full-grown individual. Agassiz had no hesitation about the analogy. "It would be easy to invent other theories that might account for the diversity of species quite as well [as Darwin's] . . . It might be assumed, for instance, that any one primary being contained the possibilities of all those that have followed, in the same manner as the egg of any animal possesses all the elements of the full-grown individual; but this would only remove the difficulty one step further back . . . Since the knowledge that we now have, that similar metamorphoses go on in the eggs of all living beings, has not yet put us on the track of the forces by which the changes they undergo are brought about, it is not likely that by mere guesses we shall arrive at any satisfactory explanation of the very origin of the beings themselves."[114]) Agassiz' argument shows clearly that he demanded of an explanation that it should indicate the causes of the changes: what internal or external phenomena led from one stage of development to another stage was the question he asked both for the individual and for the species. Darwin by-passed that question for the species, since whatever were the causes, the explanation of the direction was provided by Natural Selection. A Darwinian would therefore rather explain the development of the individual by reference to that of the species, than vice versa. Traditionalists of course preferred to have the analogy the other way round; it would then guarantee the mysteriousness of both developments. The illustration occurs again and again: "We are convinced . . . that the whole organic world arises and goes forward in an harmonious development similar to that which displays itself in the growth and action of each separate organism."[115])

[113]) *Month*, 11, 1869, 153.

[114]) *Annals and Magazine of Natural History* 6, 1860, 227.

[115]) *Month*, 11, 1869, 41. See also *Edinburgh New Philosophical Journal*, 15,

Very often the propounders of the theory of pre-determined evolution were willing to grant the efficacy of Natural Selection to a considerable extent: they only denied that it was by itself sufficient to explain Evolution. We therefore frequently meet with such expressions as "a higher law"[116] "a deeper-seated and innate principle,"[117] and the like,[118] suggested as supplements to the Natural Selection theory. The terminology well conveys both the conceptual realism and the theological implications of the anti-Darwinian position on this point.

The second method of giving substance and scientific form to the idea of a pre-determined evolution was to search for the cause of the variations in the external instead of the internal conditions of the organism. The direct adaptive influence of external conditions was a factor that Darwin himself considered seriously, and to which he was inclined to allow some weight, and increasingly as time went on. It was an ascertainable fact that many species of animals, for instance, when transferred to a colder climate, developed thicker furs, and Darwin believed that such a modification, though acquired late in life, might be subsequently transmitted to offspring. In the same manner, Darwin believed that the effects of use or disuse of organs were inheritable, and he considered this to be an important subsidiary factor in evolution. "It appears," he said in the 6th edition of the *Origin*, "that I formerly underrated the frequency and value of these latter forms of variations, as leading to permanent modifications of structure independently of natural selection."[119]

Darwin's admission of the direct influence of external conditions certainly contributed — like his references to "laws impressed on matter by the Creator"[120] — to blur the outlines of his Natural Selection theory, and thus to the misunderstandings that it gave rise to. Traditionalists were inclined to equate Natural Selection with the direct influence of external conditions, and thereby to miss the essential feature of the new theory, namely, the indefinite and random variability.

1862, 255, *Geological Magazine*, 1865, 309, (ref. La Fée) *Church Review*, 1867, Feb. 9, 135, *Contemporary Review*, 19, 1871—2, 607.

[116] *Westminster Review*, 25, 1864, 586, quoting Page.

[117] *Annals and Magazine of Natural History*, 12, 1863, 92, quot. Falconer.

[118] E. g. *Quarterly Journal of Science*, 3, 1866, 157, *Christian Observer*, 1868, 329, *Dublin Review*, 16, 1871, 486.

[119] *Origin*, 6 ed., 657.

[120] See above, Chapter 6, p. 128, 132; 7, p. 150 ff.

The following, for instance, was certainly intended as a fair exposition of the Natural Selection theory: "The theory does not suppose, like some which have preceded it, that species and genera of animals pass, in the course of successive generations, into other forms, but that certain individuals of a race, being exposed to circumstances different to the general mass, will in consequence become somewhat altered."[121]) No reader of this passage could be blamed for concluding that the Darwinian theory assumed that adaptive modifications were called forth by the external conditions. That the writer interpreted it so himself is probable; he summarized the theory in these words: "The peculiarities of every living being of necessity result from the action of certain modified external circumstances without the intervention of a superintending Providence."[122]) Other traditionalist writers were less convinced that such an interpretation was really the one intended by Darwin: "If the circumstances change, then some other corresponding modification of the organism is produced and a new variety is obtained . . . There is correlation in a very important sense, and every change acts upon every individual . . . This we understand to be the meaning and bearing of Mr. Darwin's law of natural selection, and in this sense we understand and accept his argument."[123]) On such a view there seemed indeed to be little difference between Darwin and Lamarck. Lamarck had laid down the rule, said one violently anti-Darwinian naturalist, that "Organic Forms, acquired under the presiding influence of external circumstances, are transmitted by Generation. This law involves the famous Law of Natural Selection, attributed within the last few months to Mr. Darwin."[124])

This sort of confusion between the changes wrought on a race, and on the individual is one which we have met with before.[125]) The misunderstanding as to the influence of external conditions was perhaps inevitable, for it depended not only on Lamarck's teaching, but also on what seemed to be the ordinary experience of breeders and gardeners. A clear-sighted expositor of Darwin's theory in the *Gardener's Chronicle*

---

[121]) *Edinburgh New Philosophical Journal*, 12, 1860, 234. See also: *Popular Science Review*, 1, 1861, 116.

[122]) *Edinburgh New Philosophical Journal*, 12, 1860, 239.

[123]) *Dublin University Magazine*, 56, 1860, 33.

[124]) *Annals and Magazine of Natural History*, 11, 1863, 423 (Haughton)

[125]) Above, Chapter 11, p. 239.

pointed out that the gardener was ordinarily said to make his plants vary, "and this looseness of expression betrays the inexactness of the prevalent ideas on the subject. The gardener does not necessarily or even ordinarily make the plant vary; very few useful varieties have owed their origin to design; the appearance of the variety is, in most cases at any rate, accidental."[126]

Some writers wished to make external circumstances responsible for the appearance of quite complicated organs: they were even credited with the power of creating living out of dead matter. It is a remarkable fact that many of the firmest believers in spontaneous generation in the 1860's and 1870's were to be found in the anti-Darwinian camp. Darwin himself refused to discuss the matter: "at present it seems to me beyond the confines of science,"[127] and Huxley forcibly declared that the thing had not been achieved.[128] But they naturally both refused to subscribe to the idea that it was impossible in principle. Believers in pre-determined evolution, on the other hand, were prepared to accept the inconclusive experiments that had so far been made.[129] Owen's ideas have already been referred to; the constantly anti-Darwinian *Athenaeum* gave currency to similar opinions: "The inevitable corollary of the Darwinian hypothesis seems to us to demonstrate its weakness; and to show by contrast the superiority of the Lamarckian principle of heterogenous production of the primitive types of organisms. Agreeably with this principle, we conceive that 'particles of apparently homogeneous jelly' are now, as of old, being aggregated through the operation of existing interchangeable modes of force."[130] This was a sort of spontaneous generation with which Darwin would have nothing to do, whereas anti-Darwinians wished to believe in it for the same reason as they sought to prove the occurrence of big mutations involving specific changes. The result of the external influence was to be a wonderfully complicated and well-adapted creation, which it would be impossible to dismiss as being due to chance.

While Darwin looked upon the direct action of external conditions as only a subsidiary influence, ultimately regulated by Natural Selec-

---

[126] *Gardener's Chronicle*, 1860, Jan. 21, 45.
[127] Darwin, *More Letters*, I, 273, To V. Carus, 1866.
[128] Above, Chapter 4, p. 86.
[129] *Nature* 2, 1870, 170–7, 193–201, 219–228 (H. Charlton Bastian), 473 (Huxley).
[130] *Athenaeum*, 1863: 1, 418.

tion, his opponents regarded it as an alternative explanation. Thus a biologist said at the British Association meeting in 1866 that modifications are "more likely to be the result of external influences operating upon successive generations, influencing their development, their growth and their maturity, than of 'natural selection' and 'struggle for existence'."[131]) But as long as it was not shown exactly what external conditions resulted in what modifications, such an assertion was empty of content, nor was it a scientific advance to eke it out by postulating a "latent power of adaptation"[132]) excited by the external conditions. Such a latent power was simply an *ex post facto* construction, and as such completely in line with the conceptual realism of the idealistic school, to which most of Darwin's opponents belonged. Darwin himself did not postulate a power of adaptation resident in the individual organism: the only power he postulated was that of producing *variable* offspring. The adaptation was a matter for future conditions to select: when the parent organism produced them they were no adaptations. The average offspring of any organism was no more adapted than the parent itself.

The third alternative to Natural Selection was to explain evolution as due to the action of a Vital Force, which enabled the organisms, consciously or unconsciously, to adapt themselves to changing conditions. The idea that animals could contribute to specific change by their own volition was part of the Lamarckian doctrine, and though Darwin would only admit it as far as it concerned the effect of use or disuse of organs, the general resemblance of his theory with Lamarck's permitted the popular press especially to direct against him the ridicule which had so often been the lot of his predecessor. One alleged that the theory presupposed that animals or plants could modify themselves by simply desiring to do so, or as a rhymester put it in *Blackwood's:*

> "Some, wishing to walk, manufactured a limb
> Some rigged out a fin with a purpose to swim
> Some opened an eye, some remained dark and dim
> Which nobody can deny."[133])

On this point, however, misunderstandings were less common than in the matter of the influence of external conditions, and even in periodicals which were by no means favourably disposed towards Darwin

---

[131]) *Morning Post*, 1866, Aug. 24, p. 2. reporting Humphry.
[132]) *Edinburgh Review*, 128, 1868, 444.
[133]) *Blackwood's Magazine*, 89, 1861, 615. See also: *Temple Bar*, 5, 1862, 215.

the difference between his theory and Lamarck's was recognized.[134])
There were some, however, who declared their preference for the Lamarckian view: "In Lamarck's [theory] an animal is incited to exertion by
some kind of external circumstances, and obtains an improved organization as the reward of his efforts; according to Mr. Darwin's theory, the
same advantage is gained, but (if we may use the expression) rather by
an accident of birth. We are disposed to think the less aristocratic
process somewhat the preferable one."[135])

The idea that the modification was due to the creature's own volition
could not be applied directly to the vegetable world, nor to the lower
animals. To solve the difficulty one sometimes generalised the volition
theory to comprise not only the conscious will of the individual, but
also the unconscious craving of the race, or even of Life as such. Thus
one reviewer of the *Origin* spoke of "what we might term a yearning on
the part of nature for a common pattern on which to construct the
several forms of life."[136]) The will of the individuals became the yearning
of the whole of nature, and could be assimilated with the action of a
Vital Force acting through the whole of the organic world. One naturalist
wrote: "I cannot but believe in the existence of an unconscious Organising Intelligence, an idea which Mr. J. J. Murphy has ably and logically
advocated in his 'Habit and Intelligence'. And if this inherent innate
power of change is admitted, it at once harmonises the tendency to
variation which exists in all created beings."[137]) From this point the
step was short to such ideas as Wallace's disembodied spirits directing
the evolution of Man.[138])

The unconscious urge providing the motive power of Evolution could
be conceived, as we said, as the manifestation of a Vital Force, an idea
which traditionalists regarded as "indispensable in biology", as Whewell
said. The existence of such an entity, distinguishing all living organisms
from dead matter, made it impossible for the spontaneous generation
theory to be true in the materialistic sense: vitality was something
different in kind from matter, and could not arise from any combination
of purely material molecules. Further, the entity could serve as the

[134]) *Popular Science Review*, 3, 1863—4, 372.
[135]) *Fraser's Magazine*, 61, 1860, 752.
[136]) *John Bull*, 1859, Dec. 24, 827.
[137]) *Nature*, 3, 1870, Nov. 10, 32 (Bennett).
[138]) *Saturday Review*, 29, 1870, 710.

substantive basis of those internal tendencies or powers by which evolution was often explained. It was often referred to in press discussions of Darwin's theory. The idea was implicit in the declarations that science cannot "at present hope to arrive at even the most remote approximation to the real solution of the problem of vital phenomena,"[139]) and the like.[140]) Sometimes it was quite explicit. "Life, whether animal or vegetable, is an immaterial, but a real state of existence,"[141]) wrote one theological writer. Discussion around the matter was given a strong impulse by Huxley's Edinburgh lecture on "The Physical Basis of Life", later published in the *Fortnightly*,[142]) where he attacked the idea of Vital force and vitality as a piece of conceptual realism with no scientific explanatory value.[143]) The theological-idealistic school of philosophy at once joined issue with him. "Vitality . . . bears to protoplasm a quite different relation from that which aquosity bears to water," said one commentator. and continued, "It must at one time have been possible for life to mould and vivify inert matter without being previously embodied; and it must needs have been by unembodied life that inorganic matter was first organized and animated."[144]) Whatever the quality of the logic of this argument, it is evident that views of this kind were widespread and popular. The very positive reaction showed by the press when Stokes gave support to the vitalistic position in his Presidential Adress to the British Association in 1869, and the relief felt in several quarters when Huxley declared that "biogenesis" had to be a working hypothesis in practical scientific work, are significant.[145]) Altogether, the notion of a mysterious Vital Force served as an important rallying point for all those who preferred a metaphysical to a scientific explanation of the phenomena of life — a class which included practically all anti-Darwinians. The Darwinian, empiricist answer to the challenge offered by this sort of criticism was discussed above, Chapter 9, p. 181—3.

---

[139]) *Fraser's Magazine*, 61, 1860, 742—4.

[140]) E. g. *Standard*, 1861, Sep. 12, p. 6, rep. discussion at British Association, *Edinburgh Review*, 119, 1864, 25—6.

[141]) *Friends' Quarterly Examiner*, 1867, 512.

[142]) *Fortnightly Review* February 1869.

[143]) See above, Chapter 9, p. 182.

[144]) *Contemporary Review*, 20, 1872, 671, See also *Contemporary Review*, 11, 1869, 240—63.

[145]) See Chapter 4 p. 83, 86.

CHAPTER 13.

The Case for Darwin

The Darwinian controversy, naturally enough, chiefly turned on the weak spots in the doctrine. Darwin's opponents almost exclusively dwelt on the arguments against the theory, and the Darwinians, in their turn, would reply to the criticism rather than repeat the case for Darwin, for which readers were naturally best referred to the *Origin* itself. As, moreover, Darwin's opponents in the press were much more numerous than his supporters, it is obvious that a sample of the press discussion on the theory must contain many more references to the arguments against Darwin than to the arguments for him.

The arguments that were produced on the Darwinian side — in addition to those advanced in reply to criticism, and which have been dealt with in previous chapters — were mainly concerned with the Descent theory. Darwinians felt that once the fact of transmutation had been established, the necessity of admitting a purely scientific explanation would eventually be acknowledged. This was definitely Darwin's own view, who repeatedly asserted that he wished above all to see the principle of Descent established.[1]) He did not mean, however, that he considered Natural Selection of little importance. His stubborn defence of that theory against the criticism of, for instance, his friend Asa Gray, shows that he felt deeply on this point. What he defended in Natural Selection was above all the scientific attitude towards the problems involved. He vigorously combated the non-scientific, teleological views propounded by the large majority of his opponents.

It may indeed be held that any advocacy of an empiricist philosophy of science supported, at least indirectly, Natural Selection as an explanation of Evolution. On the purely factual level there was nothing, at least before the age of statistical experiments on heredity, that could be adduced as supporting Natural Selection rather than a theory of

---

[1]) Darwin, *Letters*, II, 371, to Gray, 1863. Huxley held the same view: see Darwin, *More Letters*, I, 386, Huxley to Darwin, 1880.

directed evolution. If such a theory was acceptable at all, it could accommodate all the facts as easily, or more easily, than Natural Selection. But things were different with the Descent theory. The only real alternative to it was the theory of the separate creation of each specific type. And there were many facts which the Descent theory explained, but which received no explanation at all on the principle of separate creation.

Foremost among these facts was the observation that the organic world could be classified hierarchically in varieties, species, general, families, classes, and so on. Next came the general progression and differentiation observed in the geological record. Before Darwin, the usual explanation of both these sets of facts had been to say that such was the Creator's plan. Now the Descent theory made it intelligible why such a plan had been followed. The chief scientific reason — as distinct from theological ones — why people had not accepted the Descent theory before Darwin was that the gaps between the species had appeared too wide, and the specific characters too stable. It was also on these points that Darwin had to fight his hardest battle, as our illustrations of the missing link argument will have shown. But the very controversy on these points proves that most people were prepared to accept that *if* a complete, graduated series of organisms could be found in the records of the past, then the Descent theory was correct. Common sense recognized that similarity of structure was a valid argument for connexion by Descent. It was difficult to quarrel with Darwin when he wrote of inheritance that it was "that cause which alone, as far as we positively know, produces organisms quite like, or, as we see in the case of varieties, nearly like each other."[2] This was the foundation of the whole argument, and as it was a truism accepted on every hand it was not necessary to dwell on it. The only question to be discussed was the size of the gaps, and how to interpret them, and these discussions were presented in the foregoing chapters.

The other arguments for the Descent theory concerned more specific points which will be illustrated one by one in the following sections.

Geographical Distribution of Species

It was the facts of geographical distribution which put Darwin on the track of the Descent theory, as he says himself in the opening passage

---

[2] *Origin*, 1 ed. r., 297.

of the *Origin of Species:* "When on board H. M. S. 'Beagle', as a
naturalist, I was much struck with certain facts in the distribution
of the inhabitants of South America, and in the geological relations of
the present to the past inhabitants of that continent. These facts
seemed to me to throw some light on the origin of species."[3] He devoted
a whole chapter in his book to this problem, where he at once presented
the outline of his argument to his readers: "In considering the distribution of organic beings over the face of the globe, the first great fact
which strikes us is that neither the similarity nor the dissimilarity of
the inhabitants of various regions can be accounted for by their climatal
and other physical conditions ... A second great fact which strikes
us in our general review is that barriers of any kind, or obstacles to free
migration, are related in a close and important manner to the differences
between the productions of various regions."[4] If species were separately
created, one would expect them to be similar in regions with similar
climatic and soil conditions. But that is not generally the case. Oceanic
islands, though presenting much the same conditions all over the world
on similar latitudes, have widely different floras and faunas. Alpine
plants in different parts of the world often resemble neighbouring
lowland plants, but not each other, from a systematic point of view.
A further support for Darwin was the fact that not only the geographical
distribution of the present series, but also the paleontological record
indicated that a uniform and gradual development had taken place
within each geographical region.[5]

In the early years after the publication of the *Origin* practically
only the pro-Darwinian reviewers gave currency to the argument,[6]
while anti-Darwinians at first either ignored it or disputed its validity.
One religious organ, for instance, held that there was no difficulty for
the Creator to create whichever forms He liked wherever He liked:
far-fetched theories of migrations could thus be dispensed with.[7]
But later on, when opposition to Darwin came to turn more and more

---

[3] *Origin*, 1 ed. r., 1.

[4] *Origin*, 1 ed. r., 294—5.

[5] *Origin*, 1 ed. r., 346. See also: *London Review*, 1, 1860, 59, *Intellectual Observer*, 3, 1863, 81—3.

[6] See e. g. *Times*, 1859, Dec. 26, p. 8, *Spectator*, 1860, 380 (letter), *National
Review*, 10, 1860, 213, *London Review*, 1, 1860, 33.

[7] *London Quarterly Review*, 14, 1860, 302.

on the Natural Selection theory, the bearing of the Geographical distribution argument on the Descent theory was increasingly recognized, though hardly with good grace: one writer declared that it was "the sole argument brought forward by Darwin which seems to us to lend any countenance to the theory of a common origin and the transmutation of species."[8]

On the other hand, there were critics who pointed out that several facts of distribution were difficult to explain on the theory that all similar species were related to each other, without assuming extensive and extremely difficult migrations, or very great — and unproved — changes in the geography of large regions. Indeed, some critics declared that such migrations were impossible, since, as Agassiz put it, "most species are so narrowly confined within the limits of their natural range that even slight changes in their external relations may cause their death."[9] A half-way house between creationism and transmutationism was occupied by those critics who sought an alternative explanation in the doctrine of a limited number of centres of creation, which would explain the family likeness of the flora and fauna of large zones of the earth, while avoiding the necessity of assuming too extensive migrations. This doctrine was especially associated with the name of the English naturalist Forbes: it had been to some extent foreshadowed by Buffon.[10] Mivart, as we should expect, was favourably disposed towards this view. He suggested that some cases of similarity between species might be due to similar development in separate lines of descent. If evolution was due to an internal law of development, similar conditions ought naturally to lead to similar results. But if evolution was due to the natural selection of accidental variations, identical results in separate lines were so improbable as to be virtually unthinkable. Thus Mivart was able to allege some facts at least of the geographical distribution of animals and plants against the Darwinian doctrine, though only against its Natural Selection part.[11] This position was fairly general towards the end of the 'sixties. The facts of geographical distribution

[8] *North British Review*, 46, 1867, 316. See also: *Quarterly Journal of Science*, 3, 1866, 169, *Edinburgh Review*, 134, 1871, 198, *Month*, 11, 1869, 35.

[9] *Annals and Magazine of Natural History*, 6, 1860, 223. See also *Athenaeum*, 1859: 2, 660.

[10] *Annals and Magazine of Natural History*, 11, 1863, 424.

[11] Mivart, *Genesis of Species*, 171—2. See also *Month*, 11, 1869, 41—2.

were accepted as evidence for a Descent theory, but also used against
the peculiarly Darwinian form of it.[12])

Rudimentary Organs

The rudimentary organs very clearly indicated the points of
resemblance between the various species, and as at the same time they
could not easily be regarded as adaptive, they presented an unsolved
puzzle on the separate creation theory. Darwin wrote in the *Origin:* "The
same reasoning power which tells us plainly that most parts and organs
are exquisitely adapted for certain purposes, tells us with equal plainness
that these rudimentary or atrophied organs are imperfect and useless.
In works on natural history rudimentary organs are generally said
to have been created 'for the sake of symmetry,' or in order 'to complete
the scheme of nature,' but this seems to me no explanation, merely
a restatement of the fact. Would it be thought sufficient to say that
because planets revolve in elliptic courses round the sun, satellites
follow the same course round the planets, for the sake of symmetry,
and to complete the scheme of nature?"[13])

In the early years of the controversy, Darwin's opponents in the
press — when they took up the rudimentary organs at all — continued
to prefer the explanation in terms of a Divine plan which Darwin so
forcibly rejected. Agassiz could be cited as an authority for such views:
"The organ remains, not for the performance of a function, but with
reference to a plan."[14]) The same was true of Professor Owen, to whom
the Duke of Argyll referred in an *Edinburgh* article on the subject of
"silent members," as he called them. "Mr. Darwin, when he sees such
a member in any animal, concludes with certainty that this animal
is the lineal descendant by ordinary generation of some other animal
in which that member was not silent but turned to use. Professor
Owen, taking a larger and wider view, would say, without pretending
to explain *how* its presence is to be accounted for physically, that the
silent member has relation to a general purpose or plan which can be
traced from the dawn of life, but which did not receive its full accomp-
lishment until man was born. This is certain; the other is a theory."[15])

---

[12]) See e. g. *Nature*, 3, 1870—1, 272, *Popular Science Review*, 10, 1871, 188.
[13]) *Origin*, 1 ed. r., 384—5.
[14]) *British and Foreign Evangelical Review*, 12, 1863, 482.
[15]) *Edinburgh Review*, 116, 1862, 390.

As Argyll's article was later included in his *Reign of Law*, it gained wide publicity.[16])

Darwinians did not rate such criticism very highly. One writer quoted Argyll's saying that the rudiments might be "right either in the light of history of in the light of prophecy. They indicate either what has already been, or what may yet come to be." He commented, "This is very much akin to Mr. Paine's view, who regards fossils as the 'medals' of creation."[17])

It was clearly difficult to maintain the attitude that rudimentary organs should be regarded as an unfathomable mystery, when the Descent theory was beginning to gain ground. One writer recognized this in a somewhat naïve manner, by asking his readers to look at the fully developed rather than at the rudimentary structures in order to learn the beauty and goodness of the arrangement of the organic world: "And even if the whole of the rudimentary stage of Nature was an enigma, how could that cancel the machinery of her mature work? Whatever the introductory period may be, Nature leaves it very soon behind her, and presents to us a magnificent and consistent structure."[18])

In general therefore, the force of the Darwinian argument was acknowledged. It was a matter of course with those who were generally favourable to the new theory[19]): "When Mr. Darwin pointed out to us the persistent tips in our ears, he did more to discomfort the friends of persistent species than he did by the thousands of other facts,"[20]) wrote one reviewer of *Descent*. In the press organs which generally took an anti-Darwinian line it naturally took some time before the argument was admitted. Mivart's acceptance of it certainly hastened its recognition in the traditionalist press.[21]) Mivart, of course, used the argument for the purposes of his theory of predetermined evolution.

---

[16]) *Eclectic Review*, 12, 1867, 56—7, *Contemporary Review*, 5, 1867, 77.

[17]) *Examiner*, 1872, 603.

[18]) *Quarterly Review*, 127, 1869, 145.

[19]) See e. g. *Times*, 1859, Dec. 26, p. 8, *Spectator*, 1860, 380, (letter) *London Review*, 1, 1860, 33, *Examiner*, 1867, Nov. 30, 758, *Nature*, 3, 1870—1, 442, *Examiner*, 1871, 234, 256; 1872. 576, 602.

[20]) *Examiner*, 1871, Aug. 5. 775.

[21]) *British Quarterly Review*, 53, 1871, 566. See also: *Intellectual Observer*, 3, 1863, 81, *Month* 11, 1869, 35, *Observer*, 1871, Mar. 19, p. 3 *Edinburgh Review*, 134, 1871, 198, *Guardian*, 1871, 935, *Lancet*, 1872: 2, 297.

Embryology

The facts of embryology gave a support to the transmutation theory similar to that afforded by the rudimentary organs. The resemblances it revealed could not easily be used to prove direct adaptation and Design. Why should the embryo of a mammal, a bird, and a frog, destined to become so widely different from each other, yet resemble one another so closely as to be practically indistinguishable? To say that the resemblance depended on the similarity of the conditions under which embryos grow was hardly sufficient: moreover, a mammal embryo in the womb of its mother, a bird in an egg, or the spawn of a frog in water can hardly be said to live under similar conditions. Moreover, embryological features often had the character of rudiments, for instance, the foetal teeth of some whales, which never break the gums, and are later resorbed.

The Natural Selection theory gave an intelligible explanation of these peculiarities. The modifications which had a chance of being naturally selected were primarily those which offered the individual an advantage in the fully developed stage of existence, where the environment was much more differentiated, and in which the struggle for life was keenest, and least affected by chance influences. Therefore, if a species branched into two distinct ones in successive generations, they would resemble each other more or less exactly in the embryonal stages, in which no differentiating forces were at work, and only begin to differ in their adult forms. This was in fact the case with closely allied species of the same genus, as well as with varieties of the same species. In the end, one should expect the embryonal development of an individual to reflect, though not necessarily exactly and in every detail, the development of its forefathers throughout paleontological history. If the Darwinian Descent theory was true, one should therefore expect the embryonal development of different species to follow parallel courses for long periods, the longer the more closely allied the species were. As this was actually the case, one had here a powerful support for the Descent theory.[22] It was not, however, referred to very often in the general discussion in the press.[23]

[22] *Origin*, 1 ed. r., 372 ff.

[23] See, however, *Times*, 1859, Dec. 26, p. 8 *John Bull*, 1859, Dec. 24, 827, and in organs unfavourable to Darwin: *Athenaeum*, 1859: 2, 660, *Zoologist*, 19, 1861, 7588, *Quarterly Journal of Science*, 3, 1866, 174.

Darwin was not, of course, the first to point out the remarkable parallelism of the foetal development of different species, or even the fact that the embryo recounted the paleontological progression within the class to which the species belonged. Indeed, these facts had been strongly insisted on by Agassiz. But Agassiz did not interpret the facts as support for the Descent theory. On the contrary, the fact that no embryo of one species ever developed into an adult individual of a different species, in spite of the close parallelism in the early embryonal stages, showed instead that the Descent theory was wrong. "The most closely allied species of the same genus, or the different species of closely allied genera, or the different genera of one and the same natural family, embrace representatives which at some period or other of their growth resemble one another more closely than the nearest blood relations; and yet we know that they are only stages of development of different species distinct from one another at every period of their life ... not a single fact can be adduced to show that any one egg of an animal has ever produced an individual of any species but its own ... So that no degree of affinity, however close, can ... be urged as exhibiting any evidence of community of descent."[24]

Given an idealistic interpretation of nature, Agassiz view was not unnatural. It was of course very acceptable to anti-Darwinians, as it enabled them to say, as one of them did, "Let no one suppose that the study of embryology forces any man to believe that we have a filial relationship with the beasts that perish, or that have perished."[25] In general, however, embryology was interpreted as giving support to the Descent theory. Towards the end of the 'sixties it was so used quite frequently in the anti-Darwinian press as well. The explanation is not far to seek. The argument, as we have seen in the previous chapter, served excellently to illustrate the idea of a predetermined evolution, which was then spreading fast in opposition to Darwin's Natural Selection theory. The individual's development was clearly predetermined: hence it was natural to conclude that the development of the species was so too. It was in this sense that the argument was urged by Mivart, who made a very marked impact on the press discussion in the late

[24]) *Annals and Magazine of Natural History*, 6, 1860, 225.
[25]) *Lancet*, 1863: 2, 421.

'sixties and early 'seventies.[26]) Even pro-Darwinian writers, it seems, mainly used the argument as a general support for the Descent theory, not specifically to support Natural Selection.[27])

Mimetic Resemblance

The argument of mimetic resemblance was one of the few which Darwin himself did not advance in the *Origin*. It was brought into the discussion by the naturalist H. W. Bates, author of a book of travels, *A Naturalist on the Amazons*, and it was elaborated by A. R. Wallace, who had been Bates's travelling companion. It differed from most of the other arguments discussed in this chapter by having a direct bearing on the Natural Selection theory rather than on the Descent theory. It was not often that the Darwinians could point out the obvious advantage in the struggle for life arising from some particular, small variation. The facts of mimicry afforded such an opportunity. It seemed an obvious advantage to an innocuous fly to resemble a dangerous wasp, or for a butterfly hunted by birds to resemble a distinct species which was avoided by birds because of its nauseating taste.

The validity of the mimicry argument was acknowledged in both Darwinian and anti-Darwinian organs. By so doing, one naturally did not commit oneself to Natural Selection as an explanation of *all* evolution: mimicry was a rare phenomenon in nature. And, as we have seen, nobody maintained that Natural Selection never occurred: one only disagreed about its extent.[28])

There were, however, also dissidents. Some tried to dismiss mimicry as accidental.[29]) Others suggested that "similar conditions of food and of surrounding circumstances have acted in some unknown way to produce the resemblance."[30]) But most commonly one appealed to

---

[26]) *Month*, 11, 1869, 35, *Quarterly Review*, 127, 1869, 144, *Edinburgh Review*, 134, 1871, 198, *Guardian*, 1871, 935.

[27]) *Fortnightly Review*, 3, 1868, 614, *Nature*, 3, 1870—1, 442, *Inquirer*, 1871, 295. *Lancet*, 1872: 2, 297, *Academy*, 2, 1870—1, 11.

[28]) See above, Chapter 12. See further: *Intellectual Observer*, 6, 1864, 313, *Quarterly Journal of Science*, 1, 1864, 187, *Examiner*, 1867, Nov. 30, 758, *Month*, 11, 1869, 35, *Examiner*, 1870, May 28, 341, *Once a Week*, 4, 1869—70, 558.

[29]) *Examiner*, 1871, May 13, 492 reporting an anonymous anti-Darwinian treatise,

[30]) *Westminster Review*, 32, 1867, 34, A. R. Wallace referring to A Murray.

[31]) *Ibid.*, ref. to Westwood.

Design: "Each [mimicking] species was created a mimic for the purpose of the protection thus afforded it."[31]) An intermediate position was taken by Mivart. He admitted that mimicry provided strong support for the Natural Selection theory. Yet he would not grant, in this case any more than in others, that Natural Selection was by itself enough to achieve the mimicking effect. He adduced the uselessness of the incipient stages in support of the contention that an inherent — and designed — tendency was needed to supplement the action of Natural Selection.[32])

Analogy with Geology

Uniformitarian geology, as established in Britain by Sir Charles Lyell, had a twofold influence in preparing the public for accepting the Descent theory. In the first place, it would have been impossible to introduce, let alone to prove, a biological theory of gradual change, if geology had continued to teach that world-wide catastrophes had several times caused wholesale destruction in the world. In the second place, the uniformitarian principle that changes in the past should be explained as due to "causes now in operation," when applied to biology, implied both gradualness of evolution, and an attitude of mind which was unfavourable to the *ad hoc*, miraculous type of explanation which, in the last resort, all Darwin's opponents relied on. Darwin was fully aware that he owed a very great debt to Lyell in both respects.[33])

Several reviewers pointed out the connexion between Darwinian biology and Lyellian geology. Darwin's theory was sometimes regarded simply as a logical extension of Lyell's: "The doctrine of continuous succession, once admitted in physical geology, seems almost to necessitate the admission of the like doctrine as a corollary in regard to the succession of organic life,"[34]) wrote Carpenter in his review of the *Origin*. Such a reasoning, of course, touched only the Descent theory as such. But we also meet with statements where the naturalistic tendencies of both Lyell's geology and Darwin's biology were brought out: "Mr. Darwin's scheme . . . appears to have for its object to show that the immense

---

[32]) Mivart, *Genesis of Species*, 33—41, *Nature*, 3, 1870, 30. See also above, Chapter 12, p. 246—50.

[33]) See Darwin, *Letters*, II, 190, and Huxley, *Life*, 1 ed. 1900, I, 243.

[34]) *National Review*, 10, 1860, 210.

variety of organic forms . . . are the result of natural forces . . . We . . . mention it as exhibiting one of the many tendencies to recognise the continuous operation of uniform causes, and the powerful effects produced by a constant succession of comparatively small impulses. The growth of this idea is the culminating point of modern philosophy, and none was ever calculated to be more fertile in important practical results,"[35]) declared the positivist *Leader*.

The interdependence of geology and biology was felt to be so strong that one writer was even prepared to say that Lyell's previous belief in the permanence of species was "incompatible"[36]) with his general theory. When Lyell, at last, gave his full adherence to the Descent theory, several reviewers pointed out that this strengthened his general theory.[37]) Darwin's opponents evidently viewed the force of the analogy with some apprehension. Lyell's theory was, or had become, theologically innocuous. It was acceptable, even though not always accepted, in religious circles. But if Lyell was to prepare the ground for Darwin, the situation would become dangerous, and it was necessary to consider what steps to take: "Our purpose here is not to review the non-break, continuous theory of Sir Charles Lyell, though we are persuaded that Mr. Darwin's work will lead many to reconsider whether they have done well in accepting it,"[38]) wrote one religious organ. Not all, however, advised such a drastic course, but contented themselves with pointing out that acceptance of one theory did not necessarily imply acceptance of the other. The argument was only one of analogy. Uniformitarianism was, as one writer said, "perfectly consistent with, though by no means dependent on, a reception of the Darwinian doctrine by natural selection."[39])

Gradual development was, however, becoming the watchword of the day. Indeed, one *Edinburgh* reviewer alleged that its success was the result of a general movement of thought, thereby implying that Darwin's contribution was not very important: "It was almost inevitable that the modern scientific conceptions of unbroken continuity and

---

[35]) *Leader*, 1860, Feb. 11, 134.

[36]) *Critic*, 25 1862—3, 296.

[37]) See above, Chapter 3, p. 53 and *Gardener's Chronicle*, 1868, 521.

[38]) *North British Review*, 32, 1860, 485. *Freeman*, 1863, Apr. 8, 222. See also above, Chapter 3, p. 52—3.

[39]) *Intellectual Observer*, 6, 1864—5, 12.

progressive development of life should be applied to the highest mani-
festations of this inscrutable power [life], and that the scientific investigation of humanity as the culminating point in a great scheme of vital
evolution should be attempted."[40]) But the niggardly attitude of the
Edinburgh was not the prevalent one. Another writer declared that
"Charles Lyell and Charles Darwin are the two men who have most
deeply affected the tone of scientific thought of the present generation."[41])
The idea of continuity was expressed on every hand. It was, as we
know, the central theme of W. R. Grove's presidential address to the
British Association in 1866.[42]) And as the continuity doctrine was
neutral with regard to Natural Selection it could be adopted equally
by anti-Darwinian and Darwinian evolutionists.[43])

Analogy with Philology

The analogy between biological evolution and the development of
languages was in many respects a very apt one. Darwin himself briefly
referred to the subject in the Origin,[44]) and more elaborately in Descent.[45])
Like organic species, languages at the present time are, broadly speaking,
quite distinct from each other. There is an German language and a
French language, but not a whole series of intermediate languages
between them. Dialects may be equated with varieties. Further,
languages can be systematically classified into larger groups, in a way
resembling the classing of species into genera and orders. But what
above all interested the Darwinians was the fact that languages, unlike
species, were really proved to derive from common stocks. There was
no doubt at all that the descent theory was true in philology. Therefore,
if linguistic species could arise by the accumulation of imperceptible
changes, why could not the same be acknowledged as possible for
biological species as well? "The development of numerous specific
forms, widely distinguished from each other, out of one common stock,
is not a whit more improbable than the development of numerous

---

[40]) Edinburgh Review, 135, 1872, 89.
[41]) Student, 1, 1868, 179.
[42]) See above, Chapter 4, p. 78.
[43]) Times, 1859, Dec. 26, p. 8, Quarterly Journal of Science, 2, 1865, 194.
[44]) Origin, 1 ed. r., 359.
[45]) Descent, 137—140.

distinct languages out of a common parent language, which modern philologists have proved to be indubitably the case."[46])

Anti-evolutionists were naturally reluctant to accept this analogy, and insisted that the facts of philology were fundamentally different from those of a science like biology. One sometimes alleged that there was a difference in that, while the organic world was in the main under the sway of unconscious forces, "the impulses and efforts which have formed and which improve ... languages are not unconscious, but ... are prompted and guided at every step by human reason. The analogy, as far as it applies, not only gives no support to the theory of the production of works of Intelligence by a process of unintelligent variation and selection, but is strongly against it."[47]) There were those, indeed, who took their stand on the permanence of species to deny the analogy: "A more plausible, if not a more forcible, objection arises from the constant change which is perceptible in language. The types of nature, it is urged, are never changed".[48]) But this was simply to assume the point at issue, and in general the analogy was accepted.[49]) It was of course mainly used to support the transmutation theory in general; but there were philologists who claimed that Natural Selection also occurred in the field of philology precisely as in biology.[50]) It is evident, however, that in philology the phrase, "survival of the fittest," was even more indeterminate and difficult to define than in biology, and cautious reviewers met the argument with sound scepticism.[51])

---

[46]) *Cornhill*, 1, 1860, 445.

[47]) *British and Foreign Evangelical Review*, 20, 1871, 719.

[48]) *Edinburgh Review*, 115, 1862, 76.

[49]) *Daily News*, 1864, Oct. 21, p. 2, (reviewing Max Müller).

[50]) *Nature*, 1, 1869—70, 529. (F. W. Farrar ref. Schleicher).

[51]) *Westminster Review*, 37, 1870, 288.

CHAPTER 14

The Descent of Man

The Background

The Descent theory of evolution and the doctrine of Natural Selection were perfectly general in their application. Their domain was the whole of organic life, whether vegetable or animal. Now as man's physical organization classed him with the animals, it was inevitable that the doctrine should be applied to him as well.

For several reasons, however, it has been desirable to treat the Darwinian theory's application to Man separately from its application to the lower organic world. First, there were some problems, notably those relating to the intellectual and moral spheres, which almost exclusively concerned the human species. Second, the question of man was so closely bound up with religious and other convictions that contemporaries themselves often attempted to set it apart from the problem of evolution in general. And third, the heat and the eagerness with which the question of the theory's application to man was discussed in itself justifies a separate treatment.

In the *Origin* Darwin referred to the question of Man only at the very end of the book, where he said, "In the distant future I see open fields for far more important researches. Psychology will be based on a new foundation, that of the necessary acquirement of each mental power and capacity by gradation. Light will be thrown on the origin of man and his history."[1]) Darwin's great reserve on this subject was diplomatic. He knew that his theory would meet with much opposition, and wished to avoid hurting everybody's feelings at once. When A. R. Wallace asked him, in 1857, whether he would discuss Man in his forthcoming book, Darwin answered, "I think I shall avoid the whole subject, as so surrounded with prejudices; though I fully admit that it is the highest and most interesting problem for the naturalist."[2]) That he did not

---

[1]) *Origin*, 1 ed. r., 413—14.
[2]) Darwin, *Letters*, II, 109, to Wallace, 1857.

leave the subject altogether alone was due, as he explained in later years, to his sense of honour: "Although in the 'Origin of Species' the derivation of any particular species is never discussed, yet I thought it best, in order that no honourable man should accuse me of concealing my views, to add that by the work 'light would be thrown on the origin of man and his history.' It would have been useless and injurious to the success of the book to have paraded, without giving any evidence, my conviction with respect to his origin."[3]) That this correctly renders Darwin's feelings when he published the *Origin* is apparent from a letter to a private correspondent at that time: "With respect to man, I am very far from wishing to obtrude my belief; but I thought it dishonest to quite conceal my opinion. Of course it is open to every one to believe that man appeared by a separate miracle, though I do not myself see the necessity or probability."[4])

In any case it must have been obvious to those who followed the argument of the book that it applied to man quite as much as to the lower species. Reviewers of the *Origin* therefore seldom omitted some comment on the matter. Especially it was the "nervously anxious supporters of religious orthodoxy"[5]) who showed concern, and, as one religious organ put it, "The Darwinian theory would lose half its interest with the public if it did not culminate in a doctrine on the origin of the human species."[6]) But Darwin's judicious caution in handling the issue certainly prevented — as was intended — too violent an outcry. Indeed, there were some traditionalists who comforted themselves that evolution by descent was "a theory which the originator does not see his way clear to apply to man."[7]) The publication of the *Descent of Man* made that idea untenable, but until then, said, the *Guardian*, "Mr. Darwin's silence on the subject seemed to favour the idea that he himself . . . shrank . . . from pushing his theory to its legitimate issue."[8])

But if Darwin was cautious in order not needlessly to antagonise the religious, some of his opponents were not averse to turn the *odium theologicum* to good account for their own purposes. Thus when Owen

---

[3]) Darwin, *Letters*, I, 94, autobiographical sketch.
[4]) Darwin, *Letters*, II, 263—4, to L. Jenyns (Blomefield), 1860.
[5]) *Morning Post*, 1860, Jan. 10, p. 2.
[6]) *Literary Churchman*, 6, 1860, 393.
[7]) *Popular Science Rreview*, 2, 1862—3, 517.
[8]) *Guardian*, 1871, 935.

and the Bishop of Oxford brought up the question of man's descent at the British Association at Oxford in the summer of 1860, Huxley was probably correct in surmising that it was done with a view to arousing popular feeling against the theory. In the next few years the discussion sharpened. A good share of it was taken up by the famous *hippocampus* debate between Owen and Huxley, which was carried on both at the British Association meetings from 1860 to 1862, and in the press.[9]) A sign of the latent public interest in the question of man's descent, and also a powerful catalyst on that interest, was the tremendous popular excitement caused by the French-American traveller Du Chaillu and his gorillas. Du Chaillu lectured at the Royal Institution, published a book on his African travels and his encounters with the gorilla — a book which became a bestseller and was reviewed and commented on everywhere in the press — and appeared at the British Association meeting in 1861. Discussions on his gorillas assumed an enhanced interest from the fact that a controversy arose both on the value of specimens he had sold to the Natural History Museum, and on the veracity of his accounts from the "gorilla country." His descriptions of the gorilla's habits, especially, were sometimes alleged to be purely imaginary.[10])

As a result of all the stir, the popular press in 1861 was full of articles on the gorilla, and naturally on the resemblances and differences between the great ape and man. *Punch*, for instance, repeatedly used the unsightly beast in its satires on Ireland. A deputation of Irish "hooligans" was led by gentlemen named Mr. O'Rangoutang, Mr. G. O'Rilla, and Mr. Fitzcaliban.[11]) It is of course true that one did not go very far into the Darwinian theory in these instances, but it clearly provided an emotional background. Significantly enough, the Darwinian doctrine was usually referred to as the "ape theory" in the popular press, and the "missing link" *par préférence* was the as yet undiscovered intermediate form between ape and man. In the low-brow press, indeed, little else was said about the Darwinian doctrine than that it represented man as descended from apes. It is no wonder that Disraeli, with the politician's grasp of the mentality of the average man, found in 1864 that the great

---

[9]) See above, Chapter 3, p. 50—51, and below, p. 305—6.

[10]) Letters in the *Athenaeum*, 1861: 1, 662—3, and subsequent weeks.

[11]) *Punch*, 1861, Dec. 14, 244.

question was whether man was an ape or an angel.[12]) Nor was it surprising that he, like the majority of those who knew nothing of scientific
argumentation, decided for the side of the angels.

On the scientific level also, attention came to be focussed more and
more on the question of man's evolution. Huxley published *Man's
Place in Nature* in 1863, to some extent the fruit of researches he had
undertaken in connexion with the *hippocampus* debate. In the same
year Lyell published his great book on the *Antiquity of Man*, where a
great mass of geological evidence was marshalled in order finally to
establish a more realistic chronology of the human race. In the following
years several noteworthy publications on the subject appeared, such as
Tylor's *Early History of Mankind*, 1865, Lubbock's *Prehistoric Times*,
1865, and *Origin of Civilization*, 1870; the Duke of Argyll's *Primeval
Man* in 1869, and Wallace's *Contributions* in 1870. In these books
biological, ethnographical and anthropological questions were treated
from various points of view. Moreover, there was a continuous discussion
in the scientific and semi-scientific press, as well as at the British Association, which had a separate subsection of Anthropology from the middle
'sixties onwards.

Thus the ground was well prepared when Darwin published his
Descent of Man in 1871. It did not cause the sort of sensation produced
by the *Origin* twelve years earlier. The public were already familiar
with the argumentation on the subject. In fact, one detects a slight
tone of disappointment in many reviews of the book. Partly this was
due to the rather disproportionate amount of space that Darwin gave
to the somewhat abstruse question of Sexual Selection, partly also
to the fact that little new evidence and no radically new arguments
were produced. However, coming from Darwin, whose reputation as a
scientific authority had gone on increasing during the 60's — a reputation which had gained rather than suffered from Darwin's reluctance
to enter personally into the controversies which his theory aroused —
the book could not but exert a very great influence on the climate of
opinion.

There is one general difference between the debate on the Darwinian
doctrine's application to man, and that concerning the lower organisms:
it was much more common to make an unashamed and direct appeal

---

[12]) Disraeli's speech reported in *Times*, 1864, Nov. 26, p. 7—8.

to feelings and prejudices rather than mere facts, when man was concerned. We have indeed already seen some examples of the emotional line of argument, based specifically on religious objections to the doctrine.[13]) Many of these objections had regard to man in the first place. But in the case of man it was possible to invoke in addition the revulsion and horror that many people evidently felt in regard to a possible genetic relationship with the brute creation. To believe that man was only a developed ape was called a "mental catastrophe"[14]) and those who contributed to spreading such a belief had set as their aim to "degrade man deeply in the scale of animal existence."[15]) Ordinary folk recoiled; the question of man's origin, it was said, "touches too nearly the sanctity of the mysterious light which enshrouds our being and endows us with a responsible humanity."[16]) On this emotional level the question evidently ceased to be a factual one: "Granting even that the conclusions were fairly deducible from the facts, it is no light matter to shake the convictions of thousands ... the question here is not the truth or error of the doctrines or opinions opposed, but the feelings with which such momentous questions ought to be approached,"[17]) wrote the *Athenaeum* in 1866. The *Times* reviewer of the *Descent* tried to justify the use of an emotional argumentation by appealing to idealistic metaphysical assumptions: "There are certain instincts, which so far as we can see, are ultimate facts, and an instinct of this nature, except where it is obscured by the prejudices of speculation, impels us to a profound conviction of the essential difference between man and the rest of the animal creation. We are intimately sensible of a difference which is not one of degree, but of kind."[18])

Religious writers naturally obtained the emotional effect by appealing to the direct evidence of the Bible. "Holy Scripture plainly regards man's creation as a totally distinct class of operations from that of the lower beings,"[19]) wrote a correspondent to the *Guardian* at a time when the Descent theory was becoming generally accepted among

[13]) See above, Chapter 5, p. 100.
[14]) *Rambler*, 2, 1859—60, 368. See also: *Edinburgh New Philosophical Journal*, 19, 1864, 56, *Friends' Quarterly Examiner*, 1867, 43.
[15]) *Athenaeum*, 1863: 1, 287.
[16]) *Parthenon*, 1863, 174.
[17]) *Athenaeum*, 1866: 2, 121.
[18]) *Times*, 1871, Apr. 7, p. 3.
[19]) *Guardian*, 1871, 713.

educated people, at least in its application to the organic world below
man. Another religious organ developed the idea in more detail: "Of
the lower creation, it is said, 'Let the earth bring forth,' but of man it
is said, 'Let us make man.' Here we have a difference, not of degree,
but of kind. The laws of natural selection, development, or whatever
philosophers please to call it, may be true to any extent of all organisms
below man."[20]) In this connection critics of course also marshalled the
usual objections against Darwin's "speculative" and "hypothetical"
mode of reasoning. One orthodox naturalist, for instance, was pleased
to find that Owen, at least, "condemned any imaginary scheme by
which some anthropoid ape ... might, by Mr. Darwin's principle of
Natural Selection, become a man."[21])

But even if the purely emotional line of reasoning was not infrequent,
anti-Darwinians would not in general admit that their opposition was
based or had to be based on such grounds. "It would be absurd to
deny," said one writer, "that the adoption of Mr. Darwin's views would
materially affect many of our opinions on subjects of the highest importance; but even this should not be allowed to bias our judgment of their
truth."[22]) This latter question, however, was not so straightforward
as it may have seemed: prejudice and bias, as we have seen, came out
on a higher level, namely, in the decision on the fundamental criteria
according to which theories were to be judged.[23])

In order to illustrate the more factual debate on Man's origin, we
should distinguish not only between the arguments directed against
any Descent theory, and those directed specifically against Natural
Selection — a distinction which we have made in previous chapters —
but also between arguments based on physical, and those based on
mental characteristics. Though the Descent theory as such, of course,
had meaning only as regards man's body, it was possible to adduce
man's mental characteristics as an insuperable and absolute barrier
separating man from the animals, and hence as one of the distinctions
which made it impossible for man to be descended from them by ordinary
generation. In general, however, those who concentrated on the mental
powers of man tended to by-pass the question of man's bodily descent,

---

[20]) *Christian Observer*, 1866, 9.

[21]) *Geologist*, 1861, 529. See also: *Church Review*, 1861, Feb. p. 28.

[22]) *English Independent*, 1871, Mar. 23 273.

[23]) See above, Chapter 9.

though they were certainly not in general inclined to accept it. Their argumentation may be regarded as directed, though somewhat vaguely, against the Natural Selection theory: at any rate, if pushed somewhat further than was normally done, it could serve that purpose.

In presenting the discussion on the application of the Darwinian theory to Man, we shall first deal with the arguments employed against the Descent theory as such; i. e., the missing link arguments, based on physical or mental distinctions. We shall then discuss the arguments specifically and explicitly directed against Natural Selection. Here again both physical and mental characteristics were brought into play. Finally, we shall illustrate the argumentation concerned with setting apart the human mind or soul as a separate and unique entity. It was these latter discussions which took up the bulk of the discussion at the time.

Missing Links

The missing links were just as obvious an objection against the theory of Descent in its application to man as in its application to the lower animals. The systematic affinities of man with the higher apes were not denied, any more than one denied the general systematic affinities holding throughout the organic world. Yet the gap between man and ape was held to be too big to be bridged. All races of men were clearly men, and all the various ape species were as clearly apes. There were no intermediate forms, no graduated series where each step should not appear greater than a varietal modification. This was the more damaging to the theory, one declared, as the conditions under which Darwin supposed man had been produced were still in existence in several parts of the globe.[24] Gorillas could not be accepted as missing links. Du Chaillu and his commentators had been at pains to emphasize the differences between the beast and man. Besides, as one critic naïvely remarked, gorillas were "few and wretched."[25]

Nor was there any record of change having taken place in historical times.[26] A *Quarterly* reviewer found comfort in this fact, when he had

[24] Professor Humphry at the British Association, rep. *Morning Post*, 1866, Aug. 24, p. 2. *Manchester Guardian*, 1866, Aug. 24 p. 3. Also *Contemporary Review*, 17, 1871, 280.

[25] Leader in *Morning Advertiser*, 1866, Aug. 25, p. 4.

[26] John Crawfurd rep. *Times*, 1863, Aug. 31, p. 7, *Nonconformist*, 1863, Sep. 2, 710.

to admit the probability of a longer chronology than the traditional one for the existence of the human race on the earth: "The larger the glacial epoch, the more remarkable is the failure of proof of indefinite variation of species, the more difficult the acceptance of those consequences of 'the struggle for existence' and 'natural selection' which are offered by Darwin and his followers."[27] The *Times* reviewer of the *Descent* was more emphatic still: "It is almost incredible that no evidence should be producible of the existence of apelike creatures ... we have the undoubted and recorded experience of at least four thousand years of history, during which many races have been subjected to influences the most diversified and the most favourable to the further development of their faculties ... [yet] the earliest known examples of Man's most essential characteristics exhibit his faculties in the greatest perfection ever attained. No poetry surpasses Homer."[28]

Further, one demanded, as in the case of the lower animals, that Darwin should produce the intermediate links in the geological record. "Is it not incredible," one asked, "that not a single species of [man's] ancestors should remain, or a fossil remain of one of them to be discovered?"[29] Even a pro-Darwinian organ like the *Westminster Review* was impressed by this argument: "The gap between humanity in its most degraded physical condition, and the very highest of the apes is so great, that we may well be excused for asking for a demonstration of some of the intermediate grades, before giving an unconditional assent to the Darwinian proposition that Man has originated by the progressive development of apelike ancestors."[30]

The fossil links which were in fact produced were dismissed as not to the point. Huxley, discussing the Neanderthal and Engis skulls, admitted that though they showed some transitional features, yet they were well within the range of variation of present-day human forms, and outside that of the anthropoid apes. Such an admission was in many quarters interpreted as a near refutation of the Descent theory.

---

[27] *Quarterly Review*, 114, 1863, 411.

[28] *Times*, 1871, Apr. 8, p. 5. See also: *Journal of Sacred Literature*. 10, 1866—7, 201, *Edinburgh Review*, 132, 1870, 459.

[29] *London Review*, 1866, Aug. 25, 203. See also: *Parthenon*, 1863, 262, *Popular Science Review*, 2, 1862—3, 516, *Tinsley's Magazine*, 8, 1871, 395, *Quarterly Journal of Science*, 1, 1871, 251, *Zoologist*, 6, 1871, 2617.

[30] *Westminster Review*, 23, 1863, 584.

The Neanderthal skull, said the Duke of Argyll, referring to Huxley's authority, "might have contained the brains of a philosopher. So conclusive is this evidence against any change whatever."[31])

Other anti-Darwinians, who were more intent on saving the traditional chronology of the Bible, chose the opposite course. The fossils, they said, were not distinctly human, they were distinctly simian. "We subside into a belief that the respected Neanderthal individual was, after all, an unfortunate chimpanzee or gorilla, who came to grief in comparatively recent times," was one comment.[32]) Thus under no circumstances were the fossil skulls to be accepted as intermediate, they were apes, or they were men, for ape-men could not exist.

Two subsidiary lines of reasoning, intended more directly to prove the constancy of the species man, supported the missing link argument. One was to deny that there were any grounds for considering any of the savage races as in any way, whether physically or mentally, intermediate in the sense of more brutish. The other was to assert that the earliest men had not only been physically as highly developed as modern man, they had actually lived in a state of civilization. Now in attempting to mark off the human domain as distinctly as possible from that of the animals, the anti-Darwinians met with a dilemma. The repugnant connexion with the brutes was avoided, but it seemed at the cost of making the connexion with the savage races much closer. *Punch's* statement of the dilemma, written a few month's after Benjamin Disraeli's famous Oxford speech,[33]) can hardly be improved upon:

"The Negro's and Gorilla's shape
Comparatively scan
What kin is that anthropoid ape
To that pithecoid man?
If any, the Gorilla's proved
Our cousin some degrees removed
If none, with fellow men

---

[31]) Argyll, Primeval Man, p. 73, quot. *Dublin University Magazine*, 74, 1869, 591. See also: *Home and Foreign Review*, 2, 1863, 502, *Parthenon*, 1863, 262, *Edinburgh New Philosophical Journal*, 19, 1864, 54, *Good Words*, 9, 1868, 253 (Argyll).

[32]) *Eclectic Review*, 4, 1863, 412. See also: *Quarterly Journal of Science*, 1, 1864, 88.

[33]) See above, p. 37; 295—6.

And angels Quashee takes his stand
With Michael, Gabriel, Raphael and
Accordingly with Ben.[34])

An easy way of solving the problem was to declare that the human
races were really species, and therefore distinct both from each other
and from the animals. This solution had been advocated in America by
Agassiz, and in England it was expounded at several British Association
meetings by J. Crawfurd and W. Hunt.[35]) But in general Englishmen did
not approve of such a solution, which struck them as both uncharitable
and unorthodox. One was therefore prepared to accept the negro as
a brother.[36]) Not wholly, however. There were undoubtedly differences,
both physical and mental, between the various present human races.
Now in order to account for these differences without giving any support
to the evolution hypothesis, one turned to the opposite extreme. Civilized men had not progressed, one said: it was the savage races which
had degenerated from the perfect state of civilization in which the
whole human family had originally been placed.

This degeneration theory was evidently chiefly inspired by traditional
conceptions of the Garden of Eden and the Fall. Some extreme religious
organs brought the Biblical story directly to bear on the question. It
was suggested, for instance, that "the persistent and degraded negro
type may well have come from a sudden and violent shock — in a
word, from a supernatural punishment."[37]) Punishment for what?
"A believer in the miracles recorded in Scripture will find no difficulty
in supposing the Negro configuration to be impressed on a line of the
descendants of the unnatural and unfilial son of Noah,"[38]) or as another
journal put it, "The curse pronounced upon Canaan, the son of Ham,
would sufficiently account to us for such a variation in the skin, and
in the contour of the human face and figure, as separates the black
race from the white; it may have been the judgment of God inflicted
as a punishment for sin."[39 A])

[34]) *Punch*, 48, 1865, 160.
[35]) See above, Chapter 4, p. 75.
[36]) *Observer*, 1863, Sep. 6, p. 6, *Inquirer*, 1863, Sep. 5, 566, *Quarterly Journal
of Science*, 3, 1866, 173, *Contemporary Review*, 17, 1871, 280.
[37]) *Tablet*, 1869, Apr. 10, 780.
[38]) *Dublin University Magazine*, 74, 1869, 593.
[39 A]) *Christian Observer*, 64, 1864, 93.

In general, however, one tried to adduce factual evidence to support the contention that the savage races were degenerate, and not remnants of a previous stage of development. Now the progress of prehistoric archaeology undoubtedly showed that from the point of view of material culture, primitive society had to be equated with modern savage societies. On the material level, therefore, it became more and more difficult to uphold the degeneration theory. But even this point was conceded only very reluctantly by traditionalists.[39 B]) "We suppose we must accept modern philosophical doctrines," wrote a *Times* reviewer in 1870, "but it is not a pleasant idea to think that, for untold myriads of years, our ancestors were benighted savages."[40])

In order to accommodate such facts as these, the somewhat question-begging theory was developed that "our first parents need not have been savages, although unacquainted with modern arts."[41 A]) The view was set forth fully by the Duke of Argyll in the series of articles in *Good Words* which were later published in book form as *Primeval Man:* "What consciousness had Primeval Man of Moral Obligation, and what communion with his Creator? ... What were his innate powers of Intellect or Understanding? What was his condition in respect to Knowledge, whether as the result of intuition, or as the result of teaching?"[41 B]) In all these respects Argyll found reason to assume that early man had been "civilized". The whole question was discussed fully at the British Association meetings of 1867 and 1869,[42]) where Lubbock defended the Darwinian development position, while Wallace defended the other view, expressing his conviction that primitive man was "civilized morally".[43]) Wallace supported his thesis by referring to his experience of savage life in the East Indian archipelago, where he had found that from the ethical point of view, the savages compared

---

[39 B]) British Association reports in *Daily Telegraph*, 1865, Sep. 11, p. 3 (Rawlinson), *Morning Post*, 1869, Aug. 23, p. 2, (Morris) See also: *John Bull*, 1871, Apr. 6, 234, *Edinburgh Review*, 135, 1872, 111.

[40]) *Times*, 1870, Apr. 25, p. 4.

[41 A]) *Critic*, 1863, Apr., 295—8.

[41 B]) *Good Words*, 9, 1868, 385. See also: *Examiner*, 1869, Apr. 17, 246. A similar position expressed in British Association reports, e. g. *Daily Telegraph*, 1865, Sep. 11, p. 3.

[42]) See above, Chapter 4, p. 80—84.

[43]) Report in *Record*, 1869, Aug. 25, p. 3.

well with the Europeans.[44] This proved, one held, the possibility of an original state of civilisation in spite of material backwardness. Actual evidence of degradation was provided by the fact that many savage races possessed what could be interpreted as traces of a higher civilisation,[45] a point which Argyll developed against Lubbock.[46] One critic, referring to the brutish life of the inhabitants of the Tierra del Fuego, thought that it "points to the conclusion that man, however degraded in habits, is always man, and that he has the higher faculties, at all events latent, in his soul."[47]

Instead of focussing their attention on the absence of missing links in the record, or on the evidence of specific constancy, many writers preferred simply to stress the size of the gap as such. As a rule writers in the non-scientific press seem to have been convinced that a great gap was in itself a guarantee against the Descent theory. They did not even attempt to consider the possibility of undiscovered forms; when they had established that no sort of ape could give birth to any sort of human being, they believed they were furnished with a sufficient and direct negative to the development theory. Thus some anti-Darwinians made it appear as if a logical application of Darwin's theory implied that the existing gaps should be covered in one step. In this way it was easy work to make the theory appear absurd. "Mr. Darwin may cite instances of variation till doomsday, but the philosopher will all the while repeat the request for at least *one* actual instance of transmutation . . . We really want to be carried back to the date when he came forth, in his complete humanity, from the womb of an ape or chimpanzee. We want a record or evidence of the transmutation."[49] Another writer developed this idea further. One instance of such a change was not enough: two gorillas must almost simultaneously, and in the same region, give birth to human infants in order for the race of men to become established. This was a new version of the Adam and Eve story.[50] A speaker at the British Association, whose

---

[44] See e. g. *Times*, 1869, Aug. 25, p. 6.

[45] *Leisure Hour*, 1868, 539.

[46] *Dublin University Magazine*, 74, 1869, 599.

[47] *Contemporary Review*, 17, 1871, 280.

[48] — — —

[49] *Recreative Science*, 2, 1860—1, 273.

[50] *Dublin University Magazine*, 74, 1869, 587.

knowledge of Darwin's theory was apparently taken from such sources as those just quoted, advanced the knock-down argument that such progeny, if born to gorillas, would necessarily perish: "How was the progeny trained and elevated into man, when we, with all our education, [are] scarcely able to prevent our masses from falling back to a state rather akin to monkeys or brutes?"[51])

Man's mental capacities naturally provided the most direct and self-evident illustration of the wide and unbridgeable gulf between man and the animals. But traditionalists often tried to advance their positions by insisting on the uniqueness of purely physical characteristics. There were, for instance, the human hand and foot, differing so markedly from the corresponding organs in the apes, which were usually called *quadrumanous* — four-handed.[52]) The American naturalist Dana gave this difference a major systematic significance: "In man the anterior [limbs] are transferred from the *locomotive* to the *cephalic* series. They serve the purposes of the *head*, and are not for locomotion."[53]) On the strength of the distinction, Dana was able to put man in a separate kingdom, the *archencephalic* one.

But by far the most commonly raised physical distinction was the one associated with Professor Owen's name — the *hippocampus minor* part of the brain. Owen, in fact, laid so much stress on that distinction that he was ready to give greater weight to it than to any purely mental difference between man and ape. To state the distinction between man and ape, he said, was "the anatomist's difficulty."[54]) The subject gave rise to the famous *hippocampus* debate, which raged violently especially in 1861,[55]) partly in the press and partly at the British Association. Owen's first position had been that the *hippocampus minor* part

---

[51]) Reddie as reported in *Observer*, 1866, Aug. 26, p. 6, *Standard*, 1866, Aug. 27, p. 6,

[52]) E. g. Humphry, rep. in *John Bull*, 1861, Aug. 10, 508. See also: *Annals and Magazine*, 8, 1861, 429; other view: *Natural History Review*, 1, 1861, 296.

[53]) *Annals and Magazine of Natural History*, 11, 1863, 208.

[54]) Quoted from Owen paper of 1857 by Huxley in *Natural History Review*, 1861, Jan., p. 69.

[55]) See *Athenaeum*, 1861: 1, 395, 433, 434, 498, 536; 1861: 2, 348, 378, *Natural History Review*, 1861, Jan. 67—8, July, 296—315, *Critic*, 1861, Sep. 14, 276, *Athenaeum*, 1862: 2, 468, *Reader*, 1863, Mar. 7, 234, *British Medical Journal*, 1862, Apr. 12, 377, *Popular Science Review*, 2, 1862—3, 246, 3, 1863—4, 566, *Tablet*, 1870, Jan. 8, 39, *Examiner*, 1871, 256. See also press reports of the British Association meetings of 1861, 1862 and 1863.

of the human brain did not exist in the ape brain. When Huxley and others were able to prove that this view was incorrect, and that the difference was one of degree only, Owen and his supporters tried to save their position by holding that "in the sense used by human anatomists"[56]) the part was not to be found in apes, which was of course literally true, but also uninteresting, since nobody had asserted that the apes had a human *hippocampus*. Therefore yet another defence was worked out. The difference between man and ape was much greater, one said, than that between the apes themselves.[57]) It was indeed true, as Huxley insisted, that the difference between man and the highest ape was smaller than that between the highest ape and the lowest: yet the apes formed a continuous series of small steps, whereas no intermediate steps linked the chimpanzee with man.[58])

However, the general public were becoming impatient of the attempts to base the distinction between man and ape on "abstruse anatomical details."[59]) They were content with the bare assertion — which Owen had in fact volunteered at a British Association discussion — that there was an "impassable gulf" separating man from the lower creation.[60]) On the whole, they were content to let the distinction depend on the mental difference between man and animals. One interesting variant of the argument we have been dealing with now should, however, be mentioned. It might be argued that the physical differences between man and ape were small. But as they were accompanied by so very marked mental ones, they should be given a correspondingly great systematic importance. "The moral hiatus might be greatly out of proportion to the physical distinction,"[61]) as one orthodox naturalist put it. Indeed, from the mental point of view it was the dog, one said,

---

[56]) *Edinburgh Review*, 117, 1863, 557. (Argyll). See also *Annals and Magazine of Natural History*, 7, 1861, 458 (Owen).

[57]) *Athenaeum*, 1861: 1, 395—6, *Cassell's Illustrated Family Paper*, 7, 1861, 362.

[58]) *Edinburgh Review*, 117, 1863, 549, *Lancet*, 1868: 2, 765—6.

[59]) *Guardian*, 1863, Sep. 9, 841.

[60]) *Nonconformist*, 1861, July 17, 577. British Association report in *Daily News*, 1861, Sep. 7, *Manchester Guardian*, Sep. 5, *Standard*, Sep. 6, *Evening Star*, Sep. 7, *Record*, Sep. 6, *Patriot*, Sep. 12, *Nonconformist*, Sep. 11, 727, *Critic*, Sep. 14, p. 276; Leader in *Nonconformist* 1861, Sep. 11, 732—3.

[61]) *Daily Telegraph*, 1862, Oct. 4, p. 2, reporting Humphry and Molesworth at British Association. See also *Watchman*, 1862, Oct. 8, 331, *Good Words*, 9, 1868, 250—51, *Dublin University Magazine*, 74, 1869, 588.

rather than the monkey, that came next to man — a point of view which came natural to many people. This again showed that mental organisation had little to do with physical structure.[62])

Darwinians did not attach very much weight to the missing link argument in its specific application to man. The general homology of the human body with that of the lower animals was too obvious to require much discussion, the variability of just those parts in man and the apes where they differed most from each other,[63]) the incompleteness of the fossil record, the brevity of the time of observation — all these points held for man as they did for the lower animals. Usually, therefore, Darwin and the Darwinians relied on the general acceptance of the Descent theory for the lower creation to lead to its acceptance for man as well. They therefore tended to prefer to concentrate their efforts on the former object, since it did not so directly bring prejudices and emotion into play.[64]) And as a matter of fact, opposition to the idea of man's bodily descent from ape-like ancestors was notably weaker at the end of our period. Attention was concentrated on the origin of his mental constitution instead — a development which coincided with the increasing attention given to Natural Selection rather than Evolution as such as regards the general theory.

Natural Selection

The argument from the alleged uselessness of the incipient stages of ultimately useful structures was the chief argument against Natural Selection as applied to man. At first it appeared in a rather vague form. Man, one said, was physically weaker than most animals of comparable size, and therefore also than his supposed progenitors. His physical weakness, at the present time, is more than compensated for by his mental superiority. But in former times, when that mental superiority, according to the evolution theory, had not existed, how could his loss of strength and swiftness give him that advantage in the struggle for life which the Natural Selection theory demanded? This

---

[62]) Crawfurd at British Association, rep *Morning Post*, 1863, Apr. 15, *Daily News*, 1863, Aug. 31, *Times*, 1863, Aug. 31, *Critic*, 1863, 295—8. See also *Quarterly Review*, 114, 1863, 414.

[63]) *Natural History Review*, 1861, 510; British Association, 1863, reports of Embleton's paper (see above, Chapter 4, p. 74).

[64]) *Westminster Review*, 33, 1868, 262—3.

reasoning was advanced by Agassiz and others,[65]) and became popular
when the Duke of Argyll took it up. "Man must have had human
proportions of mind before he could afford to lose bestial proportions
of body," wrote the Duke. And if Darwinians wished to argue that the
changes might have proceeded concurrently, the Duke had an answer
at hand: "If the change in mental power came simultaneously with
the change in physical organization, then it was all that we can ever
know or understand of a new creation."[66]) A popular writer illustrated
Argyll's argument in a picturesque manner: "Place a naked high-ranking
elder of the British Association in presence of one of M. de Chaillu's
gorillas, and behold how short and sharp will be the struggle."[67]) It
appears that in most cases those who reasoned in this way consciously
or unconsciously assumed that the change must have been a sudden
and drastic one — especially the mental change, which, as we shall see
later, was often assimilated with the infusion of a soul into man's body.

A much more developed and detailed criticism of the Natural Selection
theory as applied to man was that offered by A. R. Wallace towards the
end of the 1860s, at the British Association in 1869, in an article in the
Quarterly Review in the same year, and in this *Contributions* in 1870.
"Neither natural selection nor the more general theory of evolution
can give any account whatever of the origin of sensational or conscious
life ... We may even go further, and maintain that there are certain
purely physical characteristics of the human race which are not explicable on the theory of variation and survival of the fittest,"[68]) he wrote
in the *Quarterly*, and went on to list, in addition to man's moral faculties,
his hand, his erect posture, his expressive features, the beautiful symmetry
of his features, and his power of speech, and in general the size of his
brain, as properties which could not have been developed by Natural
Selection. Instead Wallace suggested that there "seems to be evidence
of a power which has guided the action of [the laws of organic develop-
ment] in definite directions and for special ends."[69])

---

[65]) *Friends' Quarterly Examiner*, 1867, 42, *Press*, 1865, 896, ref. Guizot, *Quarterly Journal of Science*, 3, 1866, 173.

[66]) *Good Words*, 9, 1868, 253, and reviews of Argyll's book, e. g. *Examiner*,
1869, Apr. 17, 245.

[67]) *Dublin University Magazine*, 74, 1869, 589.

[68]) *Quarterly Review*, 126, 1869, 391—3.

[69]) *Quarterly Review*, 126, 1869, 393.

These remarkable views of the co-founder of the Natural Selection theory were naturally highly appreciated by anti-Darwinians,[70] while Darwinians understandably did not quite know how to treat them. Darwin contented himself with remarking, in a note in the *Descent*, that Wallace's recent views would surprise every one who had read his earlier publications on the subject.[71] Scientific writers were of course highly critical of the alternative explanation offered by Wallace.[72]

Anyhow, Wallace's arguments undoubtedly had a marked influence, especially as they coincided in many respects with the general criticism of the Natural Selection theory associated with Mivart's name: and Mivart also added further illustrations for the argument.[73] Many commentators must have felt, as one of them put it, that "Mr. Wallace's reference . . . to a Creator's will really undermines Mr. Darwin's whole hypothesis."[74]

Though several instances of purely physical characteristics were adduced by scientific writers, the most important, and also the most popular of all the arguments against the Natural Selection theory as applied to man concerned his mental powers. Wallace was among those who argued that the wonderful capabilities of the human mind were of no use to a man living under the conditions of savage life. As, on Darwin's theory, humanity must have passed through a stage of savage life, the conclusion must be that his brain could not have developed by Natural Selection alone, but would have been guided by some benevolent power; a conclusion which was welcomed in many religious organs. "We cannot conceive how, according to the Darwinian hypothesis, man might have become more crafty than the fox, more constructive than the beaver, more organized in society than the ant or the bee; but how he can have got the impulse, when he had once made

---

[70] See above, Chapter 4, p. 84.

[71] *Descent*, 73, note.

[72] *Nature*, 2, 1870, 472, *Westminster Review*, 38, 1870, 195. *Saturday Review* 29, 1870, 710.

[73] *Month*, 13, 1870, 600—3, *Guardian*, 1870, Sep. 14, 1097, *Popular Science Review*, 10, 1871, 188, *Guardian*, 1871, 935, *Academy*, 2, 1870—1, 180, (Wallace), *Standard*, 1871, Apr. 13, *Tablet*, 1871, Sep. 9, *Press*, 1871, 262—3, *Morning Post*, 1871, Aug. 8, p. 7, *Lancet*, 1871, Mar. 18, 381, *Edinburgh Review*, 134, 1871, 200, *Leisure Hour*, 1872, 364.

[74] Staniland Wake at British Association, rep. *Morning Post*, 1871, Aug. 8, p. 7.

his position on the earth secure among the animals, to follow out abstract ideas and to go working on and on, while all other creatures rested content with the sphere which they had made for themselves — this is, indeed, hard to understand ... No mere exigencies of life or struggle for existence can have given rise to the high thoughts which led to poetry and science,"[75]) wrote Sir Alexander Grant in the *Contemporary*, and a writer in the *Guardian* asked, "How would the conceptions of space and time, of form, beauty, and order, above all, of right and wrong, be of any use to a savage in his early struggle for existence?[76]) Some critics enlarged upon this idea by introducing the current theory of the loss of organs and structures by disuse: "The brain of savage man is far beyond his needs ... even if once originated, [it] ought, according to Mr. Darwin's theory, to have been lost by disuse."[77 A])

The view that the brain of the savage was too large for his needs led up to the idea that it had been intended for some future use, or else it was interpreted as evidence of a former more highly developed stage of existence. The latter tallied most satisfactorily with the deterioration theory, while the former supported the idea that savages, though they were definitely men, could not qualify as equal with the civilized races. They had a latent power of receiving the gift of civilisation, but they had not yet received it. It is necessary to keep in mind that civilization, according to the traditionalist view, was not a natural development. The accepted dogma was that no race has ever progressed by its own unaided efforts,[77 B]) a doctrine which was a natural outgrowth of the Biblical story of God's covenant with his chosen people. The savages were people who had not been offered, or had rejected, that covenant. Some such idea, at least, seems to have been implied by Grant, who concluded that "the extremely unprogressive character of savage society is an obstacle to believing that the best civilization of the world, that of the Aryan and Semitic races, can have ever taken its start from such a society [as the savage one] in the primeval ages."[78]) Grant

---

[75]) *Contemporary Review*, 17, 1871, 278, 281. See also: *Edinburgh Review*, 134, 1871, 204.

[76]) *Guardian*, 1870, Sep. 14. See also: *Standard*, 1871, Apr. 13, p. 3, *Academy*, 2, 1870—71, 183 (Wallace).

[77 A]) *Edinburgh Review*, 134, 1871, 204. See also: *Dublin University Magazine* 74, 1869, 599.

[77 B]) Archbishop Whateley's thesis; see e. g. Argyll, *Primeval Man*, 130.

[78]) *Contemporary Review*, 17, 1871, 280. See also *John Bull*, 1872, Nov. 30, 824.

clearly wished to believe that at least European man's progenitors had never led a savage life. Savages, on the other hand, had not received the Divine gift of civilization; they were, as Grant said, "the back-waters and swamps of the stream of humanity, and not the representatives of its proper and onward current."[79]) In this manner the distinction between the civilized Europeans and the uncivilized races could still be upheld as divinely ordained, while the unity of the human species was retained.

However, the progress of prehistoric archaeology made it increasingly difficult to deny that even the European races had once lived as savages now do. It then only remained to draw comfort from the supposed fact that, as Grant put it, "we know for certain that if the best races did pass through a period of communal marriages, and the like, they passed out of it early and completely."[80])

The Human Mind

Those who, like the writers quoted above, applied the argument of the uselessness of incipient structures specifically against the Natural Selection account of man's mental superiority, were only a rather small minority. Much more popular was the argument that man's mind and soul, being entities of quite another nature than his body, could not be explained in the same terms, since they did not belong to the sphere of natural science, but to that of philosophy and religion. The Natural Selection theory — indeed, any purely scientific theory — was declared to be fundamentally incapable of explaining man's mental characteristics. It was unnecessary to discuss it on its own ground and in its own terms. The theory was opposed instead on *a priori* grounds, which were largely derived from Christian religion and idealistic philosophy.

Critics of this type did not always, indeed, not usually, make a clear distinction between the human mind — a psychological entity — and his soul — a religious entity. This was natural at a time when psychology was still very largely looked upon as a branch of philosophy, and as closely bound up with religion, and when religion was in general much more intimately connected with the world of ordinary experience than it has since become. We shall here, however, discuss separately man's

---

[79]) *Fortnightly Review*, 9, 1871, 363.
[80]) *Fortnightly Review*, 9, 1871, 370.

soul, his intellectual capacities, and his moral and religious ideas; keeping in mind that usually none of these matters were sharply distinguished by writers.

The Soul

The chief aim of the argument we are now concerned with was to establish that at least the essential nature of man was something which placed him absolutely apart from the brutes — something of which the lower creation did not even possess a faint beginning, which might have served as a starting-point for a gradual development. As we have already said, this alleged fundamental distinction between the human and the animal worlds could be used as a missing link argument against the Descent theory as such, but it could also be turned directly against the Natural Selection part of the Darwinian doctrine. It was not necessary to insist on the separate and supernatural creation of man as a physical species, as long as the spiritual element in him could still be retained as a carrier of the theological interpretation.

The religious overtones of the argument were quite obvious, especially when the concept of the soul was used to express man's distinctive character. Several writers advocated an interpretation of Genesis according to which man's creation occurred when God breathed His spirit into man's animal body.[81] Man's body might be developed naturally, but his soul was in a different category: "What the law of development could do, or whatever else the law of production of species may be, seems to have terminated in the gorilla. Intellect, a moral sense, and a soul being superadded, the gorilla is converted into a man."[82]

The idea of the supernatural creation of the soul was naturally included in the fairly popular theory of evolution according to which each specific change, or at least each major evolutionary step, was due to a special Divine intervention into the course of nature. It was therefore quite in order for the Duke of Argyll to hold that "If it were proved to-morrow that the first man was 'born' from some pre-existing form of life, it would still be true that such a birth must have been in every sense of the word, a new creation."[83]

It is hardly surprising to find that the idea of the special creation

---

[81] E. g. *Recreative Science*, 2, 1860—1, 151—160.

[82] *All the Year Round*, 5, 1861, 243.

[83] *Edinburgh Review*, 116, 1862, 389.

of the human soul was advocated also by those who were inclined to consider the species below man, sometimes including the physical part of man's body, as originated by "secondary" causation. Thus while they might admit that the creation of new physical species required no new direct interposition by God, the creation of the soul did require it. The best exponent of these views was Mivart, who was well summarized by a *Tablet* reviewer: "Mivart ... is of opinion that, provided the exertion of creative power be not denied, it may be supposed to have acted through secondary causes, even in the formation of the human body, though the soul of man must, of course, have arisen from direct and immediate creation."[84]) Similar ideas, less concisely but often much more picturesquely expressed, were fairly common in the religious press. Some writers attempted to describe what the creation of the soul was like: "There is nothing shocking in the idea [of man's descent from the apes] as long as we also believe that the simian was not simply developed into a man, but was the subject matter . . . on which a creative force was exerted, producing an immortal, sinless, powerful, and wise image of Him Who is eternal, all holy, almighty, and knoweth all things."[85]) Naturally, the creation was believed to be an instantaneous event: the anti-Darwinian insistence on big mutations left room for this belief. Indeed, there can hardly be any doubt but that the theory of big mutations was largely motivated by the belief. Thus a Dissenting organ quoted a French writer with evident approval on this subject: "I assume that natural history demonstrated by solid proofs that the first man was carried in the bosom of a monkey; and I ask: What is the circumstance which set apart in the animal species a branch which presented new phenomena? What is the cause? That monkey-author of our race which one day began to speak in the midst of his brother-monkeys, amongst whom thence forward he had no fellow; that monkey, that stood erect in the sense of his dignity, that, looking up to heaven, said, My God! and that, retiring into himself, said: I! . . . What climate, what soil, what regimen, what food, what heat, what moisture . . . separated from the animal races, not only man, but human society? . . . That monkey, what shall we say of it? Do you not see that the breath of the Spirit passed over it?"[86])

[84]) *Tablet*, 1871, Feb. 25, 232.

[85]) *Church Review*, 1865, Mar. 25, 270.

[86]) *Eclectic Review*, 10, 1866, 152—3, quoting E. Naville.

It is against the background of these views that the religious were
able to save the traditional chronology of the Bible on the age of the
human race. The year 4004 B. C. was the date of the "advent of the
divine element into humanity."[87]) Thus the fossil skulls of early human
races could be summarily dismissed: they could only prove the existence
at that early age of man's earthly covering, but "there are no traces of
the existence of man — *qua* man, 'a living soul'."[88]) One wonders what
sort of evidence could have been acceptable to writers of this type.
When it became clear that these early races had at least possessed some
rudiments of culture, traditionalists only restricted their definition of
man's soul to meet the new situation. "I by no means deny," wrote
somebody in *Good Words*, "that there may have been . . . a species of
animal . . . still more like man than any of the existing ape tribes;
more like him in general organization, and with keener instincts and
more intelligence than any of those tribes now possess; and yet not
man — having none of man's higher intellectual nor a vestige of his
distinctive moral endowments."[89]) There were writers who seemed willing
to restrict the "Divine element" to the extent that even the present
savage races seemed exluded — though the unity of mankind might
be saved by the degeneration theory. The *Nonconformist*, reviewing
Sir Alexander Grant's articles, said: "Sir Alexander Grant points,
with convincing emphasis, to the fact that savage humanity is unprogressive . . . Mr. Darwin's investigation of humanity must necessarily
stop short at savage life . . . From this point the life of man is not
simply human; it is Divine, and cannot be completed without Divine
intervention, which infantile science ignores, and calls 'a break' and
leaves to be discussed 'in another place'"[90]).

It should not be thought that views such as these were confined to a
small group of religious extremists. If Sir Alexander Grant was not
himself so explicit as his commentator, his opinions were not very
widely different. And Mivart, who so successfully expressed the dominant trend of opinion on these matters in the early 70's, wrote in the
respected *Contemporary*, "If Adam was formed in the way of which

---

[87]) *Nonconformist*, 1871, 240.

[88]) *Church Review*, 1864, Mar. 26, 304.

[89]) *Good Words*, 6, 1865, 380.

[90]) *Nonconformist*, 1871, May 4, 428. See also *Quarterly Journal of Science*,
1864, 88.

I suggested the possibility [by descent from the apes], he would, till the infusion of the rational soul, be only animal vivens et sentiens, and not 'homo' at all."[91]) Further, the *Times*, quoting Froude in its review of Darwin's *Descent*, maintained that "even if our bodies could have been developed by Natural Selection to such a degree of perfection as to render them fit for our use, the soul still remains distinct from its bodily habitation."[92])

Such quite general references to the soul as a separate and distinct entity, whose creation was to be accounted for separately from that of his body, were quite common. A scientific organ like the *British Medical Journal* apparently subscribed to this idea, when it criticised those writers who only occupied themselves with the physical differences between men and animals: they miss, it said, "the true distinction between men and monkeys . . . that is to be found in man rather than in his mere carcase."[93]) An early reviewer of the *Origin*, favourably disposed towards Darwin, held that "there is something so transcendently superior in the divinity — I know not how else to express it — of man's soul, that places him, at least for the present, quite out of the argument."[94])

The soul was often said to stand in relation to life as life itself to dead matter. "Species is a mystery, life is a great mystery, the conscious rational soul is a greater mystery still,"[95]) was a typical statement.[96]) Similar ideas were advanced by Stokes in his presidential Address to the British Association in 1869, and were naturally underlined in the religious press. The *Guardian* declared in a leader that "Mind is in his judgment as much above Life, as Life itself is above Chemistry and Mechanics, and he has little expectation that Science will ever be able to unfold its laws."[97]) It may be assumed that the terms Mind, Life, and Matter were interpreted as standing for substantial entities with independent existence. Few, however, attempted to describe

---

[91]) *Contemporary Review*, 19, 1871—2, 185.

[92]) *Times*, 1871, Apr. 8, p. 5.

[93]) *British Medical Journal*, 1861, Sep. 21, 316.

[94]) *London Review*, 1, 1860, 33.

[95]) *Illustrated London News*, 1871, 243.

[96]) See e. g. *Morning Advertiser*, 1863, Sep. 4, p. 4, *London Review*, 1868, Aug. 29, 263, *Weekly Review*, 1869, Aug. 28, 816, *Dublin Review*, 17, 1871, 5, *Guardian* 1871, June 7, 682 (letter).

[97]) *Guardian*, 1869, Aug. 25, 948. See also *English Independent*, 1869, Aug. 26, 829.

more in detail what sort of entity the soul was. But the conceptual realism of most anti-Darwinians often led to the soul being discussed in terms altogether parallel to those used about the physical body. Huxley's lecture on "The Physical Basis of Life" — which caused a tremendous stir among the general public[98]) — led one writer to assert that "there may be in man a protoplasm which science cannot detect, one which he shares in common with the highest, as his physical basis rests upon the lowest."[99]) Accordingly it was also possible to extend the idea of continuous evolution to this entity: and this was also done in the more unorthodox religious organs. "It is surely not inconceivable that a germ of spiritual force might have been implanted in the race of reasoning beings, and that such germ is transmissible and capable of growth by purely natural methods."[100]) The germ itself was separately created, endowed by God with a capability of developing in a particular direction. This was altogether in line with the theories of directed evolution which traditionalists were constructing in opposition to Darwin's Natural Selection theory, towards the end of the 'sixties.[101])

The Intellect

Those writers who emphasized the intellectual capacities of man as his essential distinction had the advantage of being able to support their argument by specific factual illustrations. On the other hand, they were hampered by the fact that the intellect, unlike the soul, was not so readily apprehended as a separate entity with independent existence: conceptual realism had not here such a strong backing from religious tradition and popular thought. And without the support of such conceptual realism, it was difficult to contend that man's intellectual capacities were fundamentally and absolutely distinct from those of at least the higher animals. It is therefore not surprising that when such claims were in fact made, there was a tendency to fix the distinction by using different words for the mental capacities of man and those of animals. There was of course an old tradition for this.

Above all, there was the traditional distinction between instinct and reason. Man alone possessed the latter, while the activity of animals

---

[98]) Brown, Alan Willard, *The Metaphysical Society*, 51.

[99]) *English Independent*, 1869, Feb. 18, 151.

[100]) *Theological Review*, 9, 1872, 341.

[101]) See above, Chapter 12, p. 267—79.

was wholly guided by instinct. It is clear, however, that this assertion was based on *a priori* rather than empirical grounds. Those who made it did not in general indicate precisely how instinctive behaviour could be distinguished from reasonable activity. One writer naïvely recognized that "their instincts simulate our reason"[102]) — under such circumstances one should not expect any observable difference of behaviour. In a close argument the distinction was therefore of little value, and it was in fact primarily met with in the popular, non-scientific type of organs.[103])

Several writers tried to provide a more satisfactory definition of the faculties which were uniquely human. Argyll in the *Edinburgh*, reviewing Huxley's *Man's Place*, drew up quite an extensive list. "There is the gift of articulate language, — the power of numbers, — the powers of generalisation, — the power of conceiving the relation of man to his Creator, — the power of foreseeing an immortal destiny, — the power of knowing good from evil, on eternal principles of justice and truth."[104]) The aspects of the problem that we have distinguished — the intellectual, and the moral and religious — are here mixed together. Common to them all was their obvious dependence on an idealistic type of metaphysics: neither "immortal destiny," nor "justice," nor "truth" were considered as merely abstract constructions of the human mind; they were independent realities, "eternal principles." This was especially obvious in the case of the moral and religious concepts, as we shall illustrate more fully below. But it was also true of the purely intellectual concepts, with which we are now mainly concerned. On the empiricist view, the generalisations of abstract thinking were "logical fictions," as Mill called them. They were based on the experience of the senses: they did not furnish any knowledge beyond that of sensory experience, but only served to systematize that knowledge. Now animals, like men, adjusted their behaviour in accordance with their sensory experience; the difference was that man's adjustment was on the whole more effective, since his power of apprehending and systematizing his experience was so much greater. For this reason, empiricists like Darwin could regard the difference in intellectual capacities between men and animals as a difference of degree only.

---

[102]) *Athenaeum*, 1863: I, 288.

[103]) *News of the World*, 1862, Oct. 12, *Daily News*, 1871, Feb. 23, p. 2, *Month*, 15, 1871, 80, *Times*, 1871, Apr. 7, p. 3.

[104]) *Edinburgh Review*, 117, 1863, 567.

But if, as idealists claimed, an abstraction was something more than a logical fiction, if it gave an insight (as Whewell put it) into the "essential nature and real connexions of things," an insight which was not derived from experience alone, then it was natural to maintain that the faculty which gave rise to this insight was essentially distinct from those possessed by mere animals.

The support which the idealistic doctrine on these matters had always received from the Church and from the established philosophical schools, had won for it almost the position of a self-evident truth among the general public. It was, as the *Quarterly Review* insisted in its review of the *Descent*, "the position ... in possession — that which is commended to us by our intuitions, by ethical considerations, and by religious teaching universally. The *onus probandi* should surely therefore rest with him who, attacking the accepted position, maintains the essential similarity and fundamental identity of powers which are so glaringly diverse."[105]) The Catholic *Tablet* referred to St. Thomas Aquinas: "The souls of brutes being entirely dependent on their bodies, it follows, as S. Thomas teaches, that they can have no ideas higher than those which can be acquired by means of the senses. Man, on the other hand, is endowed with the power of subjecting the ideas obtained from his senses to analysis, and by abstraction, of arriving at the knowledge of the essence of the objects represented to him from without. This is a perfectly distinct faculty from anything to be found in the brutes."[106]) It may be due partly to Aquinas that the Roman Catholic tradition on this subject seems to have been very strong. The French naturalist, St. Hilaire, maintained that "it is by his faculties, so incomparably higher, by the addition of *intellectual* and *moral faculties* to the *faculty of sensation* and the *faculty of motion* that Man ... separates himself from the animal kingdom and constitutes above it, the supreme division of nature, the Human Kingdom.[107]) Descartes' idea of animals as automata is of course in line with this tradition.

But it would be a mistake to look upon this way of thinking as peculiarly Roman Catholic. Since it was, as the *Quarterly* said, universally sanctioned by tradition, it was also found among Englishmen with Evangelical tendencies, and a *Contemporary* reviewer expressed it as

---

[105]) *Quarterly Review*, 131, 1871, 69.

[106]) *Tablet*, 1871, Apr. 15, 456.

[107]) Quoted with disagreement in *Natural History Review*, 1862, 1.

follows: "All other animals but man seem to be under a strict limit, which they cannot pass, their faculties, however acute, and wonderful, are restricted in their direction to the finding means of bodily preservation and bodily enjoyment . . . All these tendencies in the lower animals are stopped dead, as it were, by the want of the faculty of apprehending universals."[108])

The power of apprehending universals was obviously intimately bound up with language, and it is remarkable how often Darwin's opponents adduced man's "power of articulate speech" as one of the important absolute distinctions between man and animals. In a certain sense, this was something that everybody had to agree to: no animal possessed an articulate language. But the anti-Darwinians also claimed that it was impossible for articulate language to arise by a natural development of the faculties possessed by the brutes, since man's language was based on a perception of exactly those universals to which no merely sensory experience could lead.

The chief exponent of these views in England was Max Müller, who was popularly recognized as the leading authority on the incipient science of philology. Müller's avowed opposition to the Darwinian theory on these grounds was of considerable importance, since in other respects, as we have seen, the science of philology yielded effective support to evolutionary views. His followers insisted strongly on the dependence of human language on the power of generalisation and abstraction, and contended that "the fact that words express general ideas (that is, that every word was originally a predicate) has been proved by the strictest analysis,"[109]) and accepted the view that man had been created in a perfect state, and consequently that "there was not and could not have been any period of mutism for a being whose faculty of speech was strictly natural."[110]) These views almost seem intended to compensate for the unorthodox tendencies of philology in other spheres. They were at any rate eminently acceptable to traditionalists. It is significant that Müller himself later employed his philological theory in a direct attempt to refute the Darwinian doctrine of Man's descent.[111])

---

[108]) *Contemporary Review*, 17, 1871, 277.

[109]) *Edinburgh Review*, 115, 1862, 101.

[110]) *Ibid.*

[111]) Report of Müller's lecture in *Nature*, 7, 1872—3, 145.

In religious circles, such views as Müller's were seen as a confirmation of the traditional view of language as a direct and separate Divine gift. Moreover, when philology could prove that primitive languages were at least as complex as those of modern civilized nations, and that the great linguistic families of the world were as distinct thousands of years ago as nowadays, one tended to interpret these facts as a confirmation both of the story of Babel, and of the view that man had never been different from what he is now.[112])

But it was also possible to assimilate the idealistic view of language with the evolution theory. Language was then regarded as a separate entity, developing like an organism wholly independently of all merely human wishes. But the germ of language, like the germ of any organic being, was created by God. Language, one said, "is something beyond man's control. It is . . . under the dominion of natural laws, has an existence and a vitality in common with all organic nature, and is not more the invention of man than the fuel he burns or the food he consumes. It is the gift of God to man, but then it is so only in a certain sense. It was not *communicated* to him in its perfect state, or any state at all; but it was implanted in him in such a way that he himself became the unconscious instrument of its development."[113])

Another corollary of the idealistic view of the human faculties as separate entities conferred by a special Divine act, was the widespread conviction that a fundamental distinction between man and the animals was "the stationary condition of the one and the progressiveness of the other."[114]) Animals stayed where God had placed them — or, alternatively, developed into what God willed — whereas man, with his free-will and power to understand, albeit imperfectly, the ways of God, was able to progress by his own efforts. Though superficially this view might appear as a concession to the development theory, it was not really so. On the contrary, man's ability to progress set him absolutely apart from the brutes. It was, like all his other marks of distinction, a Divine gift, "an inward impulse which led to the evolution of civilization."[115]) The inward impulse solved the problem of reconciling

---

[112]) *Dublin University Magazine*, 74, 1869, 595.
[113]) *Home and Foreign Review*, 1, 1862, 194.
[114]) *Guardian*, 1871, 1007.
[115]) *Contemporary Review*, 17, 1871, 281.

the progress theory with the equally popular one of man having been created in a state of civilization.[116])

Against the background of such specific arguments it was possible in other contexts merely to assert that the intelligence of man was of another nature than that of the animals.[117]) For it was unnecessary to develop any further an idea which was so obviously in harmony both with traditional convictions and long-established prejudices and feelings.

Morality and Religion

However much Darwin's idealistic opponents differed from him in their assessment of man's intellectual superiority over the animals, they disagreed even more strongly from him as regards man's moral nature. In this matter, religious objections obviously strongly reinforced those of idealistic metaphysics and sheer traditionalism.

According to the Darwinian theory, all the characteristics of any species had arisen by the accumulation of such chance variations as happened to favour the survival of their possessors. Only those characteristics could be selected which were beneficial in the particular environment where the organism found itself. Moreover, as Darwin himself repeatedly emphasized, there was no reason to expect any absolutely perfect adaptation, since the natural selection process was necessarily slow, and moreover, since its rate depended on the severity of the struggle for existence.[118])

Now if the Darwinian theory was to be consistently applied to man, a similarly relativistic view of human morality could not be avoided. There was no reason to expect the systems of values, whether ethical or aesthetical, which were actually to be found in human societies, to have any sort of universal and absolute validity. They had been developed in response to the particular conditions under which the different communities had found themselves. Communities adopting a moral

---

[116]) E. g. *Athenaeum*, 1863: 1, 288, *Parthenon*, 1863, 235, *Quarterly Journal of Science*, 1864, Jan., 88, *Contemporary Review*, 17, 1871, 277, *John Bull*, 1872, Nov. 30, 824.

[117]) *Daily Telegraph*, 1863, Aug. 31, p. 2, rep. Crawfurd, *Press*, 1863, 163—4, *Quarterly Journal of Science*, 3, 1866, 173, *Quarterly Review*, 126, 1869, 391, *Examiner*, 1869, Apr. 17, 245—6, *Academy*, 1, 1869—70, 236, *Inquirer*, 1871, May 20, 311, *Gardener's Chronicle*, 1871, 649, *Month*, 14, 1871, 232, *Guardian*, 1871, 1006, *Lancet*, 1872: 2, 297.

[118]) See above, Chapter 12, p. 255.

code which made them prosperous and strong would tend to proliferate. To the extent that they succeeded in inculcating their beneficial code in their descendants, or in others who came under their influence, such a moral system would tend to gain ground. But the fact that a certain moral code had established itself did not prove that it was a perfect one, even for the society which had developed it. The success of a society in the struggle for life might very well be due to other causes than its system of moral values, and even if, for the sake of the argument, this difficulty was disregarded, it could at most prove that its morality was more favourable, or more efficient in the struggle for life, than others with which it had come into competition. Just as many species in Australia had shown themselves incapable of holding their own against imported Old World forms, so also it might well be that an established moral code would prove inferior — from the point of view of survival value conferred on the community accepting it — to one developed under different conditions, or to a totally new one.

Such relativism of values was altogether unacceptable to the religious. To them the good and the right were absolute and eternal entities, and Man was endowed with a special moral intuition, which provided him with an insight into this world of eternal values. One might concede that man's intuitions were not infallible. But no doubt was allowed as to their existence, and still less about the existence of the world of absolute values which was their object. Belief in that world was enjoined not only by man's conscience, but also by the Bible and the Church. Thanks to these sources, it was possible to establish the standard according to which the values actually found in different human societies could be themselves evaluated.

Darwin, as a scientist, was not concerned with evaluating men's values according to some absolute standard, but only with ascertaining what the values were, and explaining why they had become established. But if the theory of Natural Selection was true, a possibility was provided of determining the survival value of any set of values, and if, as certainly appeared reasonable to most people, survival was recognized as something good, the Darwinian theory also offered a means of evaluating the values actually accepted by human societies. And this evaluation was completely independent of any reference to religion. Moral and aesthetical codes could be regarded as good or bad, in the sense of successful or unsuccessful, to the extent that they increased the chances of survival of the communities which accepted them. If such a standard of right and wrong was to become widely accepted, the absolute standard provided by religion could no longer be maintained.

This was a serious matter. The conflict between science and religion had already led to the exclusion of religion and religious explanations from a wide range of factual questions, so that several defenders of the faith had chosen to take the position that religion did not concern the world of factual experience at all, but only that of morality.[119]) Now if Natural Selection was applied to the development of man's ethical ideas as well, religion seemed in danger of total disintegration. "We do not see how to reconcile with our Christian faith the hypothesis ... that our moral sense is no better than an instinct like that which rules the beaver or bee; that He whom we have been accustomed to regard as the Creator of all things, is a creature of our imagination, and that our religious ideas are a development from the dreams and fears of anthropomorphous apes,"[120]) wrote one religious organ. Other writers stressed the social effect of the Darwinian account of morality. "The sense of right and wrong, according to this view, is no definite quality, but merely the result of the working together of a series of accidents controlled by natural selection for the general good. We need hardly point out that if this doctrine were to become popular, the constitution of society would be destroyed, for if there be no objective right and wrong, why should we follow one instinct more than the other, excepting so far as it is of direct use to ourselves?"[121])

What Darwin asserted of the moral ideas clearly also applied to religious ideas. They differed from community to community, and to the extent that they were of any importance in a community's struggle for existence, they were also bound to be subject to the influence of Natural Selection. Whether there was anything objective corresponding to the religious ideas that people held, — for instance, whether God existed — was a question which Darwin did not need to ask, as little as he needed ask whether anything objective corresponding to the moral ideas of people — for instance, God's will — existed. These questions could have no empirical answer. The only relevant question for Darwin was whether a community's moral and religious ideas did

---

[119]) Above, Chapter 5, p. 112.

[120]) *British and Foreign Evangelical Review*, 21, 1872, 31.

[121]) *Edinburgh Review*, 134, 1871, 217.

in fact have any survival value. If they had, they would establish themselves whether they corresponded to anything really existing or not. But since, as we said, everybody would instinctively feel that survival value was something good, it was natural to conclude that the Darwinian theory came into conflict with the current religious doctrines on these matters. It was therefore in order for one religious critic to ask, "If Christianity itself is a mere product of the human mind, how is its truth to be maintained?"[122])

Empiricists inevitably looked upon moral and religious ideas as products of the human mind. But before Darwin it had been difficult to explain both the nature and the strength of people's moral and religious beliefs on purely empiricist principles. Utilitarianism was partly an attempt to overcome these difficulties, but the very unplausibility of utilitarianism in the moral sphere had rather strengthened the hand of those who maintained that any attempt to base morality on nothing but experience was fundamentally fallacious, and who advocated instead an idealistic theory of ethics, according to which moral judgments depend on "intuitive beliefs, which certainly were in our minds as originally constituted, and were not contributed to them from without."[123]) The current idealistic doctrine was, of course, that these intuitive beliefs were implanted in the soul by the Creator. Man was endowed, one said, with a moral sense or conscience, and it was especially important to set it apart as divinely inspired — and consequently *true* — if the lower animal senses and appetites were recognized as developed by a natural process. "The moral sense or conscience," said the *Guardian*, "[is] most important to the true humanity of the individual and to the maintenance of society. If any theory tends to depose it from its spiritual throne ... and makes it but an instinct differing from others only in the greater vividness and durability ... such a theory comes home to those who care little about abstract metaphysics, and is pregnant with results which will pass beyond the school or the lecture-room, to affect the great issues of practical life."[124])

Now the Darwinian theory of Natural Selection offered a solution whereby the empiricist, utilitarian view of the origin of moral ideas

---

[122]) *British and Foreign Evangelical Review*, 17, 1868, 187.

[123]) *Friends' Quarterly Examiner*, 1867, 52.

[124]) *Guardian*, 1871, 1007. See also: *Guardian*, 1871, May 24, 624, (letter), *Fortnightly Review*, 9, 1871, 371 (Grant), *John Bull*, 1871, Apr. 6, 234.

could be made plausible, since it was no longer necessary to depend on either conscious calculations of utility, or on the effects of conscious or unconscious experience of utility, associated with certain actions during the life of the individual or the race. As the example of the bee beautifully showed, it was possible for instincts which could not possibly be experienced as useful by any of the individuals possessing them to be naturally selected, since Natural Selection affected the population as a whole, and not the individuals within it.

But the bearing of the Darwinian theory on these matters was not recognized by the large majority of the public, partly, no doubt, because at first the problems raised by the Descent theory as such caused the Natural Selection theory to be neglected, and later because anti-Darwinians were so successful in persuading the public of the insufficiency of Natural Selection as an explanation of Evolution. In addition, traditionalists were able to seek support from the current two spheres doctrine of science and religion, according to which at least the domain of morality was wholly under the jurisdiction of religion. Therefore the influential *Saturday Review* could declare in its review of the *Origin:* "To any . . . man of science who should attempt to prove to us that the moral and spiritual faculties of man have been gradually developed by the working of matter upon matter, we should reply by demurring *in toto* to the applicability of his reasoning. No conceivable amount of evidence derived from the growth and structure of animals and plants would have the slightest bearing upon our convictions in regard to the origin of conscience, or man's belief in a Supreme Being and the immortality of his own soul."[125]) Such views were very common.[126]) Naturalists might say, "With Human Progression in a moral sense, no morphological theories have anything to do; it is only with Man as an animal . . . that Zoology is concerned."[127]) And for religious people the attitude was a matter of course. "Let man be allied by specific descent to the whole animal creation, he is a 'moral and intellectual' being, and on this religion and morality depend,"[128]) wrote the *Nonconformist*, and laid it down that "Science offers as yet no certain base on which to build a new

[125]) *Saturday Review*, 8, 1859, 775.

[126]) *London Review*, 1, 1860, 33, *Morning Post*, 1860, Jan. 10, p. 2, *Spectator*, 1860, 380, (letter). See also note 131, below.

[127]) *Geological Magazine*, 1865, 309.

[128]) *Nonconformist*, 1868, May 27, 538.

theology . . . [Darwin's] now famous theory relates only to the form of life, not to the principle of life, still less to the moral principle or soul."[129])

When the bearing of the Darwinian theory on these questions was recognized, the challenge was in the large majority of cases simply met by the bare assertion that *since* man's moral sense was implanted in him by God, it was impossible to explain it naturally. "Certainly," wrote one religious critic, "the conscience is a primitive fact of man's nature, which is not formed by language, study, or habit, and which, thanks to the Father of truth, no system can ever annihilate . . . Man is endowed with a moral nature, a perception of right and wrong, and a feeling of moral obligation."[130]) Not only was it impossible to explain conscience and moral sense naturally, they also furnished a basis for a fundamental distinction between man and the animals, placing man, as Quatrefages said, in a different kingdom from the rest of the creation.[131]) It was on this basis, above all, that the theory was established that Man's creation had taken place when God had miraculously breathed a soul into man's bodily frame, which we have illustrated above.

There were also attempts to meet on its own ground the Darwinian theory of the origin of human morality. Critics sometimes went some way with Darwin in recognizing that his theory might to some extent account for the *practice* of morality. But more was involved than merely moral behaviour — "material morality".[132]) An act was moral materially, if it was objectively in accordance with moral rules, though its motives might be non-moral, or though it might be performed unconsciously. Thus even animals might exhibit a material morality. But a moral man did not only behave morally, he was also conscious of the duty so to act. Such acts as were performed with a knowledge of that duty were not only materially, but also "formally" moral, and formal morality, said Mivart and those who accepted his argument, was distinctive of man. It presupposed a conscience, a conscious knowledge of right and

---

[129]) *Nonconformist*, 1868, Sep. 2, 869. See also: *Morning Post*, 1868, Aug. 27, p. 2, where the same words occur in a report of a British Association lecture by H. B. Tristram.

[130]) *British and Foreign Evangelical Review*, 12, 1863, 486, 490.

[131]) See e. g. *John Bull*, 1861, Sep. 14, 580 (Owen), *British Medical Journal*, 1862, 591, *Times*, 1861, Apr. 3, p. 7, *Popular Science Review*, 2, 1862—3, 403, *Journal of Sacred Literature*, 10, 1866—67, 201, *News of the World*, 1871, Apr. 16, p. 6, *John Bull*, 1871, Apr. 6, 234.

[132]) Mivart, *Genesis of Species*, 220.

wrong, and often manifested itself in the feeling of remorse. Darwin's critics were convinced that such a feeling could not have arisen through Natural Selection. "No amount of the accumulation of the experiences of utility could give origin to a feeling in which utility not only had no share, but to which it was, if anything, antagonistic."[133]) The argument was obviously not to the point, for the Darwinian theory did not assume that any conscious experience of utility was needed for the feeling to be naturally selected: what was necessary was that the feeling should in fact have survival value for the race. But such misunderstandings were common in relation to the Natural Selection theory.

In many cases the basis of the argument seems to have been, that since formal morality depended on conscious judgment and insight, there was nothing in the animal world from which it could have developed, since insight into moral rules presupposed the power of forming generalisations and abstractions. When expounding the fundamental distinction between animals and men as regards the moral sense, it was therefore natural for anti-Darwinians to support their views by referring to the traditional distinction between instinct and reason. The material morality of animals might be regarded as instinctive, man's apprehension of formal morality depended on reason. Darwin, said one critic, showed no sign of metaphysical acumen by confusing the two, as when he explained the moral sense as "a mere social instinct, refined and extended by means of a superior intelligence. But instinct and intelligence are simply incompatible one with the other, instinct being essentially a blind impulse."[134]) Thus when Darwin suggested that the feeling of remorse might be caused by the conflict of two instinctive appetites, one retorted that "the conflict of the two appetites gives ... no clue to the sense of Obligation which is the distinguishing element of Conscience."[135]) The basis of that sense was, instead, the direct moral intuition which was a natural assumption for the idealistic school. A *Quarterly* reviewer expressed it, "It is a patent fact that there does exist a perception of the qualities 'right' and 'wrong' attaching to certain actions ... intellectual judgments are formed which imply the existence of an ethical ideal in the judging mind."[136])

[133]) *Edinburgh Review*, 134, 1871, 220. See also: *Month*, 11, 1869, 274.
[134]) *Tablet*, 1871, Apr. 15, 456.
[135]) *Guardian*, 1871, 935.
[136]) *Quarterly Review*, 131, 1871, 79.

Darwin's derivation of the moral sense from the social instincts was naturally unacceptable to those who held these idealistic views. "It seems to us that the moral tie or bond of any race precedes the political, not the political the moral. It is, indeed, in the position of the elementary 'accidental' (or providential) advantage from which all the future history of advantage is deduced,"[137]) wrote the *Spectator*, and an Evangelical reviewer asserted his belief that "all the higher social virtues which exist in man are a consequence of his possession of a moral sense, and not the cause of its existence."[138])

On the idealistic view there was necessarily one and only one true morality. If different nations adhered to different moral systems, the explanation must be that their moral intuitions were not equally clear. But fundamentally morality was assumed to be the same everywhere. When Wallace, at the British Association in 1869, insisted on the high moral standard of the savage races of the Malay Archipelago, it was taken as a confirmation of the view that man had been created in a state of civilization, with a moral sense specially conferred upon him. The fact that the moral ideas of many present-day savages were revolting in the extreme could be dismissed by means of the current degeneration theory: their moral sense had been perverted by their sins.

Darwin did not contradict the view that the morality of the human race was fundamentally the same everywhere. He even gave some factual support to it, for on his Natural Selection theory as well, some conformity of moral codes was to be expected. For instance, honesty and truthfulness might be assumed to be beneficial to any society whose members practiced these virtues. On the other hand, Darwin went out of his way to insist that he regarded the conformity as solely due to similarity of conditions and of constitution. Under widely different conditions a fundamentally different morality might arise. He wrote in the *Descent*: "If, for instance, to take an extreme case, men were reared under precisely the same conditions as hive-bees, there can hardly be a doubt that our unmarried females would, like the worker-bees, think it a sacred duty to kill their brothers, and mothers would strive to kill their fertile daughters; and no one would think of inter-fering."[139]) This passage exhibited Darwin's relativism quite clearly,

---

[137]) *Spectator*, 1867, 1255.

[138]) *British and Foreign Evangelical Review*, 21, 1872, 27.

[139]) *Descent*, 151—2.

and it is hardly to be wondered at that his idealistic opponents strongly objected to his illustration.[140 A])

In the parallel sphere of aesthetic judgments as well, the idealistic school found themselves in opposition to the empiricist Darwinians. Darwin explained the sense of beauty as partly dependent on purely physiological reactions, partly on habit, and saw no difference in this respect between men and animals. Indeed, he even argued that in animals the sense of beauty was sometimes more developed than in man. He further regarded the relativity of aesthetic judgments as a matter of course. His opponents, on the contrary, wished to restrict the sense of beauty to the human species, making it depend on the perception of an ideal standard, which animals were by nature incapable of apprehending. Aesthetic judgments, they further held, were not relative, but absolute. They found support for this view in the facts. A *Quarterly* reviewer of the *Descent* declared that it was not true that the standard of personal beauty, for instance, differed from nation to nation according to the type that was prevalent in each. "All cultivated Europeans, whether Celts, Teutons, or Slaves, agree in admiring the Hellenic ideal as the highest type of human beauty."[140 B])

Conclusion

As the anti-Darwinian arguments concerning man's intellectual, moral and aesthetic capacities were largely based on idealistic metaphysics, the chief Darwinian defence consisted in asserting empiricist principles of scientific inquiry. When discussing psychological questions, Darwin deliberately avoided metaphysical speculations and religious considerations, using consistently a naturalistic approach: "from the side of natural history,"[141]) as he himself put it. His psychological method may almost be called behaviouristic. As he aimed at comparing the faculties of men with those of animals, he needed a standard of comparison, and that standard was necessarily the outward behaviour, since the subjective feelings and thoughts of animals were not open to investigation.[142]) If the behaviour was similar, then the thoughts were

---

[140 A]) *Theological Review*, 1871, 167—92 (F. P. Cobbe). See also: *Academy*, 3, 1872, 230, H. Sidgwick rev. Cobbe.

[140 B]) *Quarterly Review*, 131, 1871, 63.

[141]) *Descent*, 149.

[142]) *Descent*, 126—7.

assumed to be similar also. For the purposes of the argument, thoughts were defined in terms of behaviour, which was the proper empiricist procedure. On this basis it was easy to show that the supposed fundamental difference between animal instinct and human reason, for instance, had little factual basis, and was in fact largely a product of conceptual realism. Darwin approvingly quoted Leslie Stephen as saying, "The distinctions, indeed, which have been drawn, seem to us to rest upon no better foundation than a great many other metaphysical distinctions; that is, the assumption that because you can give two things different names, they must therefore have different natures. It is difficult to understand how anybody who has ever kept a dog, or seen an elephant, can have any doubts as to an animal's power of performing the essential processes of reasoning."[143] Huxley, in his usual cutting manner, dismissed the idea that the moral sense was an "immutable and eternal" law of human nature: "I do not see how the moral faculty is on a different footing from any of the other faculties of man. If I choose to say that it is an immutable and eternal law of human nature that 'ginger is hot in the mouth', the assertion has as much foundation of truth as the other, though I think it would be expressed in needlessly pompous language."[144]

Darwinians did not of course deny that there was a great difference between animals and men in mental powers, but the difference was one of degree, not of kind.[145] One direct way of pushing this lesson home was to point to the development of human infants: in them, the acquisition of the higher faculties undoubtedly proceeded gradually.[146] On this point, however, traditionalists had various solutions ready. Some assumed the separate development of mental germs,[147] others the special providential interposition in each individual instance: "Why should not conscience, like sight, be given to each new baby as one of his faculties for the conduct of life?"[148]

---

[143] *Descent*, 120, note.

[144] *Contemporary Review*, 18, 1871, 469.

[145] See e. g. *Examiner*, 1869, Aug. 4, 519; 1871, 233, 256, *Reynolds's Newspaper*, 1871, June 11, p. 2, *Observer*, 1871, Mar. 19, p. 3, *All the Year Round*, 5, 1871, 450.

[146] *Natural History Review*, 1862, 1, *Descent*, 128.

[147] See above, p. 316.

[148] *Spectator*, 1869, 1002. See also: *Fraser's Magazine*, 67, 1863, 474, *Quarterly Review*, 133, 1872, 449.

But this solution of the difficulty was fraught with its own dangers, since it tended to blur the distinction between ordinary phenomena and miracles. The awe-inspiring mysteriousness of the miraculous would disappear.

On the whole, the more specific attempts to establish the distinctness of the human soul on the basis of factual, scientific arguments do not seem to have been very successful. It is true that the large majority did not accept the Darwinian doctrine of the development of man's mental and moral powers by purely natural selection. But neither did most people wish to commit themselves to a definite "catastrophist" view of the creation of the human soul. The various solutions which we have set forth above illustrate, so to speak, the trial and error process by which the religious world responded to the new situation which had arisen when practically the whole of the educated portion of the community had come to accept man's bodily descent from lower animal forms. For the soul, as for the body, it was the Natural Selection theory which was the chief stumbling block. Now precisely as it had been possible to reconcile the physical transmutation theory with religious feelings and theological prepossessions, provided Natural Selection was replaced by a postulate of Divine Design, so it was also possible to accept the gradual development of the soul, if it was combined with a recognition that it must have been providentially guided. The manner of that providential guidance might be obscure — hence the many speculations on the subject — but if the Natural Selection theory was rejected, it was hardly possible to deny the necessity of assuming the existence of some guiding force. It may be that the majority were not prepared to reduce their opposition to naturalistic Darwinism to these bare essentials. But they provided the foundation for all the various individual schemes of reconciling Evolution with traditional modes of thought as regards man's nature.

CHAPTER 15.

Summary and conclusion

The general public's interest in the Darwinian theory was almost wholly due to two factors which were closely bound up with each other, namely, the religious and ideological implications of the theory as such, and its bearing on traditional views concerning the history of mankind and the nature of man. It is hardly probable that ordinary people would have cared very much whether animal and plant species were transmutable or not, unless the transmutation theory had clashed with the common interpretation of the Biblical cosmogony and widely held views on the manner of God's providence. Nor would they have been greatly disturbed by the transmutation theory as such, unless it had led to a doctrine of man's descent from the lower animals as a virtually unavoidable consequence. But as the theory did have these implications, it is not surprising that it attracted very wide attention indeed. It became one of those focal points of debate on which practically everybody was compelled to have some sort of opinion. Needless to say, the opinion cannot in most cases be called an informed one. Only a small proportion of those who took sides one way or the other had even a rudimentary knowledge of the facts on which the theory was based. Attitudes were determined, as in questions of religious, political or ideological allegiance, by the established prejudices of the social group to which one belonged, rather than by a detached consideration of the factual evidence.

Now when we say that the Darwinian question interested everybody, we must bear in mind that it meant different things to different people. To the uneducated majority the question was simply whether man was descended from Adam or from the apes: and most of them seem hardly to have believed that the second alternative could be seriously entertained. On a somewhat higher educational and intellectual level, the conflict between the transmutation theory and the traditional Christian cosmogony, and thus also with important points of Christian dogma, also attracted attention. Only on the very highest level was the debate centred on the fundamental problems which the theory raised on such matters as the teleological interpretation of nature, the structure of scientific explanation, and the relation of facts to values in the moral and aesthetic spheres of human activity. Yet all these various questions were interconnected, and even though the uneducated majority were unable to appreciate the problems which occupied the minds of the country's intellectual leaders, they knew at any rate that the Darwinian question raised such problems. This gave it, even on the popular level, a greater weight and importance than it would otherwise have had. It is interesting to compare the reaction towards Chambers's *Vestiges of Creation*, in 1844, with Darwin's *Origin of Species*, in 1859, from this point of view. In a sense the *Vestiges* acted more strongly on the popular mind than the *Origin*. The book was quite as much talked about in the press during the first few years, and had undoubtedly a much wider popular readership. It appealed to the imagination by treating Evolution as concerning the whole of nature, and not just the organic world. Yet in spite of this, "Vestigianism" never reached the proportions of Darwinism as a matter of public concern. The *Vestiges* was a popular success, but no more. No scientific authority ever came forward to support its thesis. The *Origin* was sometimes taken as a more learned and less comprehensive imitation of the *Vestiges* — it was in fact described as such by the *Daily News* reviewer in 1859 — yet since it was perforce taken seriously by the intellectual *élite* of the country, the questions which it raised, whether great or small, soon took on a much deeper significance than had ever been granted to the speculations of the *Vestiges*. The broad public perhaps did not realise precisely in what way Darwin was more significant than Chambers, but the stir he caused in the intellectual world showed that he was. Darwinism concerned, as one popular commentator put it, "the tremendous issues of life."

Though the general public's interest in the Darwinian doctrine was largely due to its repercussions on religious and ideological beliefs, and though their attitudes towards it were also mainly due to these factors, yet even on the most popular level one attempted to justify one's stand either by citing the opinions of scientific authorities in support of one's own position, or by means of factual, scientific arguments. The argument most widely employed on the anti-Darwinian side — which was always in a majority, and more so among the uneducated than among the educated — was that of the missing links. As long as Darwin could produce no actual instance of the transmutation of one species into another, his doctrine was treated as a groundless speculation which did not need to be seriously considered. On a more sophisticated level, more specific arguments, such as the sterility of hybrids, and the reversion of varieties to the original type, were adduced in support of the immutability thesis. The problems raised by the Natural Selection theory, in their turn, were only discussed in the very best quality organs. It is evident that the large majority of the general public remained in ignorance of the fundamental principles of that theory. Therefore, when a Descent theory embodying a principle of providential Design was gaining ground towards the end of the 'sixties, most people failed to see that acceptance of it did not solve some of the most important problems which Darwin's naturalistic theory had raised. The fact that Darwinians had to be very much on the defensive on the question of Natural Selection, especially from about 1870 onwards, helped to confirm this misapprehension, which has in fact lingered on ever since. Though it was chiefly the Natural Selection theory which made the Descent doctrine scientifically acceptable, popular Evolutionism has by and large remained at the pre-Darwinian, unscientific stage.

On the Darwinian side, the most widely used argument was to point to the resemblances between the various forms of organic life, to the general progression in the geological strata, and to the general principles of continuity and uniformity, which appealed to the practical common sense of the ordinary man. These arguments, however, were only relevant to the Descent theory as such, not to the Natural Selection part of Darwin's doctrine. It is noticeable that the arguments specifically designed to support Natural Selection were sparse indeed. For instance, though both Darwin and Wallace explicitly acknowledged their debt to Malthus, the resemblance between the Natural Selection theory and the economic theory of *laisser faire* was not often advertised on the Darwinian side: instead, opponents used it to discredit Darwinism. The Manchester school was not popular in the 1860s.

The most effective way to support Natural Selection was to inculcate empiricist principles of scientific explanation: and on the highest intellectual level — on which alone, as we said, Natural Selection was really debated — the theory gave rise to a vigorous debate on the problems of the philosophy of science. In that debate the Natural Selection theory acted as a watershed, separating empiricist Darwinians on one hand from idealistic anti-Darwinians on the other. It is quite clear that the idealistic position was much more strongly represented among the general public — we are now speaking, of course, of the educated section — than the empiricist one. Nor is any movement of opinion towards empiricist views noticeable during the short period we have studied. Perhaps even the contrary: the challenge of the Darwinian theory caused the idealistic school to close their ranks. The revival of German post-Kantian idealism in England from around 1870 onwards should be seen as part of this movement. On the whole, the idealistic counter-offensive served its purpose. The Natural Selection theory, the spearhead of the empiricist-scientist attack, was held in check. From the viewpoint of the general educated public, at least, it seemed as if the Darwinians had had to admit that Natural Selection was insufficient as an explanation of Evolution, and that a principle of predetermined development was needed to supplement it. It is not until well into the 20th century that the progress of biology, especially in the field of genetics, has made it possible to establish the Natural Selection theory on a more secure basis.

Though the actual arguments used in the Darwinian controversy ostensibly concerned scientific points, it is quite clear that the stand taken by the disputants was ultimately determined by ideological or religious considerations. One did not on the whole disagree about the facts; one disagreed about the interpretation of the facts, and preferred the interpretation which supported the ideological position one wished to maintain. Hence there is a clear correlation between the religious and ideological views of the debaters on one hand, and their attitudes towards Darwinism on the other. Nor has the importance of the ideological factors for determining the stand on Darwinism to be inferred from this general correlation alone. There were very few reviewers indeed who refrained from offering at least some remarks on the religious implications of the theory. The importance of these implications was by no means played down: in some cases the stand was squarely based on them alone.

On the popular level it was the theory's patent contradiction of the Biblical creation story which attracted most attention when the theory's religious consequences were discussed. On a higher educational level the debate turned on the bearing of the Evolution theory on the concept

of miracles and miraculous interference. After Newton, such inter-
ferences had been virtually abandoned for celestial phenomena: the
spread of uniformitarian views in geology threatened to oust them from
the terrestrial sphere as well, at any rate as regards the material world.
Now Evolutionism bade fair to make them superfluous in their last
stronghold: the organic world. To the religious it seemed as if God was
pushed further and further away from His creation, when evolutionists
admitted Him as an original Creator and Designer, and no more.

But an even more fundamental conflict was occasioned by the Natural
Selection theory's bearing on that teleological interpreation of nature
which lies at the root of practically any sort of religious belief, and
which was of quite central importance in 19th century Britain, where
Natural Theology had become almost the whole of theology. Now on the
Natural Selection theory, almost any form of life, however complicated,
beautiful, or adapted to its function, could be explained as due to the
accumulation of purely random variations. Thanks to Darwin, the
Epicurean and Lucretian picture of a fortuitous concourse of atoms
giving rise to the world as we see it was changed from a patently absurd
speculation to an eminently plausible hypothesis. There might indeed
be room for a Creator and Designer in this theory — but there was no
need for one. It was above all this consequence which caused the
Natural Selection theory to be combated so insistently and, as we
have said, on the whole successfully.

We meet with the same concern about the Natural Selection theory in
the application of the Darwinian doctrine to man. On the popular
levels one chiefly dealt with the incompatibility of the Descent theory
with the Biblical account of the origin and history of man. But on
the higher levels one was very much concerned with the Darwinian
thesis that man's intellectual and moral nature — his soul — was
a product of the natural selection of random modifications. If that
theory was correct, the establishment of a religion or of a moral code
would have nothing to do with the question whether the beliefs associated with it were true or not. Beliefs would be established if the community
which embraced them benefited from them in the struggle for life.
One of the strongest arguments in favour of an idealistic interpretation
of the world — the intuitive nature of moral and religious beliefs, and the
difficulty of accounting for them on the basis of individual experiences
of utility, even though past experiences of the race were called in aid

— lost its force. The Natural Selection theory made it possible to give a plausible empiricist-naturalistic account of the origin of religious and other beliefs: and it was obvious that this account, which was thoroughly relativistic, was irreconcilable with the idealistic-intuitionist one, which was equally thoroughly absolutistic.

Religion was an essential ingredient in the general public's reaction towards Darwinism. But it was also a very important factor in determining the attitudes of the scientists themselves. The Darwinian controversy can probably be best characterised as one engaging religious science against irreligious science — a description already applied by C. C. Gillispie to the pre-Darwinian conflict between Genesis and Geology. The antagonism, which largely coincided with the opposition between an idealistic and a positivistic-empiricist view of the world, cropped up almost at every point. It was no coincidence that the leading Darwinians of the time — Darwin, Huxley, and Hooker — were all agnostics (a term invented by Huxley), while the leading anti-Darwinians, Owen and Mivart, were decidedly religious men.

The establishment of an evolutionary view had been virtually achieved among the educated classes before the end of the first decade after the publication of the *Origin of Species*. That was the first and most palpable change that the Darwinian theory worked in our outlook on man and the world. But it was not necessarily the most important. Some contemporaries felt keenly, others obscurely, that the extension of purely scientific methods of reasoning to subjects which had hitherto lain outside the scope of science, was the most explosive and revolutionary element in the new doctrine. One could not foresee where this development would lead: but the threat to an established body of beliefs was apparent. One therefore felt the need to strengthen them: and this inevitably led to inquiries into their foundations. In the ensuing debate, no final solution of the problems that arose in this connection was reached: none could be expected. Some of those problems are still with us; indeed, a century during which the scientific spirit has unrelentingly advanced into wider and wider fields has made them even more pressing. Hence the Darwinian debates of a hundred years ago still has an interest for us, for though the solutions that were then suggested may perhaps often seem irrelevant and naïve to a later age, the debates ranged both widely and deeply, and above all, were carried on with an earnestness and force of conviction which commands respect.

Statistical Analysis of the Press Reaction

Discussion of Methods

The purpose of this analysis is to map out, by means of a statistical technique, the attitudes taken towards Darwinism in various classes of periodicals, as well as the amount of attention given to the subject over the years, from 1859 to 1872. When classifying the periodicals I have given special attention to the characteristics of their readerships. The chief classifications are according to political and religious opinions, and according to educational standard. The latter evidently largely overlaps that of social status.

The periodicals studied for the purposes of this investigation are those listed and described in Appendix II. Of the 115 titles given there, eight have been excluded from this survey since they contain no mention whatever of the Darwinian theory. The statistics are therefore based on a material of 107 periodicals and newspapers (Table I). Review organs, which have been specially attended to, make up more than half the total number, the rest being shared equally between newspapers proper on the one hand, and magazines and weekly journals on the other. In terms of circulation the dominance of the Reviews disappears, the popular journals and Sunday newspapers accounting for roughly three fourths of the total. The aggregate circulation of the whole material is between two and three million copies per issue, which is a substantial portion of what was published in Great Britain in those days. It has been calculated that 546 million copies of newspapers were sold in England and Wales in 1864; my newspaper sample covers c. 189 million, or about one third of this, and includes almost all the important London papers.

TABLE 1.

General survey of periodicals and their readerships.

	PERIODICALS			CIRCULATION				
	Num-ber	% of total	% of sub-group	Aggregate circul. 1000's			Average all periods	
				I	II	III	% of total	% of subgr
Periodicals read	115							
Periodicals in statistics	107							
Period I (1859—1863)	92			1938				
Period II (1864—1869)	102				2635			
Period III (1870—1871)	87					2940		
General classification	107	100		1938	2635	2940	100	
Newspapers, Daily–Tri-weekly	13		12.1	227	459	621		17.4
Newspapers, Weekly	12		11.2	417	873	1107		31.8
Reviews, Weekly	24		22.4	68	79	84		3.1
Reviews, Fortnightly-Quarterly	35		32.7	80	101	76		3.4
Monthly Magazines	10		9.3	171	135	102		5.4
Weekly Journals............	13		12.1	975	988	950		38.6
Reviews, all classes	59	55		148	180	160	6.5	
Reviews, Religious	26		44.1	36	40	36		22.9
Reviews, Scientific	14		23.8	33	49	45		26.0
Reviews, General	19		32.2	79	91	79		51.2
Political classification	69	64.5		1165	1782	2083	66.7	
Conservative	17		24.6	106	161	204		9.4
Liberal-conservative	5		7.2	173	185	185		10.8
Liberal	36		52.2	559	513	687		35.0
Liberal-Radical	5		7.2	260	688	723		33.3
Radical...................	6		8.7	67	235	284		11.7
Religious classification	45	42.1		316	359	374	14.4	
High Church	11		24.4	80	89	90		24.6
Low Church	10		22.2	147	170	197		49.0
Broad Church	3		6.7	48	48	34		12.4
Roman Catholic	4		8.9	5	7	6		1.7
Methodist.................	3		6.7	15	24	28		6.4
Congreg., Baptist, Presbyt., Quaker	10		22.2	16	16	13		4.3
Unitarian	4		8.9	5	5	6		1.5
Educational classification	107	100		1938	2635	2940	100	
High	44		41.1	97	126	122		4.6
Middle	51		47.6	861	1014	1113		39.7
Low......................	12		11.2	980	1495	1705		55.6

The magazine material is less efficiently covered. About one sixth of the aggregate monthly circulation of 6,094,000 in 1864 are included in the sample. But it may be safely inferred that the majority of the magazines would yield little or no material for the purposes of this investigation; in fact most of the eight titles excluded from the statistics in this chapter were magazines.

I have no independent evidence as to the aggregate circulation of all Review organs, but there is every reason to believe that this group is covered more efficiently than any other in my material. All the general Reviews of any importance — as judged from references in the contemporary press — have been included, and also all the main organs of the various religious groups. I think therefore that my sample can be relied upon as fairly representative of the British press, as well as of the British reading public, in the 1860s. Some support for this contention may also be gathered from the distribution of the publications in the sample over the various groups of the public, in terms of aggregate circulation. The distribution of political views, as well as the educational and social stratification, is what we should expect in the Mid-Victorian reading public. For the periodicals classed according to religious opinions, however, the figures for aggregate circulation do not mean much, since the group is so inhomogeneous. Some denominations are represented by Review organs only, others by newspapers and weekly magazines. The circulation figures here are included in the table for completeness' sake.

All circulation figures are estimates only; some are no more than guesses. The over-all picture, however, is probably not far wrong. For a discussion of the circulation figures, and the data on which they are based, see my paper on "The Readership of the Periodical Press in Mid-Victorian Britain", Acta Universitatis Gothoburgensis, 63, 1957, no. 3.

As we are here concerned with the state of information and opinion within *classes* of periodicals, taken to represent *classes* of the public at large, the statements contained in each particular periodical have been classified, so that individual statements may be compared with each other by referring to some common standard. A quantitative scale is obviously desirable, and I have constructed such a scale both for the amount of information provided, and for the attitudes taken with regard to the Darwinian theory.

In order to estimate the amount of information that the various periodicals offered on Darwinism, that concept has been divided into three main parts: — 1, the Evolution theory as such; 2, the Descent theory in its application to Man; and 3, the theory of Natural Selection. These three parts have been treated separately. For each part, the amount of information can be considered from two points of view, one taking into account the frequency of information — the number of references to the idea in question, — the other taking into account the depth of the information — the degree of thoroughness with which the idea was developed. I have in the main chosen the second point of view. The procedure has been as follows.

For each periodical studied, each reference to any part of the Darwinian theory (as defined above) has been noted. (In a material covering something like 1 million pages, it is obviously impossible to avoid overlooking some, or indeed many, references, though it can reasonably be expected that no major references should be left out. There is little reason to expect that inevitable omissions have had any appreciable effect on the statistical argument, since the material has been gone through in a uniform manner). Each such reference has been graded according to the amount of information contained in it. A four-point scale has been used: the value 1 standing for a bare mention of the relevant part of the theory, without any discussion of evidence or implications, 2 standing for information on at least some pieces of evidence or some implications, and 3 standing for information on all the chief evidence and implications. Zero on the scale naturally stands for no information whatever on the point in question.

The next step has been to indicate the *highest* score for each part of the theory that the periodical has obtained *during each successive year*. These scores are given for each individual periodical in Appendix II. Most periodicals did not contain references to the theory every year: for the purposes of the statistical treatment it is therefore convenient to use a longer period than one year in order to arrive at an estimate of the depth of information that the periodical provided for its public. This is justifiable, since we have reason to believe that the publics of the periodicals remained fairly stable over the years. The time-span 1859—1872 has been divided into three periods, the first covering the years 1859—1863, the second 1864—1869, and the third 1870—1872, as explained above, Ch. 2. The *highest* score obtained by each periodical *within each period* — irrespective of the number of references — has then been taken to indicate the depth of information during that period.

The attitude towards the Darwinian theory has been expressed in terms of the degree of acceptance that each periodical was willing to extend to the new doctrines. For this purpose the scales described above, Chapter 2, p. 30—32, have been used: a five-point scale for the Evolution theory in general, and a three-point scale for the theory in its application to Man.

Applying these scales, I have noted, for each periodical, each instance where it has expressed or endorsed any opinion on any of the two portions of Darwin's theory relevant in this context: the general Evolution theory, and that theory applied to man. The scores given are indicated for each periodical and year in Appendix II. It should be borne in mind that many of the expressions of opinion were not very explicit, so that it is often not possible to say exactly which position a writer was prepared to accept. Sometimes, indeed, the writer himself indicated a range of alternative choices, and sometimes it is not clear whether the writer was willing to take any stand at all. A subjective element inevitably enters when the scoring is concerned. I have indicated ranges of alternative positions both in the cases where the writer's opinion is vaguely expressed, and where the writer himself indicates alternatives. For these opinion scorings, naturally, I have only taken into account such contributions as may be taken to represent the editorial line of the periodical in question: leaders, editorial comments, book reviews, and, in the Reviews, main articles. The reader will be able to some extent to judge for himself whether my scorings are adequate, by means of the Index of Periodicals, Appendix II. The Index contains references to all instances where a periodical is cited in this book. Unless otherwise stated in the relevant note, or in the accompanying text, the citation concerns the kind of matter just indicated. Thus most newspaper references, for instance, are to leading articles.

Information on Darwin's Theories

The periodicals have been divided into three groups according to their educational standard. Table 2 indicates, for each group, the number of periodicals whose *fullest* reference to the subject of Evolution during

Diagram 2 A. Distribution of scorings for depth of information, in % of total scorings, in three educational groups.

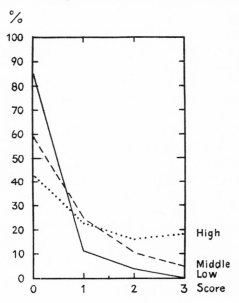

the year in question was rated by me at the value indicated at the top of the column. As is natural, score 3, i. e. a full discussion of the theory, was attained most commonly in the highest educational group, less often in the middle group, and never in the lowest group. Conversely, score 0, i. e. no mention at all, was most common in the lowest group, and least common in the highest. The difference appears clearly from Diagram 2 A which shows the distribution of the four scores, in percent of the total number of scorings during the whole period of 13 years, in the three educational groups. The diagram is constructed from the figures in the bottom row of Table 2.

Diagram 2 B represents the fluctuations in depth of information from year to year in the three educational classes. It has been derived from Table 2 by calculating, for each group and year, the average score, defined as the arithmetical mean of the individual scores. In this diagram as well the very much greater depth of information provided in the highbrow organs is apparent, but the chief use of the diagram is to show when Darwinism was most in the public eye, and when least.

TABLE 2.

Information on Evolution, year by year, in three educational classes.

YEAR	HIGH					MIDDLE					LOW				
	0	1	2	3	Average	0	1	2	3	Average	0	1	2	3	Average
1859	23	1	1	5	0.60	30	1	2	5	0.53	8	0	0	0	0
1860	5	3	3	20	2.23	18	9	4	9	1.10	6	1	1	0	0.38
1861	12	12	5	4	0.97	27	7	7	2	0.63	7	2	0	0	0.22
1862	18	11	5	0	0.62	24	13	5	2	0.66	9	0	0	0	0
1863	12	8	10	5	1.23	21	18	4	2	0.71	7	3	0	0	0.30
1864	21	6	6	3	0.75	35	7	3	0	0.29	9	1	0	0	0.10
1865	19	7	6	2	0.74	25	13	6	1	0.62	9	1	0	0	0.10
1866	17	12	2	3	0.74	20	12	10	3	0.91	8	2	0	0	0.20
1867	14	12	3	5	0.97	31	12	1	0	0.30	10	0	1	0	0.18
1868	11	7	7	9	1.41	21	13	9	3	0.87	9	3	0	0	0.25
1869	14	8	3	7	1.09	28	14	3	0	0.45	10	2	0	0	0.17
1870	14	7	7	3	1.03	28	14	2	0	0.41	12	0	0	0	0
1871	4	4	8	15	2.22	15	19	4	5	0.98	8	2	2	0	0.50
1872	11	8	7	4	1.14	33	5	3	1	0.33	11	0	1	0	0.17
Total scorings	195	106	73	85		356	157	63	33		123	17	5	0	
% of total scorings	42.6	23.1	15.9	18.5		58.7	25.8	10.4	5.4		85	11.7	3.4	0	

Diagram 2 B. Changes in average depth of information on Evolution, in three educational classes, year by year.

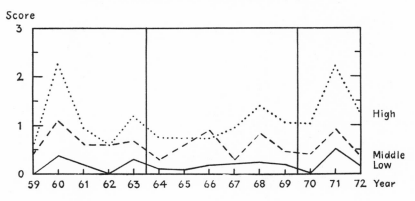

On the whole, the lower groups reflect the marked changes occurring in the highest one, but on a smaller scale. Not surprisingly, the great peaks were reached in 1860 and in 1871, indicating the impact of the *Origin of Species* (1859) and the *Descent of Man* (1871). Two smaller peaks occurred in 1863 and 1868, one connected above all with the publication of Sir Charles Lyell's *Antiquity of Man* (1863), the other with Darwin's *Animals under Domestication* (1868), as well as with Darwinian debates at the British Association. The latter is also the explanation of the peak in the middle group for 1866.

Diagram 2 C, also constructed from Table 2, is, and ought to be, the converse of Diagram 2 B. It indicates the percentage of periodicals, out of the total in each year, which contained no mention at all of Evolution during the year in question.

In order to show up more clearly the differences between the three educational classes, I have in Table 3 included their scores for all the three parts of the doctrine, giving the average values for each in Diagram 3 A—C. In this instance I have not followed the development year by year, but have only considered the highest scores obtained by each periodical in each complete period, i. e. I: 1859—1863, II: 1864—1869, and III: 1870—1872. The differences between the educational classes as regards depth of information on Darwin stand out as very marked for all three parts, and it is also apparent that the Natural Selection part of the doctrine was much less referred to than the rest. In the

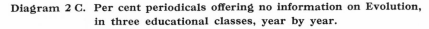

Diagram 2 C. Per cent periodicals offering no information on Evolution, in three educational classes, year by year.

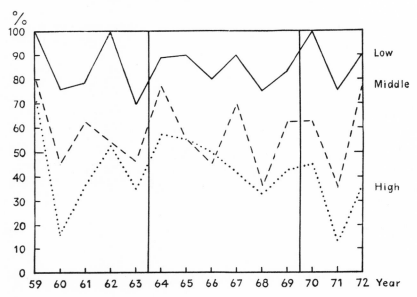

"lowbrow" press it was hardly referred to at all, and in the middle group it was seldom discussed at any length: score 1 sometimes implies no more than a mere mention of the phrase Natural Selection whose meaning may have been obscure to both writer and reader.

A comparative study of the graphs for the development of man on one hand, and Evolution in general on the other, yields interesting results. The high score for Man in period III is natural in view of the stir caused by the *Descent of Man* in 1871. More remarkable is the fact that in the middle and low educational classes the subject of man was not much behind that of evolution in general, even in period I. In the educated class conditions were markedly different.

Attitudes towards Darwinism

Table 4 is intended to show the development of attitudes towards the Darwinian doctrine over the years 1859—1872. As explained above, the scoring for attitude is not quite as simple as for information, where

TABLE 3.

Information on Darwin's theory, in three educational groups, over three periods.

CLASS	PERIOD	EVOLUTION					MAN					NATURAL SELECTION				
		0	1	2	3	Av.	0	1	2	3	Av.	0	1	2	3	Av.
HIGH	I	1	5	5	25	2.53	4	14	15	3	1.47	9	3	16	8	1.64
	II	5	7	9	20	1.58	8	16	13	4	1.32	16	7	12	6	1.19
	III	2	4	7	18	2.32	2	9	6	14	2.04	7	6	14	4	1.48
MIDDLE	I	4	17	10	15	1.78	5	25	15	1	1.26	22	10	12	2	1.08
	II	12	17	15	5	1.27	16	25	7	1	0.86	31	11	7	0	0.51
	III	12	22	4	6	1.09	12	20	7	5	1.11	30	10	3	1	0.66
LOW	I	5	4	1	0	0.60	5	5	0	0	0.50	10	0	0	0	0
	II	7	4	1	0	0.50	8	4	0	0	0.33	12	0	0	0	0
	III	7	2	3	0	0.67	5	6	1	0	0.67	11	1	0	0	0.08
ALL	I	10	26	16	40	1.92	14	44	30	4	1.25	41	13	28	10	1.07
	II	24	28	25	25	1.49	32	45	20	5	0.97	59	18	19	6	0.71
	III	21	28	14	24	1.45	19	35	14	19	1.37	48	17	17	5	0.75

Diagram 3 A—C. Amount of information on various parts of the Darwinian doctrine, in three educational groups, over three periods.

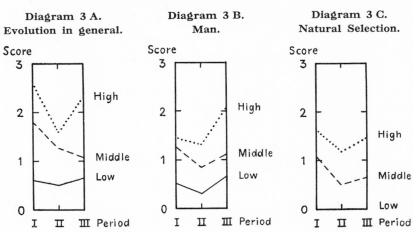

Diagram 3 A. Evolution in general.	Diagram 3 B. Man.	Diagram 3 C. Natural Selection.

TABLE 4.

Attitude towards Darwin's theories year by year (whole material).

	POSITION		YEAR 59	60	61	62	63	64	65	66	67	68	69	70	71	72
EVOLUTION	1	max. min.	1 3	10 13	6 9	2 8	9 14	2 4	2 5	6 9	4 10	2 6	1 3	5 6	9 13	1 4
	2	max. min.	2 5	7 16	6 10	5 5	6 12	2 3	4 3	4 10	5 5	6 6	4 3	1 3	3 9	2 5
	3	max. min.	1 0	8 4	3 3	4 0	5 0	3 2	4 1	3 3	5 4	2 7	2 4	2 4	10 9	6 4
	4	max. min.	5 2	12 6	7 1	3 2	8 3	4 4	4 6	13 5	6 3	11 10	6 5	6 2	15 15	6 4
	5	max. min.	3 2	5 3	1 —	1 —	3 2	6 4	3 —	1 —	4 2	11 3	6 4	4 3	10 1	6 4
		Total	24	84	46	30	62	34	32	54	48	64	38	36	94	42
		Average	3.08	2.58	2.22	2.23	2.31	3.15	2.84	2.56	2.65	3.33	3.42	2.89	2.96	3.31
MAN	1	max. min.	2 3	14 17	13 16	8 11	25 32	5 6	9 9	11 13	8 9	6 8	4 5	7 9	15 26	8 9
	2	max. min.	1 —	4 1	3 —	3 —	7 1	3 2	— 1	4 2	1 1	3 2	2 1	4 3	16 13	1 2
	3	max. min.	— —	— —	1 1	1 1	6 5	3 3	3 2	— —	1 —	2 1	2 2	4 3	15 7	5 3
		Total	6	36	34	24	76	22	24	30	20	22	16	30	92	28
		Average	1.16	1.14	1.20	1.29	1.39	1.77	1.46	1.20	1.20	1.50	1.69	1.70	1.80	1.68

only one single score value was awarded for each year. The attitude scoring, on the other hand, often takes the form of, say, 1—4, meaning that the periodical's position has been interpreted as 1 at the lowest, and 4 at the highest. In Table 4 I give, for each position and year, one figure denoting the number of periodicals taking the position as the highest ('max.' in the table), and another denoting the number of periodicals taking it as the lowest position ('min.' in the table). This means that periodicals taking one position definitively, and without any alternative, contribute both to the "max." and to the "min." row, for that position. This is as it should be, since in this manner each periodical expressing an opinion on Darwin contributes two, and always two figures to the total. The actual number of periodicals taking a stand at all on Darwinism is therefore exactly half the aggregate sum of the highest and lowest positions taken. ("Total" row in Table 4).

From Table 4 I have constructed two diagrams. Diagram 4 A indicates for each year the number of periodicals expressing an opinion on Darwinism, whether on Evolution in general or on Man. The graph

Diagram 4 A. Number periodicals expressing an opinion on Darwin's theories, year by year.

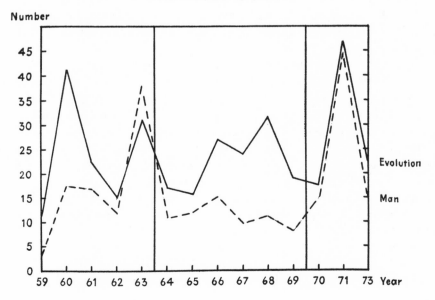

Diagram 4 B. Average position on Darwinism in whole material, year by year.

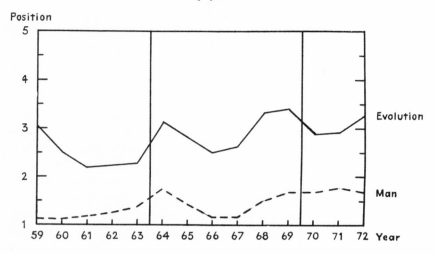

is obviously to some extent an index of the *amount* of attention given to the theory in each year, and should be compared with Diagram 2 B, which shows the yearly changes in *depth* of information provided. As we should expect, there is a good general agreement between the two.

Diagram 4 B attempts to show the yearly changes in the attitude towards Darwin. The average position for each year is the arithmetical mean of all the highest and lowest positions taken during the year. (The difference between the average as calculated in this way, and in other possible ways, e. g. the one used below, p. 356, is negligible). The graph represents the whole of the material of 107 periodicals, without taking into account the wide differences between these periodicals as regards circulation, intellectual and educational standard, importance, etc. These matters will be dealt with shortly. Meanwhile, some comments may be ventured on the present diagram.

The comparatively high rating — indicating a comparatively favourable attitude towards Evolution — for the year 1859 reflects what was said above, Chapter 2, of the first reviews of the *Origin*. In 1860 the attitude hardened. The continued fall in 1861, however, may be due chiefly to the fact that the periodicals expressing an opinion in that year tended to be of a different type than those of the preceding years.

Moreover, the material for many of the years in the 'sixties is rather small. The marked peak for 1864 therefore may not mean much, but there is no doubt that on the whole the attitude towards Evolution was gradually getting more favourable from 1863 to 1868, at least. For both these years there is ample material, as shown by Diagram 4 A. The increasing opposition to Darwinism in 1870—1871 is probably also more than a fortuitous variation due to the heterogeneity of the sample.

The two diagrams, 4 A and 4 B, provide some of the material that is needed for discussing the problem of how to subdivide the time-span covered in the investigation into periods. As we said above, the material is not large enough for statistics based on yearly intervals to yield significant results. It appears convenient to operate instead with three periods. In order to have each period reasonably homogeneous as regards attitude, and similar as regards amount of information, Diagram 4 B would seem to indicate 1859—1863 as the early period, 1864—1869 as the middle one, and 1870—1872 as the last. According to Diagram 4 A, the early period would then include two high peaks of interest in Darwinism, and the late period the years around the last high peak.

Development of opinion on Darwin in three educational classes

Each of the periodicals in my material has been referred to one of three educational classes, which may, for convenience, be called highbrow, middlebrow, and lowbrow. In order to assess the difference between these classes with regard to their attitude to Darwin, I have used a similar procedure to the one employed for constructing Table 4, with the difference that I now count the highest and lowest ('max.' and 'min.' in the table) position taken during each period, instead of during each year. The results are set out in Table 5. Diagrams 5 A—B indicates the differences between the three groups over the three periods. It appears that on the whole the highbrow class was most favourable to Darwin, and the lowbrow class least so. This holds both for the general Evolution part of the doctrine, and for its application to man. The lowbrow group being very small, the fact that it reached, in period II, a higher average score than the middle group in regard to Evolution can probably be disregarded as a random fluctuation. The remarkably high score for the middle group in the first period, however, is not

Diagram 5 A. Average position on Evolution in three educational classes, over three periods.

Diagram 5 B. Average position on Man in three educational classes, over three periods.

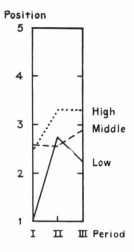

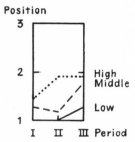

without significance. It is a reflection of the many quite favourable early reviews of Darwin in the daily and weekly papers. Many of these later changed their attitudes, when several socially important leaders of opinion — in the press and elsewhere — had deployed their forces against the new doctrines. (See above, p. 36).

From Table 5 it is also possible to get a clearer picture about which of the positions were most popular within the various groups over the periods. Diagrams 5 C—F show the relative frequency, in percent of the total of highest and lowest positions taken, of each position over the three periods. Comparing the highbrow and the middlebrow groups — the lowbrow group having been left out of account because of its small size — we notice in Diagrams 5 C and 5 D a decline for positions 1 and 2, and a rise for positions 3 and 4, which is natural, since the two pairs are correlative. The attitude toward Darwin was on the whole getting more favourable. The development with regard to position 5, implying full acceptance of the Darwinian theory of Evolution, is remarkable. In the highbrow group it was quite common in period II, but dropped again sharply in III. In the middlebrow group, it declined slightly all the time.

TABLE 5.
**Highest and lowest positions taken,
over three periods, in three educational classes.**

CLASS	PERIOD		EVOLUTION						MAN			
			1	2	3	4	5	Av.	1	2	3	Av.
HIGH	I	max.	3	7	8	12	4		18	8	4	
		min.	15	12	2	2	3		26	—	4	
		Average						2.46				1 40
		%	26.5	28	14.7	20.6	10.3		73.4	13.3	13.3	
	II	max.	1	5	4	11	13		8	7	10	
		min.	9	10	5	5	5		14	4	7	
		Average						3.26				1.90
		%	14.7	22	13.2	23.5	26.5		44	22	34	
	III	max.	1	1	9	9	7		9	9	11	
		min.	6	4	8	8	1		16	6	7	
		Average						3 26				1.90
		%	13	9.3	31.4	31.4	14.8		36.7	36.7	26.5	
MIDDLE	I	max.	4	6	4	9	6		16	4	3	
		min.	14	8	2	4	1		21	1	1	
		Average						2.60				1.28
		%	31	24.2	10.3	22.4	12.1		80.5	10.9	8.7	
	II	max.	6	6	3	9	6		13	2	1	
		min.	15	6	4	4	1		15	—	1	
		Average						2.56				1.19
		%	35	20	11.6	21.7	11.6		87.6	6.2	6.2	
	III	max.	5	2	4	10	4		8	8	9	
		min.	7	7	3	8	0		14	8	3	
		Average						2.86				1.80
		%	24	18	14	36	8		44	32	24	
LOW	I	max.	5	—	—	—	—		3	—	—	
		min.	5	—	—	—	—		3	—	—	
		Average						1				1
		%										
	II	max.	1	—	—	—	1		3	—	—	
		min.	1	—	—	1	—		3	—	—	
		Average						2.75				1
		%										
	III	max.	2	—	1	1	—		4	1	—	
		min.	2	—	1	1	—		4	1	—	
		Average						2.24				1.20
		%										

Diagram 5 C. Distribution of positions taken on Evolution in highbrow class, over three periods.

Diagram 5 D. Distribution of positions on Evolution in middlebrow class, over three periods.

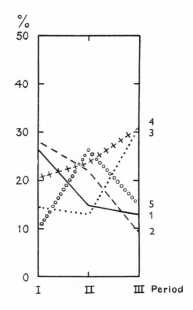

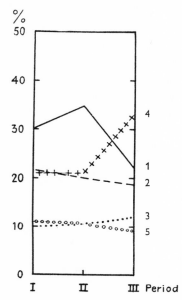

The development of opinion as regards man is shown in Diagrams 5 E and 5 F. Evolution was steadily gaining ground in both educational groups. The decline for position 3, which implied acceptance of evolution for both body and soul, was apparently connected with the decline for Natural Selection. The connection was not a logical one, for it was quite possible to accept 3 for man while rejecting 5 for Evolution in general. But there seems to have been a psychological connection, since few were prepared to accept 3 for man while stopping at 4 for Evolution, and nobody who accepted 5 for Evolution failed to accept 3 for man.

Correlation between information and opinion on Darwin

In order to follow up the results obtained from the previous tables, and also indirectly to test the correctness of my classification into educational groups, I have studied the correlation between depth of information, as defined above, and opinion on Darwin. For this purpose

Diagram 5 E. Distribution of positions on Man in highbrow class,
over three periods.

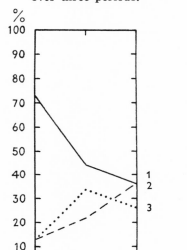

Diagram 5 F. Distribution of positions on Man in middlebrow
class, over three periods.

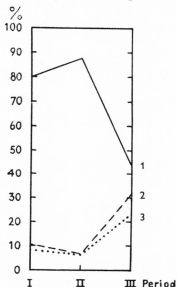

I have divided my material into three groups, one consisting of periodicals having 1 as the highest score for information, whether for Evolution,
Man, or Natural Selection, during the relevant period, and the other
two groups consisting of periodicals similarly scoring 2 and 3 as highest,
respectively.

In this, as in most subsequent calculations, I have not calculated
only the overall average position scored by each group in each period,
but have indicated both the average highest and the average lowest
positions taken ('max.' and 'min.' in the table). The average highest
position is defined as the arithmetical mean of the highest positions
scored by the periodicals during the period in question. The average
lowest position is calculated correspondingly. Each periodical evidently
scores both a maximum and a minimum — which are identical if the
periodical consistently scored only one position during the period,
without either changing or admitting alternatives. The advantage
of this notation is that it indicates the average range of variation for
the individual periodicals. If we calculate the overall average as the

TABLE 6.
Correlation information-attitude.

INFORMATION	PERIOD		EVOLUTION							MAN				
			0	1	2	3	4	5	Av.	0	1	2	3	Av.
1	I	max.	11	6	4	2	1	0	1.85	11	13	—	—	1
		min.	11	9	4	—	—	—	1.31	11	13	—	—	1
	II	max.	14	4	3	2	4	3	2.94	17	12	1	—	1.08
		min.	14	9	3	2	2	—	1.81	17	13	—	—	1
	III	max.	14	15	1	3	9	—	2.88	14	8	7	3	1.72
		min.	14	6	4	3	5	—	2.39	14	10	6	2	1.56
2	I	max.	7	3	3	4	3	2	2.87	10	9	2	1	1.33
		min.	7	9	5	—	—	1	1.60	10	11	—	1	1.16
	II	max.	4	4	3	2	9	8	3.53	13	8	3	6	1.88
		min.	4	10	5	3	4	4	2.50	13	11	1	5	1.65
	III	max.	4	2	1	1	3	2	3.22	2	6	—	5	1.91
		min.	4	3	1	2	3	—	2.56	2	7	3	1	1.45
3	I	max.	—	2	6	7	17	8	3.58	9	15	10	6	1.71
		min.	—	15	11	4	7	3	2.30	9	26	1	4	1.29
	II	max.	2	—	5	3	7	9	3.83	12	4	5	5	2.07
		min.	2	6	8	4	4	2	2.50	12	8	3	3	1.65
	III	max.	—	1	2	9	9	9	3.77	—	7	11	12	2.17
		min.	—	6	7	6	10	1	2.77	—	17	6	7	1.67

arithmetical mean of the maximum and minimum averages we obtain substantially the same result as by the method used above, p. 350.

Table 6 and Diagrams 6 A—C show that there is a clear correlation between depth of information and attitude. The periodicals which went most deeply into the subject were more likely to be favourably inclined towards it. This is not surprising. The fullest information was naturally to be found in the high-brow publications, and these, as we have seen, tended to be more favourable towards Darwin.

Diagram 6 A. Develop-
ment of attitude to-
wards Darwinism in
periodicals giving 1 as
highest information.

Diagram 6 B. Develop-
ment of attitude to-
wards Darwinism in
periodicals giving 2 as
highest information.

Diagram 6 C. Develop-
ment of attitude to-
wards Darwinism in
periodicals giving 3 as
highest information.

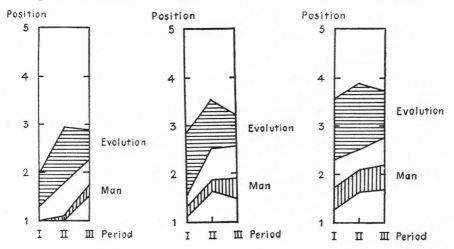

Development of opinion in Review organs

The Review group, including the whole of the "highbrow" class, and in addition 15 organs classed as "middlebrow", has been subdivided into three main groups, one consisting of organs of a decidedly religious character, another of organs giving special attention to scientific questions, and a third omnibus group consisting chiefly of literary and political Reviews. The distribution of the groups over the three periods, together with the estimated aggregate circulations of each, were set out in Table 1.

In order to assess the average position towards Darwinism in each group I have used the same procedure as for Table 6, estimating the average highest ('max.') and the average lowest ('min.') positions with regard to both Evolution in general, and the development of Man, the overall average being defined as the arithmetical mean of these two values. In Table 7 and 8 account has been taken not only of the number of periodicals concerned, but also of their circulations. In the upper

TABLE 7.

Periodicals and circulation:
Highest and lowest positions taken in three classes of Reviews, over three periods.

CLASS		PERIOD		Position EVOLUTION							Position MAN				
				0	1	2	3	4	5	Av.	0	1	2	3	Av.
GENERAL	Periodicals	I	max.	1	1	2	2	7	4	3.69	4	4	5	4	2.0
			min.	1	3	10	—	1	2	2.31	4	10	—	3	1.46
		II	max.	2	2	—	2	3	5	3.75	6	2	1	5	2.38
			min.	2	2	4	3	1	2	2.75	6	3	2	3	2.0
		III	max.	3	—	—	2	5	3	4.10	1	1	4	7	2.50
			min.	3	—	2	2	5	1	3.50	1	3	3	6	2.25
	Circulation (1000's)	I	max.	5	2	2	15	28	27	4.02	12	14	18	35	2.32
			min.	5	20	29	—	20	5	2.38	12	40	—	27	1.80
		II	max.	28	11	—	7	17	28	3.81	43	11	4	33	2.46
			min.	28	11	11	17	2	22	3.20	43	15	3	30	2.32
		III	max.	14	—	—	10	29	26	4.24	10	1	21	47	2.67
			min.	14	—	11	10	40	4	3.57	10	12	13	44	2.46
RELIGIOUS	Periodicals	I	max.	2	3	6	6	5	0	2.65	5	17	—	—	1.0
			min.	2	13	5	1	1	—	1.50	5	17	—	—	1.0
		II	max.	6	1	5	3	10	1	3.25	9	12	4	1	1.35
			min.	6	9	7	2	2	—	1.85	9	16	—	1	1.12
		III	max.	2	3	1	4	9	—	3.11	1	9	8	1	1.55
			min.	2	7	3	5	2	—	2.12	1	13	4	1	1.33
	Circulation (1000's)	I	max.	3	3	8	13	9	—	2.85	7	29	—	—	1.0
			min.	3	20	10	2	1	—	1.52	7	29	—	—	1.0
		II	max.	7	2	8	3	19	1	3.27	11	17	11	1	1.45
			min.	7	15	13	2	3	—	1.78	11	28	—	1	1.07
		III	max.	4	4	2	7	19	—	3.27	2	14	19	1	1.61
			min.	4	11	5	12	4	—	2.28	2	26	7	1	1.26
SCIENTIFIC	Periodicals	I	max.	—	—	3	2	4	2	3.46	—	5	4	2	1.73
			min.	—	6	2	1	1	1	2.0	—	9	—	2	1.36
		II	max.	2	—	2	—	2	8	4.33	6	1	3	4	2.38
			min.	2	3	2	1	3	3	3.08	6	3	2	3	2.0
		III	max.	2	—	—	3	2	4	4.11	2	2	1	6	2.44
			min.	2	1	—	4	4	—	3.22	2	4	3	2	1.78
	Circulation (1000's)	I	max.	—	—	19	4	8	2	2.78	—	8	23	2	1.82
			min.	—	24	6	1	1	1	1.45	—	31	—	2	1.12
		II	max.	11	—	16	—	3	19	3.66	34	1	5	9	2.53
			min.	11	4	16	3	8	7	2.95	34	3	5	7	2.27
		III	max.	4	—	—	20	2	19	3.98	4	5	15	21	2.39
			min.	4	1	—	21	19	—	3.42	4	21	12	8	1.68

portion of the table for each class ("Periodicals") each periodical has been counted as just one unit; in the lower portion ("Circulation") each periodical has contributed as many units as its estimated circulation, in thousands of copies. For example, the first row in Table 7 corresponds to the seventh row, the second to the eighth, etc. There is in the first row one periodical omitting to take a stand (position 0), and this periodical, according to the seventh row, has a circulation of 5,000. Continuing, the single periodical taking position 1 has a circulation of 2,000, two periodicals taking position 2 have an aggregate circulation of 2,000, two taking position 3 have an aggregate of 12,000, etc. The averages in the "circulation" portions express the average per reader, so to speak. Each part of the table is accompanied by a diagram showing the results graphically (Diagrams 7 A—F).

Comparing the results as obtained from the "unweighted" periodicals (Diagrams 7 A, C, E) with those obtained from the same periodicals when weighted for circulation (Diagrams 7 B, D, F) it appears that the large-circulation publications tended on the whole to be more favourable

Diagrams 7 A—F. Development of attitude towards Darwinism in three Review classes.

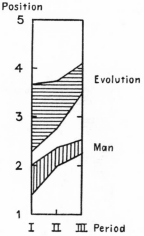

Diagram 7 A. General class, periodicals.

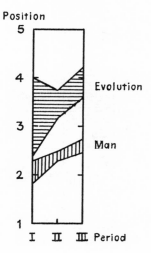

Diagram 7 B. General class, circulations.

**Diagrams 7 A—F cont. Development of attitude towards Darwinism in
three Review classes.**

Diagram 7 C. Religious class,
periodicals.

Diagram 7 D. Religious class,
circulations.

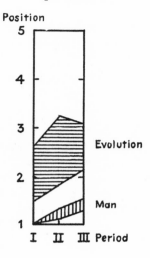

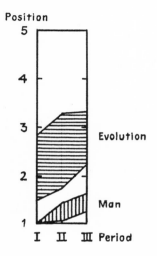

Diagram 7 E. Scientific class,
periodicals.

Diagram 7 F. Scientific class,
circulations.

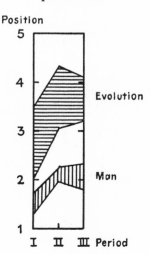

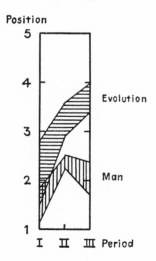

TABLE 8.
Periodicals and circulation.
Highest and lowest position taken in two political groups, for two classes of periodicals, over three periods.

Type of periodical	Political class	PERIOD		POSITION											
				EVOLUTION							MAN				
				0	1	2	3	4	5	Av.	0	1	2	3	Av.
REVIEWS	RIGHT	I	max.	1	2	—	3	1	1	2.86	1	6	1	—	1.14
			min.	1	5	2	—	—	—	1.29	1	7	—	—	1
		II	max.	1	2	2	—	2	1	2.71	1	5	—	2	1.57
			min.	1	5	—	1	—	1	1.86	1	5	—	2	1.57
		III	max.	2	1	—	3	—	1	3	—	4	2	1	1.57
			min.	2	3	—	1	1	—	2	—	5	1	1	1.43
		I	max.	2	2	—	11	10	2	3.40	2	17	8	—	1.32
			min.	2	22	3	—	—	—	1.12	2	25	—	—	1
		II	max.	1	3	3	—	9	18	4.10	1	7	—	26	2.57
			min.	1	7	—	8	—	18	3.67	1	7	—	26	2.57
		III	max.	3	2	—	12	—	20	4.07	—	7	10	20	2.35
			min.	3	6	—	8	20	—	3.23	—	9	8	20	2.29
	LEFT	I	max.	—	1	5	3	9	4	3.46	6	9	2	5	1.75
			min.	—	8	9	—	2	3	2.22	6	12	—	4	1.50
		II	max.	5	—	2	4	8	6	3.90	10	6	4	5	1.93
			min.	5	4	9	3	2	2	2.45	10	10	2	3	1.53
		III	max.	2	1	1	1	12	2	3.76	1	2	9	7	2.28
			min.	2	1	4	5	6	1	3.11	1	6	6	6	2
		I	max.	—	2	6	14	25	26	3.81	11	18	8	36	2.29
			min.	—	14	32	—	21	6	2.56	11	34	—	28	1.90
		II	max.	31	—	4	9	23	12	3.90	47	9	14	9	2
			min.	31	8	22	10	3	5	2.48	47	23	3	6	1.47
		III	max.	5	1	2	2	45	6	3.95	2	3	28	28	2.42
			min.	5	1	15	12	24	4	3.45	2	22	12	25	2.05
MIDDLESBROW ORGANS	RIGHT	I	max.	4	3	1	2	3	2	3	8	5	2	—	1.29
			min.	4	5	3	2	1	—	1.91	8	7	—	—	1
		II	max.	4	3	4	1	2	1	2.45	8	7	—	—	1
			min.	4	8	1	2	—	—	1.45	8	7	—	—	1
		III	max.	6	2	2	3	1	—	2.37	5	6	3	—	1.33
			min.	6	4	4	—	—	—	1.50	5	9	—	—	1
		I	max.	51	8	10	103	6	57	3.50	165	12	58	—	1.83
			min.	51	21	104	58	1	—	2.21	165	70	—	—	1
		II	max.	106	7	8	65	88	20	3.56	279	15	—	—	1
			min.	106	80	80	28	—	—	1.73	279	15	—	—	1
		III	max.	45	4	10	206	70	—	3.18	42	20	273	—	1.93
			min.	45	10	280	—	—	—	1.96	42	293	—	—	1
	LEFT	I	max.	8	1	3	2	5	3	3.43	10	9	1	2	1.21
			min.	8	7	4	—	3	—	2.64	10	11	—	1	1.17
		II	max.	9	2	2	1	6	3	3.43	16	5	1	1	1.43
			min.	9	5	4	2	3	—	2.21	16	6	—	1	1.28
		III	max.	7	3	—	—	7	3	3.54	7	2	4	7	2.39
			min.	7	3	1	2	7	—		7	4	6	3	1.92
		I	max.	193	10	4	7	44	180	4.50	274	56	80	28	1.84
			min.	193	50	15	—	180	—	3.27	274	144	—	20	1.24
		II	max.	222	26	3	1	37	177	4.39	301	12	3	150	2.84
			min.	222	31	15	3	195	—	3.49	301	15	—	150	2.83
		III	max.	303	32	—	—	24	148	4.25	315	7	7	178	2.89
			min.	303	32	2	2	168	—	3.49	315	59	116	17	1.78

towards Darwin, which is most marked in the general group. On the assumption of a broad correlation between press and public, this means that press opinion in the sense of the average opinion of all writers in the press, irrespective of the size and importance of their organ, tended to be less advanced as regards Darwin than public opinion. This only means, of course, that the numerous writers contributing to in small, out of the way publications tended to be more hostile to Darwin than writers in established organs.

Turning now towards the difference between the groups themselves, it is obvious that the religious organs were considerably behind the others in accepting the Darwinian doctrines. We also note that the scientific ('popular scientific') group lagged behind the general one during the first period.

Correlation between political opinion and attitude towards Darwin

The periodicals classed by me for political opinion have been divided into two groups, one consisting of the Review organs, the other of all periodicals classed as middle-brow, which includes all daily newspapers and most magazines and weekly journals. There is a small, and in terms of aggregate circulation, negligible overlap between these groups. The "Right" group includes conservative and liberal-conservative organs; the "Left" group liberal and radical ones. The results are set forth in Table 8.

The positions in the Reviews is graphically represented in Diagrams 8 A—D. If the periodicals alone are considered, disregarding their circulation, the correlation between political opinion and Darwinian position is clearly brought out. It is consistent all through the periods, both for the Evolution theory in general and for its application to man. But if circulations are taken into consideration, the picture is not the same. The difference still holds for the first period, but not for the later two. Too much should not be made of this, since the high scores in the conservative group are almost wholly due to one single paper, the *Saturday Review*, which cannot be taken as a typical conservative organ, since with its dominating position in the world of letters at the time, it was willing to accommodate both liberal and conservative writers. As the paper's historian has written, the editor "employed Liberals to write on the matters where they were most conservative, and the Conservatives on topics which they could treat liberally."

Diagrams 8 A—H. Development of attitude towards Darwinism in two political groups and two classes of periodicals.

Diagram 8 A. Reviews, right, periodicals.

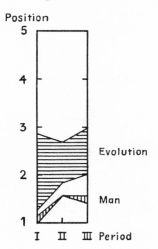

Diagram 8 B. Reviews, right, circulations.

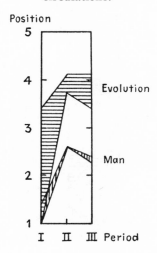

Diagram 8 C. Reviews, left, periodicals.

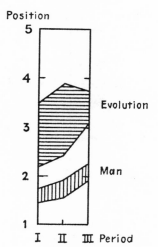

Diagram 8 D. Reviews, left, circulations.

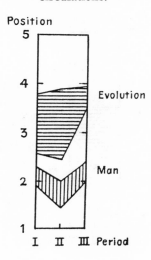

Diagrams 8 A—H cont. Development of attitude towards Darwinism in two political groups and two classes of periodicals.

Diagram 8 E. Middlebrow organs, right, periodicals.

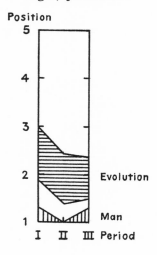

Diagram 8 F. Middlebrow organs, right, circulations.

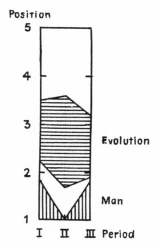

Diagram 8 G. Middlebrow organs, left, periodicals.

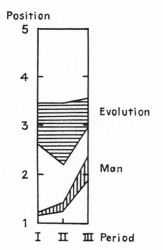

Diagram 8 H. Middlebrow organs, left, circulations.

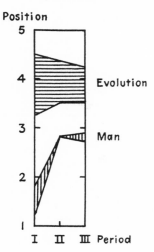

TABLE 9.

Highest and lowest positions taken in periodicals classed according to religious views.

GROUP		PERIOD	POSITION									
			EVOLUTION						MAN			
			I	I	3	4	5	Av.	I	2	3	Av.
HIGH CHURCH	Max.	I	2	1	4	2	1		7	3	—	
		II	—	2	2	4	—		3	2	1	
		III	1	1	3	1	—		5	3	—	
		Total	3	4	9	7	1	2.96	15	16	3	1.42
	Min.	I	6	3	1	—	—		10	—	—	
		II	4	2	2	—	—		5	—	1	
		III	3	1	2	—	—		7	1	—	
		Total	13	6	5	—	—	1.67	22	1	1	1.12
LOW CHURCH	Max.	I	4	2	—	1	1		7	—	—	
		II	3	1	3	1	—		6	2	—	
		III	2	1	2	2	—		6	2	—	
		Total	9	4	5	4	1	2.31	19	4	—	1.17
	Min.	I	5	2	—	1	—		7	—	—	
		II	5	3	—	—	—		8	—	—	
		III	3	3	1	—	—		7	1	—	
		Total	13	8	1	1	—	1.57	22	1	—	1.04
BROAD CHURCH	Max.	I	—	—	—	1	1		—	—	2	
		II	—	—	—	1	—		—	—	—	
		III	—	—	—	3	—		—	1	2	
		Total	—	—	—	5	1	4.17	—	1	4	2.60
	Min.	I	—	1	—	1	—		1	—	1	
		II	—	1	—	—	—		—	—	—	
		III	—	1	—	2	—		1	—	2	
		Total	—	3	—	3	—	3.0	2	—	3	2.20
ROMAN CATHOLIC	Max.	I	—	1	1	—	—		2	—	—	
		II	—	1	—	1	—		1	—	—	
		III	—	—	2	1	—		1	2	—	
		Total	—	2	3	2	—	3.0	4	2	—	1.33
	Min.	I	1	—	1	—	—		2	—	—	
		II	1	—	—	1	—		1	—	—	
		III	2	—	1	—	—		3	—	—	
		Total	4	—	2	1	—	2	6	—	—	1

TABLE 9. (cont.)

GROUP		PERIOD	POSITION									
			EVOLUTION						MAN			
			1	2	3	4	5	Av.	1	2	3	Av.
METHODISTS	Max.	I	2	—	1	—	—		3	—	—	
		II	3	—	—	—	—		2	—	—	
		III	1	—	1	—	—		1	—	—	
		Total	6	—	2	—	—	1.5	6	—	—	1.0
	Min.	I	3	—	—	—	—		3	—	—	
		II	3	—	—	—	—		2	—	—	
		III	2	—	—	—	—		1	—	—	
		Total	8	—	—	—	—	1.0	6	—	—	1.0
CONGREGATIONALISTS, BAPT., PRESBYTARIANS, QUAKERS	Max.	I	—	3	1	3	—		7	—	—	
		II	—	3	2	4	—		6	2	—	
		III	1	1	—	4	—		2	4	—	
		Total	1	7	3	11	—	3.09	15	6	—	1.29
	Min.	I	6	1	—	—	—		7	—	—	
		II	5	4	—	—	—		8	—	—	
		III	1	2	2	1	—		3	3	—	
		Total	12	7	2	1	—	1.64	18	3	—	1.14
UNITARIANS	Max.	I	—	—	—	3	—		—	—	1	
		II	—	—	—	2	1		—	—	2	
		III	—	—	—	3	—		—	1	2	
		Total	—	—	—	8	1	4.11	—	1	5	2.84
	Min.	I	—	2	—	1	—		—	—	1	
		II	—	—	1	2	—		—	1	1	
		III	—	—	1	2	—		—	1	2	
		Total	—	2	2	5	—	3.34	—	2	4	2.67

(M. M. Bevington, *The Saturday Review*, 35). On the whole, however, it seems likely that among the Review readers — a small and intelligent *élite* of the population — the difference between conservatives and liberals was fairly small from the middle 60s onwards.

In the middlebrow group of periodicals the position was markedly different in this respect, as appears from Diagrams 8 E—H. The difference between the right and the left is very striking, whether individual periodicals or circulations are taken into account.

Correlation between religious opinion and attitude towards Darwinism

Table 9 sets out the attitudes towards Darwinism in the religious press. I have not here differentiated between periodicals of different educational standard, since the material is too small for further sub-division. For this reason there is no point in weighting the periodicals for size of circulation. The opinions arrived at, therefore, may be said to represent press opinion in the sense of the opinions of a sample of press writers. In view of the small number of periodicals in each group, as well as the heterogeneity of the material, the averages shown must be interpreted with caution. But considering that within each group there was little divergence from the line taken by the leading organs of the group, some reliance at least can be put on the results. Again because of the small size of each sample, I have not considered it worth while to calculate the development of opinion over the three periods. The diagram represents the pooled maximum and minimum averages (see above, p. 356) for all three periods together.

Diagram 9. Attitude towards Darwinism in periodicals classed according to religious views.

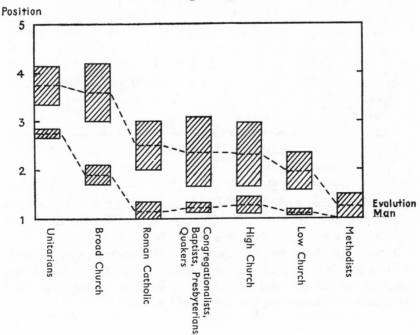

APPENDIX II

The Periodicals: Classification, Coverage of Darwinism, Index of References

E x p l a n a t i o n. The periodicals are listed in alphabetical order. The figures after the name indicates the years studied (59 = 1859, etc.) Many of the periodicals were in existence only during part of the period studied (1859—1872), and some of the files available to me (chiefly in the British Museum) have been incomplete. The classified description employs the classes used in the statistics.

First: general classification. Abbreviations: d = daily, d(e) = daily evening, d(tri) = tri-weekly, m = monthly, q = quarterly, w = weekly. Gen = General, Sci = Scientific, Rel = Religious.

Second: political classification. Abbreviations: N = neutral or unclassified, C = conservative, L = liberal, R = radical.

Third: religious classification. Abbreviations: N = neutral or unclassified, B = Baptist, Br = Broad Church, C = Congregationalist, H = High Church, L = Low Church, M = Methodist, P = Presbyterian, Q = Quaker, R = Roman Catholic, U = Unitarian.

Fourth: educational classification. Abbreviations: 1 = Highbrow, 2 = Middlebrow, 3 = Lowbrow.

Fifth: estimated circulation. Figures indicate thousands. First figure: period 1859—1863. Second figure: period 1864—1869. Third figure: period 1870—1872.

The periodical's coverage of Darwinism is indicated year by year (59 = 1859, etc.) When there is no figure for any year, the implication is that the periodical did not deal with the question during that year. There are two sets of figures, one for amount of information (Inf.), the other for attitude (Att.).

Inf. The amount of information is graded as explained above, p. 341, on a scale 0—3, 0 standing for nothing, 3 for a full discussion. For each year, there are always three figures, the first for Evolution in general, the second for Man, the third for Natural Selection. Thus

"69: 110" means: "In 1869, the periodical briefly referred to, but did not discuss, the general evolution theory and its application to man, but did not even mention the natural selection theory.'

Att. The attitude towards Darwinism is graded as explained above, p. 342, on a 5-point scale for Evolution in its general application, (E), and on a 3-point scale for the theory as applied to Man (M). Hyphenated figures indicate range or alternatives (see above, p. 342). Figures for Evolution are given first, then those for Man. Thus "E: 71: 4—5" means, "In 1871, the periodical was prepared to accept the general Evolution theory fully, and perhaps even Natural Selection as an explanation of Evolution."

Ref. Under *Ref.* are given references to chapters and notes where the periodical is mentioned. I have preferred to use note numbers instead of page numbers, since the large majority of the mentions occur only in the notes, and moreover most mentions in the text itself are accompanied by a corresponding note.

As explained above, p. 342, a reference is always to matter which can be taken to represent the editorial line of the paper or periodical, unless otherwise indicated in the note or the corresponding text.

For further information on the periodicals, see my article on "The Readership of the Periodical Press in Mid-Victorian Britain", Acta Universitatis Gothoburgensis, 63, 1957, no. 3.

Academy, 69–72

Review m Sci	Polit N	Relig N	Éduc 1	Circ. –, 8, 8.

Inf. **69**:110, **70**:120, **71**:332, **72**:220
Att. E: **71**:4–5, **72**:5 M: **71**:2–3
Ref. **2**:11 **3**:102, 114, **5**:29, **6**:94, **13**:27, **14**:73, 76, 117, 140.

All the Year Round, 59–69

Journal w	Polit L	Relig N	Educ 2	Circ. 80, 80, 50.

Inf. **60**:312, **61**:020, **66**:200, **68**:010, **71**:330
Att. E: **60**:4–5, **71**:4–5, M: **61**:1–2, **71**:1–3
Ref. **2**:7, 11 **3**:67, **6**:59, **10**:23, **14**:82, 145

Annals and Magazine of Natural History, 59–72

Review m Sci	Polit N	Relig N	Educ 1	Circ 2, 2, 2.

Inf. **60**:313, **61**:332, **62**:200, **63**:303, **64**:300, **65**:321, **66**:100, **68**:111, **70**:200, **71**:201
Att. E: **60**:1–2, **61**:1–2, **62**:1–2, **63**:1–4, **64**:1–2, **65**:4 M: **60**:1
Ref. **1**:9, 10, **2**:8 **3**:28, 29, 30, 72, 74, 76, 84, 102, **6**:25, 26, 61, 61, 61, 62, 77, **7**:32, 33, **8**:57, 66, 82, **9**:62, **10**:3–11, 28, 42, 46, 47, 49, 52, **11**:4, 5, 14, 26, 29, 34, 36, 39, 44, 46, 51, 61, 65, 76, 79, 80, 81, 83, 86, 87, 110, **12**:20, 22, 23, 34, 41, 41, 41, 43, 72, 92, 99, 112, 114, 117, 124, **13**:9, 10, 24, **14**:52, 53, 56.

Argosy, 66–72
Mag m   Polit N   Relig N   Educ 3   Circ –, 20, 20
Not included in statistics

Athenaeum, 59–72.
Review w Sci   Polit N   Relig N   Educ 1   Circ. 15, 15, 15.
Inf. **59**:312, **60**:302, **61**:321, **62**:120, **63**:333, **64**:110, **65**:220, **66**:311, **68**:322, **69**:321, **70**:232, **71**:232, **72**:120
Att. E: **59**:1–2, **60**:1–2, **61**:2, **63**:2, **66**:2, **69**:2, **71**:3 M: **59**:1, **63**:1–2, **70**:1, **71**:1–2
Ref. **2**:8, 11, 11, **3**:33, 40, 58, 63, 64, 85, 95, 96, 100, **4**:21, 23, 29, **5**:20, **7**:10, 52, **8**:62, 75, **9**:52, 79, 106, **10**:31, 50, **11**:7, 52, 58, 77, 86, **12**:72, 98, 106, 130, **13**:9, 23, **14**:10, 15, 17, 55 55, 55, 57, 102, 116.

Belgravia, 67–72.
Mag m   Polit N   Relig N   Educ 3   Circ –, 20, 18.
Not included in statistics.

Bentley's Miscellany, 59–68.
Mag m   Polit N   Relig N   Educ 2   Circ 5, 3, –.
Inf. **60**:101
Att. –
Ref. –

Blackwood's Magazine, 59–72.
Mag m   Polit C   Relig N   Educ 2   Circ 10, 8, 7.
Inf. **61**:210, **62**:200, **67**:002, **71**:110
Att. E: **62**:1–2, **67**:3–4 M: –
Ref. **2**:16, **3**:11, **5**:16, 55, **6**:47, **9**:64, **11**:58,106, **12**:41, 71B, 73, 133.

British and Foreign Evangelical Review, 59–72
Review q Rel   Polit N   Relig L   Educ 1   Circ 1, 1, 1.
Inf. **60**:300, **61**:110, **63**:220, **66**:300, **67**:110, **68**:110, **69**:110, **70**:101, **71**:110, **72**:232.
Att. E: **60**:2, **61**:2, **63**:2, **66**:2, **68**:2, **70**:2, **71**:2–4, **72**:3 M: **61**:1, **63**:1, **68**:1, **71**:1, **72**:1.
Ref. **2**:8, 11, **3**:38, 107, **5**:35, **6**:61, 77, **7**:62, **8**:10, 38, 44, 62, 65, 80, **10**:12, **11**:21, 66, 68, 80, 85, **12**:41, 47, 69, **13**:14, 47, **14**:120, 122, 130, 138.

British Medical Journal, 59–72.
Review w Sci   Polit N   Relig N   Educ 1   Circ 3, 3, 4.
Inf. **61**:120, **62**:120, **63**:020, **66**:120, **68**:100, **71**:211
Att. E: **62**:1–2, **71**:3 M: **61**:1–2, **62**:1–2, **71**:1
Ref. **4**:44, **14**:55, 93, 131.

British Quarterly Review, 59–72.
Reviel q Rel.   Polit L   Relig CB   Educ 1   Circ 2, 2, 2.
Inf. **60**:312, **61**:121, **62**:210, **63**:323, **65**:111, **66**:100, **67**:111, **68**:201, **69**:110, **70**:110. **71**:332, **72**:110
Att. E: **60**:2, **61**:2–4, **63**:2, **65**:2, **67**:1–2, **68**:1–2, **69**:1–2, **71**:2 M: **60**:1, **62**:1, **63**:1, **65**:1, **67**:1, **70**:1.
Ref. **2**:8, 11, **3**:27, 59, 87, 94, **5**:25, 42, **6**:54, **7**:46, 59, 62, **8**:28, 31, **9**:52, **10**:39, 39, 59, **11**:6, 21, 58, **12**:4, 18, 25, 70, 72, 73, **13**:21.

British Standard, 59–66.

Newspaper w Rel　　　Polit L　　　　Relig CB　　Educ 2　　　Circ 1, 1, –.

Inf. **63**:010, **64**:010, **65**:100

Att. E: **65**:1–2. M: **63**:1.

Ref. –

Broadway, 67–72.

Mag m　　　　　　　Polit N　　　　Relig N　　　Educ 3　　　Circ –, 15, 15.

Not included in statistics.

Cassell's Illustrated Family Paper, Cassell's Magazine, 59, 72.

Journal w　　　　　　Polit N　　　　Relig N　　　Educ 3　　　Circ 200, 200, 200.

Inf. **61**:020

Att. E: – M: **61**:1

Ref. **14**:57

Chambers's Journal, 59–72.

Journal w　　　　　　Polit N　　　　Relig N　　　Educ 2　　　Circ 80, 70, 60.

Inf. **59**:302, **60**:100, **61**:100, **62**:200, **67**:100, **71**:100

Att. E: **59**:4, **62**:4, **71**:4 M: –

Ref. **1**:1, **2**:7, **3**:8, 65, **4**:19, **6**:59, **11**:64, 68, 70, **12**:15.

Christian Observer, 59–72.

Review m Rel　　　　Polit N　　　　Relig L　　　Educ 1　　　Circ 1, 1, 1.

Inf. **60**:313, **61**:100, **66**:111, **68**:101, **70**:202, **71**:312.

Att. E: **60**:1, **61**:1, **66**:1–4, **68**:3–4, **70**:1,–**71**:1 M: **60**:1, **66**:1, **71**:1.

Ref. **2**:8, **3**:12, **5**:43, **7**:15, **8**:32, 42, 58, 61, 70, **9**:70, 71, **10**:39, **11**:6, 31, 76, **12**:72, 72, 72, 118, **14**:20, 39

Christian Remembrancer, 59–68.

Review q Rel　　　　Polit N　　　　Relig H　　　Educ 1　　　Circ 2, 1, 1.

Inf. **60**:110, **67**:100

Att. E: **60**:1–2, **67**:3 M: **60**:1

Ref. **6**:31, **7**:55, 62, **10**:39, **12**:25, 70.

Churchman. See English Churchman.

Church Review, 61–69, 73.

Review w Rel　　　　Polit C　　　　Relig H　　　Educ 1　　　Circ 1, 1, 2.

Inf. **61**:110, **62**:210, **63**:010, **64**:010, **65**:010, **66**:010, **67**:100, **68**:100, **73**:010

Att. E: **61**:1, **62**:1, **67**:1–2, **68**:2–4 M: **61**:1, **62**:1, **63**:1, **64**:1, **66**:1, **73**:1.

Ref. **7**:20, **8**:37, 38, 63, 68, 72, **9**:71, 79, 85, **12**:115, **14**:21, 85, 88.

Church Times, 63–72.

Newsp. w Rel　　　　Polit C　　　　Relig H　　　Educ 2　　　Circ 5, 5, 5.

Not included in statistics.

Contemporary Review, 66–72.

Review m Gen　　　　Polit L　　　　Relig L　　　Educ 1　　　Circ –, 4, 4.

Inf. **66**:222, **67**:202, **68**:311, **69**:120, **70**:302, **71**:232, **72**:203.

Att. E: **66**:2, **67**:2, **68**:2–3, **69**:2, **70**:2–4, **72**:3–4, M: **66**:1, **68**:1–2, **69**:1, **71**:1–2

Ref. **2**:11, **5**:36, 51, **6**:10, 23, 24, 44, 61, 77, **7**:3, 5, 6, 7, 8, 37, 57, 58, 65, 65, 67, **8**:81, **9**:33, 74, 79, 82, 82, 86, 96, **10**:42, 69, **11**:21, 56, 81, **12**:33, 48, 49, 51, 70, 71B, 79B, 107, 115, 144, 144, **13**:16, **14**:24, 36, 47, 75, 78, 91, 108, 115, 116, 144.

Cornhill Magazine, 60–72.

Mag m   Polit L   Relig N   Educ 2   Circ 80, 30, 18
Inf. **60**:302, **62**:322.
Att. E: **60**:4, **62**:4–5 M: –
Ref. **2**:7, **3**:4, **10**:22, 40, **13**:46.

Critic, 59–63

Review w Gen   Polit N   Relig N   Educ 1   Circ 1, –, –.
Inf. **59**:312, **60**:200, **61**:120, **62**:110, **63**:120.
Att. E: **59**:2–4, **62**:2, **63**:2–3, M: **59**:1–2, **63**:1–2.
Ref. **2**:8, **3**:5, 41, 57, **6**:59, **10**:58, **11**:3, 66, **13**:36, **14**:41, 55, 60, 62.

Daily News, 59–72.

Newsp. d   Polit L   Relig N   Educ 2   Circ 5, 10, 90.
Inf. **59**:302, **61**:222, **62**:120, **63**:120, **64**:101, **65**:210, **66**:300, **67**:010, **68**:301, **69**:100, **71**:110, **72**:120.
Att. E: **59**:2–3, **65**:2–3, **66**:2–4, **71**:4–5 M: **71**:2–3
Ref. **2**:6, 11, **3**:102, **4**:5, 8B, 11, 15, 16, 27, 37, 44, 49, 58, 59, 64, 72, 80, 81, 90, 96, 123, **5**:43, **6**:59, **7**:70, **8**:55, **9**:64, 65, 87, 94, **10**:39, 42, **11**: 21, 35, 47, 53, 107, **12**:17, 19, 36, **13**:49, **14**:60, 62, 103.

Daily Telegraph, 59–72.

Newsp. d   Polit LR   Relig N   Educ 2   Circ 70, 150, 190
Inf. **60**:100, **62**:120, **63**:120, **64**:100, **65**:010, **66**:100, **67**:010, **68**:211, **69**:110, **71**:111, **72**:100.
Att. E: **64**:4, **66**:4, **68**:4–5 M: **68**:3.
Ref. **3**:4, **4**:5, 15, 15, 35, 53, 61, 64, 78, 80, 90, 103, 125, 134, 138, **5**:43, **6**:59, **12**:102, **14**:39B, 41B, 61, 117.

Dublin Review, 59–72.

Review q Rel   Polit N   Relig R   Educ 1   Circ 2, 2, 2.
Inf. **60**:313, **63**:010, **71**:322, **72**:201.
Att. E: **60**:3, **71**:3, M: **60**:1, **63**:1, **71**:1–2
Ref. **2**:8, **5**:6, 40, **8**:5, 41, 49, 65, 68, **9**:84, **10**:41, 48, **11**:72, 99, **12**:55, 107, 118, **14**:96

Dublin University Magazine, 59–72.

Mag m   Polit C   Relig N   Educ 2   Circ 3, 2, 2.
Inf. **60**:302, **65**:110, **69**:131.
Att. E: **60**:3, 65:1–2, **69**:1 M: **65**:1, **69**:1
Ref. **2**:7, **3**:92, **5**:26, **6**:58, 77, **8**:68, **9**:94, **11**:80, **12**:9, 123, **14**:31, 38, 46, 50, 61, 67, 77, 112

Ecclesiastic, 59–68.

Review m Rel   Polit C   Relig H   Educ 1   Circ 1, 1, –.
Inf. **60**:211
Att. E: **60**:2–3 M: **60**:1.
Ref. **2**:8 **6**:43, 60, **7**:72, **8**:64

Echo, 68–72.

Newspaper d(c)   Polit R   Relig N   Educ 3   Circ –, 80, 80.
Inf. **69**:100, **71**:111
Att. E: **69**:4–5, **71**:4 M: **71**:2
Ref. –

Eclectic Review, 59–68

Review m Rel   Polit L   Relig CB   Educ 1   Circ 1, 1, —.

Inf. **60**:312, **61**:010, **63**:120, **64**:100, **65**:100, **66**:010, **67**:111,**68**:211

Att. E: **60**:1, **63**:1–2, **64**:3, **65**:2–3, **67**:2–3, **68**:2 M: **60**:1, **61**:1, **63**:1, **66**:1, **68**:1

Ref. **2**:8, **3**:48, **6**:12, 27, 30, 77, **7**:11, 24, 65, 69, **8**:6, 18, 25, 30, 38, 40, 43, 47, 48, 83,
    10:39 51, 59, **11**:6, 19, 37, 45, 58, 73, 87, 106, **12**:24, 33, 35, 66, 71, 71B, 71B,
    72, 72, 72, **13**:16, **14**:32, 86

Economist, 59–72.

Review w   Polit L   Relig N   Educ 1   Circ 4, 4, 4.

Not included in statistics.

Edinburgh New Philosophical Journal, 59–65.

Review q Sci   Polit N   Relig N   Educ 1   Circ 1, 1, –.

Inf. **60**:312, **61**:301, **62**:212, **63**:210, **64**:330

Att. E: **60**:1, **61**:2–3, **62**:2–3, **64**:1–2 M: **60**:1, **62**:1, **64**:1

Ref. **2**:8, **3**:15, **5**:7, **6**:49, 72, 73, **7**:21, 59, **8**:1, 20, 40, 74, **10**:48, **11**:22, 30, 34, 67, 68,
    84, 106, 107, **12**:41, 44, 68, 72, 72, 109, 115, 121, 122, **14**:14, 31.

Edinburgh Review, 59–72

Review q Gen   Polit L   Relig N   Educ 1   Circ 7, 7, 7.

Inf. **60**:302, **62**:111, **63**:230, **68**:303, **70**:200, **71**:332, **72**:221

Att. E: **60**:2–3, **62**:2–3, **63**:2–3, **68**:3–4, **71**:3–4, **72**:2–4. M: **62**:1–2, **63**:1–2, **71**:2,
    72:1

Ref. **2**:8, 11, **3**:10, 21, 36, 49, 57, 70, 82, 93 A, 104, 110, **5**:12, **6**:61, 66, 72, **7**:59, **9**:50,
    52, 60, 71, 91, **10**:14, 17, 26, 70, **11**:16, 28, 39, 58, 91, 93, **12**:36, 41, 42, 51, 69,
    71B, 71 B, 80, 89, 90, 96, 104, 132, 140, **13**:8, 15, 21, 26, 40, 48, **14**:28, 39B, 56,
    58, 73, 75, 77, 83, 104, 109, 110, 121, 133

English Churchman, Churchman, 59–72.

Review w Rel   Polit C   Relig H   Educ 2   Circ 1, 1, 2.

Inf. **59**:100, **61**:010, **62**:110, **63**:110, **65**:010, **66**:200, **67**:100, **69**:100, **70**:110

Att. E: **63**:1, **67**:1–2, **70**:1, M: **62**:1, **63**:1, **65**:1, **70**:1

Ref. **2**:8, 18 B, **4**:38, 77, 87, 111, 115, **6**:61, **7**:71, **8**:9

English Independent, 67–72.

Review w Rel   Polit L   Relig CB   Educ 2   Circ –, 1, 2.

Inf. **67**:010, **68**:110, **69**:110, **70**:100, **71**:110

Att. E: **71**:2–4, M: **67**:1, **68**:1, **71**: 1–2.

Ref. **2**:11, **4**:78, 83, 110, 115, 130, **5**:16, **6**:61, **7**:65, **8**:62, **9**:82, **14**:22, 97, 99

Evening Star. See **Star.**

Examiner, 59–72.

Review w Gen   Polit 7   Relig N   Educ 1   Circ 4, 2, 2.

Inf. **59**:302, **61**:211, **63**:110, **67**:311, **68**:303, **69**:330, **70**:210, **71**:333, **72**:312.

Att. E: **59**:2–4, **61**:2–3, **67**:3–4, **68**:4–5, **70**:4–5, **71**:5, **72**:5, M: **71**:3

Ref. **2**:8, 11, **3**:6, 32, 58, 102, **4**:90, 124 A, 124 B, 135, **6**:74, **7**:76, **9**:59, 59, 82, 106,
    10:39, **11**:27, 80, 93, **12**:94, 105, **13**:17, 19, 19, 19, 19, 20, 28, 28, 29, **14**:41B,
    55, 66, 117, 145, 145

Family Herald, 59–72

Journal w   Polit N   Relig N   Educ 3   Circ 200, 200, 200.
Inf. **60**:110, **61**:100, **62**:010, **63**:010, **65**;110, **67**:210, **71**:210
Att. E: **60**:1, **67**:1, **71**:1, M: **60**:1, **61**:1, **63**:1, **65**:1, **67**:1, **71**:1
Ref. **3**:77, 96, **5**:13, 14, **6**,:14B, **7**:59, 74, **8**:38, **10**:32, 39, **11**:6, 6, 58, 106, 108.

Fortnightly Review, 65–72.

Review m Gen   Polit LR   Relig N   Educ 1   Circ –, 3, 3.
Inf. **65**:121, **67**:100, **68**:302, **71**:222.
Att. E: **65**:2–3, **68**:4–5, M: **71**:2–3
Ref. **5**:55, **7**:4, 62, **8**:79, **10**:69, **11**:6, 21, 42, 99, **12**:23, 55, 142, **13**:27, **14**:79, 80, 124.

Fraser's Magazine, 59–72.

Rev.-mag m Gen   Polit L   Relig Br   Educ 2   Circ 8, 8, 6.
Inf. **60**:312, **63**:230, **68**:001, **70**:110, **72**:110
Att. E: **60**:2–4, **63**:4, **72**:4, M: **60**:1–2, **63**:3, **72**:3
Ref. **2**:7, **3**:89, 109, **7**:17, 18, **8**:73, **9**:86, 88, 89, **10**:13, 16, 54, **11**:18, 80, **12**:2, 7, 135, 139, **14**:148

Freeman, 59–72.

Review w Rel   Polit L   Relig B   Educ 2   Circ 2, 2, 1.
Inf. **60**:202, **62**:010, **63**:010, **65**:100, **66**:210, **68**:200, **69**:010, **70**:110, **71**:001
Att. E: **60**:2–3, **63**:1, **66**:1–3, **68**:3–4, **70**:3–4. M: **66**:1, **70**:2.
Ref. **2**:8 **3**:58B, **6**:61, 61, 77, 77, **7**:54, **8**:12, 32, 38, 60, **11**:50, 58, **12**:71B, 72, **13**:38.

Friend, 59–72.

Review m Rel   Polit L   Relig Q   Educ 2   Circ 2, 2, 2.
Inf. **61**:211, **63**:010, **68**:110
Att. E: **61**:1–2, **68**:1–2. M: **61**:1, **63**:1, **68**:1.
Ref. **3**:73, 80, 86, **8**:45

Friend's Quarterly Examiner, 67–72.

Review q Rel   Polit L   Relig Q   Educ 1   Circ 2, 2, 2.
Inf. **67**:221, **69**:102, **71**:110.
Att. E: **67**:1–4, **69**:3–4, **71**:3–4. M: **67**:1–2, **71**:2
Ref. **5**:15, **6**:61, **8**:81, **12**:141, **14**:14, 65, 123.

Fun, 61–72.

Journal w   Polit N   Relig N   Educ 2   Circ 10, 20, 20.
Inf. **60**:010, **64**:010, **65**:010
Att. –
Ref. –

Gardener's Chronicle, 59–72

Newsp. w   Polit N   Relig N   Educ 2   Circ 5, 5, 5.
Inf. **59**:300, **60**:303, **62**:110, **63**:020, **64**:100, **65**:100, **67**:100, **68**:202, **69**:200, **71**:230, **72**:202
Att. E: **60**:5, **64**:5, **68**:5, **71**:4–5. M: **63**:2–3, **71**:2–3
Ref. **2**:8, 11, **3**:71, **10**:23, 63, **11**:33, 34, **12**:2, 29, 55, 62, 126, **13**:37, **14**:117

Gentleman's Magazine, 68–72.

Mag m   Polit N   Relig N   Educ 2   Circ –, 10, 10.
Inf. **70**:010, **71**:110
Att. –
Ref. –

Geological Magazine, Geologist, 59–72.

Review m Sci   Polit N   Relig N   Educ 1   Circ 1, 2, 2.
Inf. **60**:302, **61**:111, **62**:020, **63**:220, **64**:200, **65**:212, **68**:210
Att. E: **60**:3–4, **65**:4–5. M: **61**:1, **63**:1, **65**:2–3.
Ref. **2**:8, **3**:17, **6**:60, **11**:66, **12**:72, 115, **14**:21, 127

Geologist. See **Geological Magazine.**

Globe, 59–72.

Newsp d(e)   Polit C   Relig N   Educ 2   Circ 3, 3, 7.
Inf. **62**:010, **63**:110, **64**:220, **66**:100, **67**:010, **68**:110, **69**:110, **71**:110.
Att. E: **68**:1–2, **71**:2. M: **67**:1, **71**:1.
Ref. **2**:5, **4**:55B, 76, 125, **5**:52, **7**:42, **8**:59, **9**:52, 79, 98, **11**:38.

Good Words, 60–72.

Journal w Rel   Polit N   Relig L   Educ 2   Circ 30, 70, 100.
Inf. **61**:010, **62**:301, **65**:320, **68**:220.
Att. E: **62**:1–2, **65**:1–2, **68**:2–3. M: **61**:1, **65**:1, **68**:1–2.
Ref. **3**:79, **5**:10, **7**:29, 65, **8**:7, **10**:42, 57, **11**:39, 59, 74, 76, **12**:18, 41, 55, 71B, 71B
       101, **14**:31, 41B, 61, 66, 89, 90A.

Guardian, 59–72.

Review w Rel   Polit L   Relig H   Educ 1   Circ 5, 5, 6.
Inf. **60**:212, **61**:110, **62**:110, **63**:110, **65**:210, **66**:111, **67**:101, **68**:201, 69:120, **70**:331,
       71:322, **72**:101.
Att. E: **60**:2–3, **61**:2, **65**:2–3, **67**:2–4, **68**:3–4, **69**:3–4, **70**:3–4, **71**:3, **72**:3–4. M: **63**:1,
       69:1–2, **70**:1–2, **71**:2.
Ref. **2**:8, 11, **3**:20, 35, 106, 108, **4**:8A, 10, 20, 43, 51, 60, 68, 70, 78, 89, 93, 94, 98, 110,
       5:24, 48, 49, 50, **6**:44, 61, 61, 67, 67, **7**:27, 44, 62, **8**:12, 29, 32, 38, 43, 50, 72, 81,
       9:63, 67, 72, 86, **10**:52, **11**:8, 85, 97, 98, 98, **12**:2, 23, 33, 41, 71B, 72, 79B, **13**:21,
       26, **14**:8, 19, 59. 73, 73, 76, 96, 97, 114, 117, 124, 124, 135.

Home and Foreign Review. See **Rambler.**

Illustrated London News, 59–72.

Journal w   Polit LC   Relig N   Educ 2   Circ 100, 80, 70.
Inf. **60**:100, **62**:020, **63**:110, **64**:200, **65**:111, **66**:200, **68**:100, **69**:010, **70**:201, **71**:312,
       72:100.
Att. E: **60**:2–3, **64**:2–3, **66**:3–4, **71**:2–4, **72**:2–4. M: **71**:1–2.
Ref. **2**:19, **3**:8, 111, **4**:6, **7**:64, **9**:100, **10**:42, **11**:11, 27, 35, 60, **14**:95.

Inquirer, 59–72.

Review w Rel   Polit L   Relig U   Educ 2   Circ 1, 1, 1.
Inf. **60**:110, **61**:110, **63**:110, **65**:010, **66**:200, **68**:200, **71**:311.
Att. E: **60**:2–4, **61**:3–4, **68**:3–4, **71**:3–4. M: **71**:2.
Ref. **2**:11, **3**:101, **4**:21, 50, 89, 118B, 130, **6**:29, **8**:50, 77, **12**:76, **13**:27, **14**:36, 117.

Intellectual Observer, See **Recreative Science.**

John Bull, 59–72.

Newsp. w   Polit C   ReligH   Educ 2   Circ 2, 2, 3.

Inf. **59**:302, **60**:100, **61**:121, **62**:101, **63**:311, **66**:100, **67**:110, **68**:100, **69**:110, **71**:321, **72**:220.

Att. E: **59**:4, **60**:2–4, **61**:2–4, **63**:2–4, **67**:1–2, **68**:1–2, **71**:1–2, **72**:1–2. M: **61**:1, **63**:1, **67**:1, **71**:1, **72**:1.

Ref. **2**:6, 7, 11, **3**:8, 39, 57, **4**:15, 21, 33, 86, 133, **6**:11, 41, 59, **9**:52, 64, 68, 71, 82, 82, **10**:23, 42, 60, **11**:6, 21, 21, 48, 58, 85, **12**:44, 72, 136, **13**:23, **14**:39B, 52, 78, 116, 124, 131, 131.

Journal of Sacred Literature, 59–68.

Review q Rel   Polit N   Relig H   Educ 1   Circ 1, 1, –.

Inf. **60**:100, **64**:110, **66**:110, **67**:110.

Att. E: **64**:2–4, **66**:2–4, **67**:2–3. M: **64**:2, **66**:2, **67**:1.

Ref. **5**:44, **8**:19, 81, **14**:28, 131.

Lancet, 59–72.

Review w Sci   Polit N   Relig N   Educ 1   Circ 3, 3, 3.

Inf. **59**:100, **61**:100, **63**:110, **65**:200, **66**:100, **68**:321, **70**:102, **71**:321, **72**:320.

Att. E: **59**:2, **61**:2–3, **66**:4, **68**:4–5, **71**:4–5, **72**:4–5. M: **63**:1, **71**:2–3, **72**:2–3.

Ref. **2**:11, **3**:47, 93B, **10**:39, 52, 69, **11**:58, 85, **12**:18, **13**:21, 25, 27, **14**:58, 73, 117.

Leader, 59–60.

Review w Gen   Polit R   Relig N   Educ 1   Circ 1, –, –.

Inf. **60**:302.

Att. E: **60**:5, M: –.

Ref. **3**:66, **11**:41, **12**:98, **13**:35.

Leisure Hour, 59–72.

Journal w Rel   Polit N   Relig L   Educ 3   Circ 100, 80, 80.

Inf. **63**:110, **66**:010, **67**:010, **68**:010, **72**:210.

Att. E: **63**:1, **72**:2. M: **63**:1, **66**:1, **67**:1, **68**:1, **72**:1.

Ref. **3**:26, 78, **5**:41, **6**:42, **7**:59, **8**:38, 39, **10**:19, 38, **11**:62, **14**:45, 73.

Literary Churchman, 59–72.

Review w Rel   Polit N   Relig H   Educ 1   Circ 1, 1, 1.

Inf. **60**:110, **62**:101, **65**:010, **71**:110, **72**:100.

Att. E: **60**:1, **62**:1–3, **71**:1–3, **72**:2–3. M: **60**:1, **71**:1.

Ref. **3**:14, 24, **5**:32, **6**:60, 61, **7**:29, 60, **8**:11, **9**:79, **14**:6.

Lloyd's Weekly London Newspaper, 59–72.

Newsp. w   Polit LR   Relig N   Educ 3   Circ 150, 500, 500.

Inf. **62**:010, **65**:010, **66**:100, **68**:100.

Att. E: – M: **65**:1.

Ref. –.

London Journal, 59–72.

Journal w   Polit N   Relig N   Educ 3   Circ 300, 200, 150.

Not included in statistics.

Ref. **2**:14, **8**:34.

London Quarterly Review, 59–72.

Review q Rel   Polit C   Relig M   Educ 1   Circ 2, 2, 2.

Inf. **60:**312, **61:**100, **63:**120, **64:**100, **65:**100, **70:**110, **71:**211, **72:**310.

Att. E: **60:**1, **61:**1, **63:**1–3, **66:**1, **70:**1–3, **71:**1, **72:**1. M: **63:**1, **66:**1, **71:**1, **72:**1.

Ref. **2:**8, 11, **3:**57, 103, **6:**21, 36, 40, **7:**16, 23, 37, 51, 59, **8:**17, 19, 27, 33, 66, 67, 68, 83, **9:**86, 94, 94, 95, **10:**39, 42, 42, **11:**9, 21, 27, 53, 58, 63, 81, 98, **12:**2, 18, 36, 72, 106, **13:**7.

London Reader, 63–72.

Journal w   Polit N   Relig N   Educ 3   Circ 75, 75, 75.

Inf. **69:**010.

Att. –

Ref. –.

London Review, 60–69.

Review w Gen   Polit L   Relig N   Educ 1   Circ 1, 1, –.

Inf. **60:**312, **61:**110, **62:**111, **63:**110, **64:**101, **65:**120, **66:**110, **67:**100, **68:**311, **69:**010

Att. E: **60:**3–4, **62:**2–4, **66:**2–4, **68:**4–5. M: **60:**1–2, **66:**1–2, **68:**2–3.

Ref. **3:**42, 43, 90, **4:**34, 53, **6:**59, **10:**40, 64, **11:**25, 91, 93, 114, **12:**72, **13:**5, 6, 19, **14:**29 94, 96, 126.

London Society, 62–72.

Mag m   Polit N   Relig Br   Educ 2   Circ 20, 20, 20.

Inf. **68:**102, **69:**100, **70:**110, **71:**111.

Att. E: **68:**2–4, **70:**2–4, **71:**2–4. M: **70:**1–2, **71:**1–2.

Ref. **2:**11, **5:**43, **6:**61, 61, **11:**107, **12:**106.

Macmillan's Magazine, 59–72.

Rev.-mag. m   Polit L   Relig Br   Educ 2   Circ 20, 20, 8.

Inf. **59:**202, **60:**302, **61:**312, **70:**100, **71:**322.

Att. E: **59:**5, **60:**4–5, **61:**4, **70:**4, **71:**4. M: **61:**3, **71:**3.

Ref. **2:**7, 7, 11, **3:**22, 23, 69, **4:**91, **5:**9, 55, **6:**59, 80, **8:**78, **9:**58, **12:**74.

Manchester Guardian, 59–72.

Newsp. d   Polit LR   Relig N   Educ 2   Circ 20, 25, 30.

Inf. **61:**210, **62:**010, **63:**010, **64:**100, **65:**100, **66:**110, **67:**100, **68:**100, **69:**100, **70:**100, **71:**100.

Att. E: **64:**4—5, **66:**4, **68:**4–5. M: –.

Ref. **4:**7, 9, 14, 17, 26, 58, 90, 97, 108, 119, 126, **5:**43, **9:**66, 75, **11:**58, **12:**76, **14:**24 60.

Methodist Recorder, 61–72.

Newsp. w Rel   Polit L   Relig M   Educ 2   Circ 10, 20, 25.

Inf. **63:**110, **64:**100, **66:**200, **70:**100, **71:**100.

Att. E: **63:**1, **66:**1, **70:**1. M: **63:**1.

Ref. **3:**34, **4:**8 B, 67, 112, 115, 118B, **5:**11, **6:**61, **8:**71, **9:**82, **11:**6.

Month, 64–72.

Review m Rel   Polit N   Relig R   Educ 1   Circ –, 2, 2.

Inf. **66:**100, **69:**303, **70:**020, **71:**220, **72:**010.

Att. E: **69:**4, **71:**1—4. M: **70:**1, **71:**1, **72:**1.

Ref. **2:**11, **6:**11, 34D, 70, **8:**59, **10:**42, 52, **11:**6, 27, 99, 107, **12:**23, 55, 71B, 81, 92, 113, 115, **13:**8, 11, 21, 26, 28, **14:**73, 103, 117, 133.

Morning Advertiser, 59-72.

Newsp. d Polit L Relig L Educ 2 Circ 6, 6, 6.

Inf. 62:010, 63:110, 65:200, 66:210, 67:010, 68:100, 70:110, 71:110.

Att. E: 66:1, 67:1, 70:1, 71:1. M: 63:1, 66:1, 70:1, 71:1.

Ref. 4:5, 50, 52, 79, 87, 112, 122, 131, 7:41, 8:15, 40, 46, 71, 9:79, 11:21, 14:25, 96.

Morning Post, 59-72.

Newsp. d Polit C Relig H Educ 2 Circ 3, 3, 4.

Inf. 60:312, 61:110, 62:010, 63:120, 64:010, 65:200, 66:220, 67:010, 68:211, 69:220, 71:010, 72:010.

Att. E: 60:4, 63:1. M: 60:2, 63:1, 72:1.

Ref. 2:6, 4:5, 8B, 15, 70, 75, 88, 92, 133, 8:38, 79, 9:76, 99, 10:42, 11:21, 12:131, 14:5, 24, 39B, 62, 73, 74, 126, 129.

Morning Star. See Star.

National Review, 59-64.

Review q Rel Polit L Relig U Educ 1 Circ 1, 1, -.

Inf. 60:313.

Att. E: 60:4. M: -.

Ref. 2:8, 5:37, 45, 6:78, 9:106, 10:69, 11:90, 13:6, 34.

Natural History Review, 61-64.

Review q Sci Polit N Relig N Educ 1 Circ 1, 1, -.

Inf. 61:220, 62:020, 63:322, 64:320.

Att. E: 63:4-5, 64:5. M: 62:3, 63:3, 64:3.

Ref. 3:56, 14:52, 54, 55, 55, 63, 107, 146.

Nature, 69-72.

Review w Sci Polit N Relig N Educ 1 Circ -, 5, 5.

Inf. 69:222, 70:222 71:333, 72: 201.

Att. E: 69:5, 70:5, 71:4-5, 72:4-5. M: 69:3, 70:3, 71:3.

Ref. 2:11, 3:31, 99, 105B, 4:17, 113, 11:80, 101, 12:23, 26, 27, 45, 55, 55, 129, 137 13:12 19 27, 32, 50, 14:72, 111.

New Monthly Magazine, 59-72.

Mag m Polit LC Relig N Educ 2 Circ 3, 2, 2.

Inf. 61:110, 71:100.

Att. E: 71:1, M: -.

Ref. -.

News of the World, 59-72.

Newsp. w Polit L Relig N Educ 3 Circ 80, 60, 50.

Inf. 60:200, 62:010, 71:010.

Att. E: 60:1. M: 62:1, 71:1.

Ref. 2:6, 10:34, 59, 11:71B 14:103, 131.

Nonconformist, 59-72.

Review w Rel Polit L Relig CBMP Educ 2 Circ 3, 3, 3.

Inf. 60:210, 61:210, 62:010, 63:201, 64:100, 65:110, 66:211, 67:110, 68:210, 69:110, 71:110, 72:001.

Att. E: 60:1, 63:2-4, 66:2, 68:4, 69:4, 71:4. M: 60:1, 61:1, 63:1, 66:1, 68:2, 69:2, 71:2.

Ref. **2**:11, **3**:4, **4**:26, 28, 58, 82, 118B, 129, **5**:43, 53, **6**:61, 61, **7**:56, 78, **8**:16, 50, 68, 68, 72, 74, 79, **9**:73, 77, 78, 94, **11**:6, 6, 20, 40, 68, 79, 85, 98, **12**:72, 77, **14**:26, 60, 60, 60, 87, 90B, 128, 129.

North British Review, 59–71.

Review q Gen   Polit L   Relig L   Educ 1   Circ 2, 3, 2.
Inf. **60**:302, **67**:303, **69**:110, **70**:110.
Att. E: **60**:1, **67**:2–3, **70**:3. M: **70**:2.
Ref. **2**:8, **3**:13, 91, **5**:17, 38, 53, 54, **6**:17, **7**:28, 30, **8**:24, 35, 38, 50, 56, 76, 83, **10**:12, 20, 33, 42, 42, 55, **11**:58, 58, 98, 106, **12**:5, 8, 11, 35, 71B, **13**:8, 38.

Observer, 59–72.

Newsp. w   Polit L   Relig N   Educ 2   Circ 6, 3, 3.
Inf. **60**:110, **62**:110, **63**:120, **65**:010, **66**:110, **68**:100, **71**:220.
Att. E: **71**:4. M: **71**:3.
Ref. **2**:11, **3**:105A, **4**:15, **5**:47, **8**:50, **13**:21, **14**:36, 51, 145.

Once a Week, 59–72.

Journal w   Polit L   Relig N   Educ 2   Circ 60, 60, 40.
Inf. **61**:020, **62**:100, **69**:002, **71**:010, **72**:011.
Att. –.
Ref. **3**:104, **13**:28.

Pall Mall Gazette, 65–72.

Newsp. d(e)   Polit L   Relig N   Educ 2   Circ –, 2, 8.
Inf. **65**:110, **66**:100, **67**:100, **68**:202, **69**:010, **70**:100, **71**:120.
Att. E: **66**:3–4, **68**:4–5, **71**:4–5. M: **71**:2–3.
Ref. **2**:11, **3**:98, **4**:12, 61, 98, **6**:28, **7**:62, **10**:27, **11**:109, 114, **12**:71B.

Parthenon, 62–63.

Review w Gen   Polit N   Relig N   Educ 1   Circ 1, –, –.
Inf. **63**:220.
Att. E: **63**:2–4. M: **63**:1–2.
Ref. **12**:41, **14**:16, 29, 31, 116.

Patriot, 59–68.

Review w Rel   Polit L   Relig C   Educ 2   Circ 2, 1, –.
Inf. **60**:301, **61**:010, **62**:210, **63**:210, **66**:100.
Att. E: **60**:2–4, **63**:1, **66**:2–4. M: **63**:1.
Ref. **2**:8, **3**:46, **4**:13, 38, 50, 60, 66, **5**:43, **6**:51, **7**:12, 41, **8**:13, 68, 71, **9**:93, **11**:3, 49, 88, **14**:60.

Popular Science Review, 61–72.

Review q Sci   Polit N   Relig N   Educ 1   Circ 3, 3, 3.
Inf. **61**:201, **62**:110, **63**:320, **64**:222, **65**:202, **66**:201, **67**:300, **68**:201, **69**:300, **70**:100, **71**:332, **72**:301.
Att. E: **61**:2–4, **63**:2–3, **64**:3–5, **65**:4–5, **66**:4–5, **67**:5, **68**:5, **69**:5, **71**:4–5, **72**:5. M: **63**:1–2, **64**:2, **71**:3.
Ref. **2**:11, **3**:41, 44, 90, **5**:15, 39, **6**:38, 59, **7**:73, **9**:86, 106, **10**:24, 25, 39, 52, **11**:21, 21, 66, 67, 71, 80, 99, **12**:23, 55, 71B, 90, 121, 134, **13**:12, **14**:7, 29, 55, 55, 73, 131.

Press, 59–72.
Review w Gen   Polit C   Relig L   Educ 2   Circ 2, 1, 1.
Inf. **59**:202, **60**:210, **61**:010, **62**:110, **63**:110, **64**:010, **65**:110, **66**:100, **69**:010, **71**:020.
Att. E: **59**:2–5, **60**:2–4, **63**:2–4, **65**:1, **66**:1. M: **61**:1, **63**:1, **64**:1, **65**:1, **71**:1.
Ref. **2**:8, 11, **3**:4, 68, **4**:18B, **7**:41, **8**:14, 68, **9**:64, **12**:72, **14**:65, 73, 117.

Progressionist, 63.
Review m Gen   Polit N   Relig N   Educ 2   Circ 1, –, –.
Inf. **63**:110.
Att. E: **63**:2. M: **63**:1.
Ref. –.

Public Opinion, 60–72.
Review w Gen   Polit N   Relig N   Educ 2   Circ 5, 10, 10.
Inf. **62**:010, **63**:220, **65**:110, **66**:020, **67**:110, **68**:010, **69**:111, **70**:111, **71**:220, **72**:220.
Att. E: **65**:1. M: **65**:1, **66**:1.
Ref. **2**:11, **9**:79.

Punch, 59–72.
Journal w   Polit L   Relig N   Educ 2   Circ 40, 40, 40.
Inf. **61**:111, **62**:010, **63**:120, **64**:010, **65**:010, **68**:010, **69**:010, **71**:111, **72**:010.
Att. –.
Ref. **3**:16, 18, 45, **8**:68, **9**:83, 83, **12**:9, **14**:11, 34.

Quarterly Journal of Science, 64–72.
Review q Sci   Polit N   Relig N   Educ 1   Circ –, 1, 1.
Inf. **64**:221, **65**:300, **66**:332, **67**:303, **68**:200, **69**:200, **71**:221.
Att. E: **64**:2–3, **65**:3–4, **66**:3–4, **67**:3–5, **68**:3–4, **69**:3, **71**:3–4. M: **64**:1–2, **66**:1–2,
71:1–3.
Ref. **2**:11, **3**:90, **6**:35, 61, 61, **10**:39, 66, **11**:21, 66, 100 **12**:118 **13**:8 23, 28, 43,
14:29, 32, 36, 65, 90, 116, 117.

Quarterly Review, 59–72.
Review q Gen   Polit C   ReligH   Educ 1   Circ 8, 8, 8.
Inf. **60**:312, **63**:220, **65**:102, **67**:100, **69**:332, **71**:333, **72**:100.
Att. E: **60**:1, **63**:1–3, **69**:3–4, **71**:3. M: **60**:1, **63**:1–2, **69**:3, **71**:2.
Ref. **2**:8, 11, **3**:61, 103, **4**:99, **5**:8, **6**:18, 33, 53, 61, 64, 65, **7**:22, 35, 38, 49, **8**:50, 66,
71, **9**:49, 64, **10**:30, 36, 42, 59, 62, **11**:17, 35, 76, 80, 80, 91, **12**:72, 92, 93, **13**:
18, 26, **14**:27, 62, 68, 69, 105, 117, 136, 140B, 148.

Rambler, Home and Foreign Review, 59–64.
Review q Rel   Polit L   Relig R   Educ 1   Circ 1, 1, –.
Inf. **59**:100, **60**:313, **63**:120.
Att. E: **59**:1, **60**:2. M: **60**:1, **63**:1.
Ref. **2**:8, **3**:25, 38, **5**:19, 28, **7**:29, **8**:4, **10**:43, **11**:85, **14**:14, 31, 113.

Reader, 63–66.
Review w Gen   Polit R   Relig N   Educ 1   Circ 1, 1, –.
Inf. **63**:220, **64**:202, **65**:020.
Att. E: **63**:5, **64**:5. M: **63**:3, **64**:3.
Ref. **3**:50, **14**:55.

Record, 59–72.

Newsp. d(tr) Rel   Polit C   Relig L   Educ 2   Circ 4, 4, 3.

Inf. **61**:010, **62**:110, **63**:110, **65**:100, **66**:110, **67**:100, **68**:100, **69**:110, **70**:210, **71**:100, **72**:300.

Att. E: **62**:1, **63**:1, **66**:1, **67**:1, **70**:1, **71**:1, **72**:1–3. M: **62**:1, **63**:1, **66**:1, **70**:1.

Ref. **4**:38, 66, 71, 73, 100, 110, 115, 121, **5**:5, **6**:61, **8**:47, 62, 68, **9**:64, 69, 79, 81, 82. **11**:6, 21, 58, 107, **12**:18, 19, 23, 72, 107, **14**:43, 60.

Recreative Science, Intellectual Observer, Student, 59–71.

Review m Sci   Polit N   Relig N   Educ 2   Circ 2, 3, 1.

Inf. **60**:010, **61**:311, **62**:200, **63**:320, **64**:202, **65**:100, **66**:201, **67**:200, **68**:302, **69**:202, **70**:130.

Att. E: **61**:1, **62**:2–4, **63**:2–4, **64**:4, **65**:4, **66**:4, **67**:4, **68**:4, **69**:4–5, **70**:3–4. M: **60**: 1–2, **61**:1, **63**:1–2, **70**:2–3.

Ref. **2**:8, **3**:51, 75, 90, 93B, **5**:11, **6**:34D, 77, 79, 88–90, **8**:8, 49, 68, **9**:51, 90,, **10**:12, 20, 40, 42, 61, **11**21, 49, 75, 80, 85, 87, 96, 107, **12**:72, 103, **13**:5, 21, 28, 39, 41, **14**:49, 81.

Reynolds's Newspaper, 59–72.

Newsp. w   Polit R   Relig N   Educ 3   Circ 60, 150, 200.

Inf. **61**:010, **71**:010.

Att. –.

Ref. **14**:145.

Saint James' Chronicle, 59–66.

Newsp. d(tri)w   Polit C   Relig L   Educ 2   Circ 1, 1, –.

Inf. **60**:312, **61**:010, **62**:100.

Att. E: **60**:4. M: –.

Ref. **2**:6.

Saint James' Magazine, 61–72.

Mag m   Polit C   Relig N   Educ 3   Circ 20, 20, 20.

Inf. **61**:100, **63**:100.

Att. E: **61**:1, **63**:1, M: –.

Ref. –.

Saint Paul's Magazine, 67–72.

Mag m   Polit N   Relig N   Educ 2   Circ –, 20, 20.

Not included in statistics.

Saturday Review, 59–72.

Review w Gen   Polit LC   Relig N   Educ 1   Circ 10, 18, 20.

Inf. **59**:312, **60**:312, **61**:202, **62**:100, **63**:200, **64**:112, **66**:012, **68**:202, **69**:201, **70**:221, **71**:332, **72**:110.

Att. E: **59**:1–4, **60**:2–3, **61**:2–4, **64**:5, **68**:5, **69**:5, **70**:5, **71**:4–5. M: **59**:1, **64**:3, **70**:3, **71**:3.

Ref. **2**:8, 11, **3**:3, 19, 37, 62, 102, **9**:97, **11**:76, 100, **12**:53, 75, 76, 138, **14**:72, 125.

Spectator, 59–72.

Review w Gen   Polit L   Relig U   Educ 1   Circ 3, 2, 4.

Inf. **59**:202, **60**:312, **61**:210, **62**:200, **63**:221, **64**:201, **65**:010, **66**:110, **67**:213, **68**:300 **71**:322.

Att. E: **60**:2–4, **63**:4, **64**:4, **67**:4, **68**:4, **71**:4. M: **63**:3, **67**:2–3, **71**:3.

Ref. **2**:8, 11, **3**:9, 57, 81, 97, **4**:104, 105, **5**:18, **6**:13, 34, **8**:43, **9**:48, 97, **11**:8, 15, 27, 29, 41, 78, 79, 82, 105, **12**:12, 67, 70, 71B, **13**:6, 19, **14**:126, 137, 148.

Standard, 59–72.

Newsp. d   Polit C   Relig N   Educ 2   Circ 40, 100, 140.

Inf. **61**:010, **62**:110, **66**:320, **67**:110, **68**:100, **71**:131.

Att. E: **71**:2–3. M: **71**:1–2.

Ref. **4**:5, 55B, 59, 71, 73, 88, 96, 132, **7**:77, **9**:71, **11**:98, 107, **12**:140, **14**:51, 60, 73, 76.

Star, 59–69.

Newsp. d   Polit LR   Relig N   Educ 2   Circ 20, 10, –.

Inf. **60**:201, **61**:010, **62**:010, **63**:121, **65**:200, **66**:320, **67**:010, **68**:301.

Att. –.

Ref. **4**:5, 18, 23, 45, 46, 54, 58, 62, 96, 102, **14**:60.

Student. See Recreative Science.

Sunday Times, 59–72.

Newsp w   Polit LC   Relig N   Educ 2   Circ 5, 20, 30.

Inf. **65**:010, **68**: 100, **69**:010, **71**:010.

Att. E: **68**:3–5. M: –.

Ref. **8**:12.

Tablet, 59–72.

Review w Rel   Polit C   Relig R   Educ 1   Circ 2, 2, 2.

Inf. **68**:100, **69**:310, **70**:020, **71**:331, **72**:211.

Att. E: **64**:1, **69**:1–2, **71**:1–3, **72**:3. M: **64**:1, **70**:1, **69**:1, **71**:1–2, **72**:1.

Ref. **2**:11, **3**:96, 108, 113, **6**:31, 61, 67, **7**:22B, 75, **8**:74, **10**:42, 42, **11**:21, **12**:2, 9, 13, 23,A, 23B, 70, 72, **14**:37, 55, 73, 84, 106, 134.

Tait's Edinburgh Magazine, 59–61.

Review m Gen   Polit R   Relig N   Educ 2   Circ 1, –, –.

Inf. **60**:110.

Att. E: **60**:2. M: **60**:1.

Ref. **8**:69, **11**:107.

Temple Bar, 61–72.

Mag m   Polit L   Relig N   Educ 2   Circ 30, 20, 13.

Inf. **61**:211, **62**:210, **65**:211, **72**:010.

Att. E: **61**:3–4, **62**:1–3, **65**:4. M: **62**:1, **72**:2–3.

Ref. **3**:8, 18, 88, **5**:16, **7**:19, 68, **9**:106, **11**:39, 41, **12**:133.

Theological Review, 64–72.

Review q Rel   Polit L   Relig U   Educ 1   Circ –, 1, 1.

Inf. **65**:212, **67**:100, **69**:100, **72**:111.

Att. E: **65**:4–5, **67**:4–5, **69**:4, **72**:4. M: **65**:3, **72**:3.

Ref. **5**:36, **6**:77, **14**:100, 140.

Times, 59–72.

Newsp. d   Polit LC   Relig H   Educ 2   Circ 55, 65, 63.

Inf. **59**:303, **60**:100, **61**:110, **62**:110, **63**:121, **64**:010, **65**:200, **66**:100, **67**:102, **68**:220, **69**:121, **70**:110, **71**:231, **72**:122.

Att. E: **59**:5, **61**:3–5, **66**:1–3, **67**:1–3, **71**:2–3, **72**:2–3. M: **61**:1–2, **62**:1–2, **71**:1–2, **72**: 1–2.

Ref. **2**:6, 8, **3**:58, **4**:5, 31, 32, 39, 48, 65, 92, 96, 101, 114, 120, 125, 137, **5**:15, 21, 26, **6**:54, **7**:3, 25, 40, **9**:71, 71, 82, 94, **10**:23, 40, 53, **11**:21, 21, 39, 89, **12**:71 B, 73, **13**:6, 19, 23, 43, **14**:12, 18, 26, 28, 40, 44, 62, 92, 103, 131.

Tinsley's Magazine, 67–72.

| Mag m | Polit N | Relig N | Educ 3 | Circ –, 20, 10. |

Inf. **71**:220.

Att. E: **71**:1. M: **71**:1.

Ref. **8**:68, **11**:107, **14**:29.

Tomahawk, 67–72.

| Journal w | Polit N | Relig N | Educ 2 | Circ –, 10, 10. |

Inf. **67**:110.

Att. –.

Ref. –.

Universe, 60–72.

| Newsp w Rel | Polit L | Relig R | Educ 3 | Circ 5, 10, 20. |

Not included in statistics.

Vanity Fair, 68–72.

| Journal w | Polit N | Relig N | Educ 2 | Circ –, 3, 5. |

Inf. **71**:100.

Att. E: **71**:2–3. M: –.

Ref. –.

Watchman, 59–72.

| Newsp w Rel | Polit C | Relig M | Educ 2 | Circ 3, 2, 1. |

Inf. **61**:200, **62**:010, **63**:111, **64**:010, **66**:110, **68**:101.

Att. E: **61**:1, **63**:1, **64**:1, **66**:1, **68**:1. M: **62**:1, **64**:1.

Ref. **4**:26, **9**:52, 52, 92, **10**:37, **11**:55, **14**:61.

Weekly Dispatch, 59–72.

| Newsp w | Polit L | Relig N | Educ 3 | Circ 20, 10, 140. |

Inf. **63**:100, **64**:100, **66**:100, **68**:100, **71**:100.

Att. –.

Ref. –.

Weekly Review, 62–72.

| Review w Rel | Polit L | Relig P | Educ 2 | Circ 1, 1, 1. |

Inf. **62**:110, **63**:110, **64**:100, **65**:110, **66**:110, **67**:010, **69**:100, **70**:100, **71**:110.

Att. E: **63**:1–2, **66**:2–3, **69**:2–3, **71**:1. M: **63**:1, **65**:1, **67**:1, **69**:1, **71**:1.

Ref. **5**:44, **9**:79, **11**:111, **14**:96.

Weekly Times, 59–72.

| Newsp w | Polit L | Relig N | Educ 3 | Circ 75, 100, 150. |

Inf. **68**:100, **69**:100, **71**:010.

Att. –.

Ref. –.

Westminster Review, 59–72.

Review q Gen   Polit R   Relig N   Educ 1   Circ 4, 4, 4.

Inf. **60**:303, **61**:100, **62**:101, **63**:220, **64**:200, **65**:111, **66**:100, **67**:302, **68**:100, **69**:312, **70**:322, **71**:112, **72**:030.

Att. E: **60**:5, **63**:5, **67**:5, **69**:5, **70**:5, **72**:5. M: **63**:3, **65**:3, **70**:3, **72**:3.

Ref. **2**:8, 11, 17, **3**:2, 60, 90, 90, 112, **5**:27, **6**:59, **8**:23, 50, 52, 53, **9**:104, **10**:23, 40, 65, **11**:58, 93, **12**:10, 45, 85, 86, 116, **13**:30, 31, 51, **14**:30, 64, 72.

Zoologist, 59–72..

Review m Sci   Polit N   Relig N   Educ 1   Circ 1, 1, 1.

Inf. **58**:303, **59**:302, **61**:312, **62**:100, **66**:110, **67**:100, **68**:300, **71**:310, **72**:100.

Att. E: **61**:1–2, **66**:2–4, **67**:1, **68**:1, **71**:1, **72**:1–3. M: **61**:1, **66**:2, **68**:1, **71**:1.

Ref. **2**:1, 8, 11, **3**:103, **6**:13, 15, 16, 22, 71, 77, **7**:43, **8**:12, 60, **9**:47, **10**:39, 45, 56, **11**:6, 27, 57, 71, 80, 87, **12**:3, 25, 43, 54, 111, **13**:23, **14**:29.

Index of Names, with Biographical Notes

Bibliography of Citations

For periodicals quoted, see Appendix II, p. 368—384.

Argyll, Duke of; see Campbell.

Bates, Henry William. *The Naturalist on the Amazons*. London, 1863.   p. 288.

Bevington, Merle Mowbray. *The Saturday Review, 1855—1868*. New York, 1941.
     p. 366.

British and Foreign Medico-Chirurgical Review.   p. 8, 11.

British Association for the Advancement of Science, *Report*, 1831, in progress.
(Quoted as *Report*.)   p. 62—94 *passim*.

Brown, Alan Willard. *The Metaphysical Society*. New York, 1947. p. 104, 141,
     316.

Buckland, William. *Bridgewater Treatises*. Treatise 6. *Geology and Mineralogy
considered with reference to Natural Theology*. London, 1836.   p. 16.

Campbell, George Douglas, 8th Duke of Argyll. *Primeval Man*. London, 1869.
     p. 52, 84, 301, 303.

— *The Reign of Law*. London, 1867. (Quoted from 18 ed., 1884.) p. 124, 130,
     132, 151, 251, 285, 296.

Chambers, Robert. *Vestiges of the Natural History of Creation*. London, 1844.
(Quoted from 6 ed., London, 1847.) p. 11, 12, 14, 16, 32, 41, 44, 47, 268, 333.

Colenso, John William. *The Pentateuch and the Book of Joshua Critically Examined*.
7 parts, London, 1862—1879.   p. 27, 106.

Darwin, Charles Robert. *The Descent of Man*. London, 1871. (Quoted from
Popular ed., London, 1901 as *Descent*.)   *Passim*.

— *On the Origin of Species*. London, 1859. (Quoted as *Origin*, 1 ed. 1859.) Reprinted, London, 1950. (Quoted as *Origin*, 1 ed. r.) 6 ed. 1872, Popular Impression,
London, 1900 (Quoted as Origin, 6 ed.)   *Passim*.

— *The Variation of Animals and Plants under Domestication*. London, 1868.
     p. 28, 139, 222, 345.

— *On the various Contrivances by which Orchids are Fertilized by Insects*. London,
1862.   p. 27, 72, 188.

— *Journal of the Proceedings of the Linnaean Society*, London, 1858. p. 18, 52,
     185.

Darwin, Francis. *Life and Letters of Charles Darwin*. 3 vols., London, 1888.
(Quoted as *Letters*.)   *Passim*.

— *More Letters of Charles Darwin*. 2 vols., London, 1903. (Quoted as *More Letters*.)
     Passim.

Daubeny, Charles. "Remarks on the Final Causes of the Sexuality of Plants."
London, 1860.   p. 69.

Dobzhansky, Theodosius. *Genetics and the Origin of Species*. 3 ed., New York, 1951.
                                                                          p. 215.
Du Chaillu, Paul Belloni. *Explorations and adventures in Equatorial Africa*. London, 1861.   p. 43.
Ellegård, Alvar. "The Readership of the Periodical Press in Mid-Victorian Britain",
    Acta Universitatis Gothoburgensis, 63, 1957, no. 3.   p. 20, 340, 369.
— "The Darwinian Theory and Nineteenth Century Philosophies of Science",
    Journal of the History of Ideas, 18, 1957, 362—393.   p. 175.
— »Public Opinion and the Press: Reactions to Darwinism», *Journal of the History
    of Ideas*, 19, 1958, 379—387.   p. 20.
Elliott-Binns, L. E. *English Thought, 1860—1900: The Theological Aspect*. London, 1956.   p. 106, 156, 172.
Essays and Reviews. London, 1860.   p. 27, 106, 156, 158.
Fisher, R. A. *The Genetical Theory of Natural Selection*. Oxford, 1930. p. 215.
Freke, H. *On the Origin of Species through Organic Affinity*. London, 1861.
                                                                          p. 272.
Gillispie, Charles Coulston. *Genesis and Geology*. Harvard Historical Studies, 58,
    Cambridge, Massachusetts, 1951.   p. 16, 95, 102, 108, 118, 159, 337.
Glover, Willis B. *Evangelical Nonconformists and Higher Criticism in the 19th
    Century*. London, 1954.   p. 95, 105, 156, 172.
Gray, Asa. "Natural Selection not inconsistent with Natural Theology." London,
    1861.   p. 137—8.
Hall, Everett W. *Modern Science and Human Values*. New York, 1956.
                                                                     p. 112, 173.
Hofstadter, Richard. *Social Darwinism in American Thought*. *1860—1915*.
    Philadelphia and London, 1944.   p. 254.
Howarth, O. J. R. *The British Association*. London, 1922.   p. 64.
Hume, David. *Dialogues concerning Natural Religion*. 1779. (Quoted from ed.
    N. Kemp Smith, London, 1947.)   p. 115, 116.
Huxley, Leonard. *Life and Letters of Sir J. D. Hooker*. 2 vols., London, 1918.
                                                                          p. 78.
Huxley, Leonard. *Life and Letters of Thomas Henry Huxley*. 3 vols. London,
    1900. (Quoted as Huxley, *Life*.)   p. 49, 92, 183, 289.
Huxley, Thomas Henry. *Collected Essays*. II, London, 1893.   p. 140.
— *Hume*. London, 1881.   p. 140, 180, 195.
— *Essays upon Controversial Subjects*. London, 1892.   p. 181, 183.
— *Lay Sermons*. London, 1870.   p. 82, 180, 182, 214, 237, 238.
— *Lectures and Essays*. London, 1908.   p. 180, 182, 187, 195.
— *Man's Place in Nature*. London, 1863.   p. 28, 50, 73, 74, 154, 296, 317.
Journal of the Transactions of the Victoria Institute. London, 1866.   p. 104.
London, Edinburgh and Dublin Philosophical Magazine.   p. 8.
Lubbock, John. *Origin of Civilization*. London, 1870.   p. 296.
— *Prehistoric Times*. London, 1865.   p. 296.
Lyell, Charles. *Antiquity of Man*. London, 1863. p. 28, 52, 73, 128, 160, 192,
                                                                     296, 345.

— *Principles of Geology.* London, 1830—33. p. 51, 53, 159.
— *Life, Letters and Journals of Sir Charles Lyell,* edited by his sister in law. 2 vols., London, 1881. p. 125.
Mill, John Stuart. *A System of Logic.* 2 vols., London, 1843. (Quoted from 9 ed., London, 1875, as Mill.) p. 175—185 *passim.*
Mivart, St George. *Genesis of Species.* London, 1870. (Quoted from 2 ed., 1871). p. 28, 60, 79, 248, 249, 256, 266, 271, 272.
Mozley, James Bowling. *Eight Lectures on Miracles.* London, 1865. p. 141.
Murphy, Joseph John. *Habit and Intelligence.* London, 1869. p. 278.
Newton, Sir Isaac. *Philosophiae Naturalis Principia Mathematica.* 1687 (Quoted from 3. ed., 1726.) p. 114.
Nordenskiöld, Nils-Erik. *Biologins Historia.* 1—3, Stockholm, 1920—24. p. 13.
Owen, Richard. *Anatomy of the Vertebrates.* 3 vols. London, 1866—68. p. 136, 266, 271.
Owen Richard. *Life of Richard Owen.* London, 1894. p. 12.
Paley, William. *A View of the Evidences of Christianity,* 2 ed., London 1794. p. 115.
— *Natural Theology.* 3 ed., London, 1803. p. 115.
Roberts, Windsor Hall. *The Reaction of American Protestant Churches to the Darwinian Philosophy, 1860—1900.* (Abstract of dissertation., Chicago 1936 (1938)). p. 113.
Rouché, Max. "Herder précurseur de Darwin." *Publications de la faculté de Lettres de l'Université de Strasbourg,* Fasc. 94, 1940. p. 35.
Simonsson, Tord. *Face to Face with Darwinism.* A Critical Analysis of the Christian Front in Swedish Discussion of the Later Nineteenth Century. Doctoral diss. Lund, 1958. p. 113.
Spencer, Herbert. *First Principles.* 6 ed., London, 1900. p. 255.
Strauss, David Friedrich. *Leben Jesu.* English translation, *Life of Jesus,* London, 1846. p. 105.
Thomson, William. "On the Secular Cooling of the Earth", *Transactions* of the Royal Society of Edinburgh, 1861, 157. p. 236.
The Times. *History of The Times.* 3 vols. London. p. 128.
Todhunter, Isaac. *William Whewell.* London, 1876. p. 185
Tylor, Edward Burnett. *Researches into the Early History of Mankind.* London, 1865. p 296.
Voltaire, Jean François Marie Arouet de. *Singularités de la Nature;* in *Oeuvres,* ed. Genève, 1758—1757, *Mélanges Philosophiques,* Tome 4. p. 96.
Wallace, Alfred Russell. *Contributions to the Theory of Natural Selection.* London, 1870. p. 28, 84, 296.
— *Journal of the Proceedings of the Linnaean Society,* London, 1858. p. 18, 52, 185.
— *My Life.* 2 vols., London, 1905. p. 28.
Whewell, William. Bridgewater Treatises. 3. *Astronomy and General physics considered with reference to Natural Theology.* London, 1833—36. p. 128.

— *History of the Inductive Sciences.* 3 vols. 2 ed., London, 1847.   p. 14.

— *Indications of the Creator.* London, 1847.   p. 14.

— *The Philosophy of the Inductive Sciences.* 2 vols., London, 1840. p. 14, 15, 175—185 *passim.*

Wood, H. G. *Belief and Unbelief since 1850.* Cambridge, 1955.   p. 95, 172.

Zeuner, Frederick E. *Dating the Past.* London, 1946.   p. 238.